AF769086

Lecture Notes in Physics

Editorial Board

R. Beig, Wien, Austria
J. Ehlers, Potsdam, Germany
U. Frisch, Nice, France
K. Hepp, Zürich, Switzerland
W. Hillebrandt, Garching, Germany
D. Imboden, Zürich, Switzerland
R. L. Jaffe, Cambridge, MA, USA
R. Kippenhahn, Göttingen, Germany
R. Lipowsky, Golm, Germany
H. v. Löhneysen, Karlsruhe, Germany
I. Ojima, Kyoto, Japan
H. A. Weidenmüller, Heidelberg, Germany
J. Wess, München, Germany
J. Zittartz, Köln, Germany

Springer-Verlag
Berlin Heidelberg GmbH

ONLINE LIBRARY
Physics and Astronomy
http://www.springer.de/phys/

Editorial Policy

The series *Lecture Notes in Physics* (LNP), founded in 1969, reports new developments in physics research and teaching -- quickly, informally but with a high quality. Manuscripts to be considered for publication are topical volumes consisting of a limited number of contributions, carefully edited and closely related to each other. Each contribution should contain at least partly original and previously unpublished material, be written in a clear, pedagogical style and aimed at a broader readership, especially graduate students and nonspecialist researchers wishing to familiarize themselves with the topic concerned. For this reason, traditional proceedings cannot be considered for this series though volumes to appear in this series are often based on material presented at conferences, workshops and schools (in exceptional cases the original papers and/or those not included in the printed book may be added on an accompanying CD ROM, together with the abstracts of posters and other material suitable for publication, e.g. large tables, colour pictures, program codes, etc.).

Acceptance

A project can only be accepted tentatively for publication, by both the editorial board and the publisher, following thorough examination of the material submitted. The book proposal sent to the publisher should consist at least of a preliminary table of contents outlining the structure of the book together with abstracts of all contributions to be included.
Final acceptance is issued by the series editor in charge, in consultation with the publisher, only after receiving the complete manuscript. Final acceptance, possibly requiring minor corrections, usually follows the tentative acceptance unless the final manuscript differs significantly from expectations (project outline). In particular, the series editors are entitled to reject individual contributions if they do not meet the high quality standards of this series. The final manuscript must be camera-ready, and should include both an informative introduction and a sufficiently detailed subject index.

Contractual Aspects

Publication in LNP is free of charge. There is no formal contract, no royalties are paid, and no bulk orders are required, although special discounts are offered in this case. The volume editors receive jointly 30 free copies for their personal use and are entitled, as are the contributing authors, to purchase Springer books at a reduced rate. The publisher secures the copyright for each volume. As a rule, no reprints of individual contributions can be supplied.

Manuscript Submission

The manuscript in its final and approved version must be submitted in camera-ready form. The corresponding electronic source files are also required for the production process, in particular the online version. Technical assistance in compiling the final manuscript can be provided by the publisher's production editor(s), especially with regard to the publisher's own Latex macro package which has been specially designed for this series.

Online Version/ LNP Homepage

LNP homepage (list of available titles, aims and scope, editorial contacts etc.):
http://www.springer.de/phys/books/lnpp/

LNP online (abstracts, full-texts, subscriptions etc.):
http://link.springer.de/series/lnpp/

I. P. Williams N. Thomas (Eds.)

Solar and Extra-Solar Planetary Systems

Lectures Held at the Astrophysics School XI
Organized by the European Astrophysics
Doctoral Network (EADN) in The Burren,
Ballyvaughn, Ireland, 7-18 September 1998

Springer

Editors

Iwan P. Williams
University of London
Astronomy Unit, School of Mathematical
Sciences, Queen Mary a. Westfield College
Mile End Road
E1 4NS London, United Kingdom

Nicolas Thomas
Max-Planck-Institut für Aeronomie
Max-Planck-Straße 2
37191 Katlenburg-Lindau, Deutschland

Cover Picture: The true relative sizes of the planets and the Sun, taken from The new Cosmos -an Introduction to Astronomy and Astropyhsics, A. Unsöld/B. Baschek, 5th Edition 2001, ISBN 3-540-67877-8.

Library of Congress Cataloging-in-Publication Data.

Die Deutsche Bibliothek - CIP-Einheitsaufnahme

Solar and extra-solar planetary systems : lectures held at the Astrophysics School XI, in The Burren, Ballyvaughn, Ireland, 7 - 18 September 1998 / I. P. Williams ; N. Thomas (ed.). Organized by the European Astrophysics Doctoral Network (EADN).
(Lecture notes in physics ; Vol. 577)
(Physics and astronomy online library)
ISBN 978-3-662-14310-0 ISBN 978-3-540-44807-5 (eBook)
DOI 10.1007/978-3-540-44807-5

ISSN 0075-8450
ISBN 978-3-662-14310-0

This work is subject to copyright. All rights are reserved, whether the whole or part of the material is concerned, specifically the rights of translation, reprinting, reuse of illustrations, recitation, broadcasting, reproduction on microfilm or in any other way, and storage in data banks. Duplication of this publication or parts thereof is permitted only under the provisions of the German Copyright Law of September 9, 1965, in its current version, and permission for use must always be obtained from Springer-Verlag Berlin Heidelberg GmbH.
Violations are liable for prosecution under the German Copyright Law.

http://www.springer.de

© Springer-Verlag Berlin Heidelberg 2001
Originally published by Springer-Verlag Berlin Heidelberg New York in 2001
Softcover reprint of the hardcover 1st edition 2001

The use of general descriptive names, registered names, trademarks, etc. in this publication does not imply, even in the absence of a specific statement, that such names are exempt from the relevant protective laws and regulations and therefore free for general use.

Typesetting: Camera-ready by the authors/editors
Camera-data conversion by Steingraeber Satztechnik GmbH Heidelberg
Cover design: *design & production*, Heidelberg

Printed on acid-free paper
SPIN: 10792099 54/3141/du - 5 4 3 2 1 0

Preface

The 11th pre-doctoral school of the European Astrophysics Doctoral network (EADN) took place during 7 – 18 September in 1998. The Burren, Ballyvaughn, Ireland, with the participation of eight lecturers and 17 students from six countries.
The subject of the school "Solar and Extra-Solar Planetary Systems" was selected in view of the recent discoveries of the first extra-solar planets and the high level of activity in space exploration at the current time.
Following the tradition of the EADN summer schools, the subject matter of the lectures was both theoretical and observational. All students gave a short presentation of their recently started research projects.
This volume contains the lectures presented at the school but not the student seminars.
We acknowledge with grateful appreciation financial support from: the Training and Mobility of Researchers programme of the European Union, the Granholm Foundation of Sweden, the European Space Agency, and the Max-Planck Institute für Aeronomie, Katlenburg-Lindau.
Special thanks are due to Eimhear Clifton and Hilary O'Donnell for all the local arrangements and to Andrea Macke for generating the LaTeX version of the book.
We also thank Tom Ray, secretary of the EADN, for his help in getting the school started and acting as "local" host, and Carl Murray, for the design of the poster.
Finally, we wish to thank the teachers for their excellent lectures, and staff and students alike for braving the cold Atlantic weather (their only comfort being the local pubs).

October 2001

Iwan Williams
Nicolas Thomas

Table of Contents

List of Contributors

T. Encrenaz
Observatoire de Paris
92195 Meudon, France

A. Fitzsimmons
Department of Pure
and Applied Physics
Queens University
Belfast BT7 1NN, Northern Ireland

I. Mann
Max-Planck-Institut für Aeronomie
Max-Planck-Str. 2
37189 Katlenburg-Lindau, Germany

C. D. Murray
Astronomy Unit
Queen Mary
University of London
Mile End Road
London E1 4NS, UK

R. P. Nelson
Astronomy Unit
Queen Mary
University of London
Mile End Road
London E1 4NS, UK

T. P. Ray
School of Cosmic Physics
Dublin Institute for Advanced
Studies
5 Merrion Square
Dublin 2, Ireland

N. Thomas
Max-Planck-Institut für Aeronomie
Max-Planck-Str. 2
37189 Katlenburg-Lindau, Germany

I. P. Williams
Astronomy Unit
Queen Mary
University of London
Mile End Road
London E1 4NS, UK

Plates

Plate 1: A 3-colour composite image of HH 30 based on data from HST observations. Note that the image has been rotated so that the blueshifted jet is oriented upwards. Here blue light represents continuum emission, which in this case is scattered starlight from the "top" and "bottom" of the disk. The source itself is highly embedded and the disk can be seen in silhouette as the dark lane bisecting the nebular cusps. Red light is [SII]$\lambda\lambda$6716,6731 emission and green represents Hα. Note the asymmetry in excitation conditions in the flow and the counterflow. Image reconstruction done by C.R. O'Dell and S.V.W. Beckwith.

Plate 2: Both Enceladus (left) and Miranda (right) show both heavily cratered and relatively smooth regions indicating different ages for the surface units (photo: NASA).

Plate 3: Three full-disk colour views of Jupiter's volcanic moon Io as seen by NASA's Galileo spacecraft are shown in enhanced colour to highlight details of the surface. Note the absence of craters. Major changes between observations acquired by Voyagers 1 and 2 and those from Galileo were seen (photo: NASA).

Plate 4: Images of the nuclei of comet Halley (left: acquired by the HMC instrument on board ESA's Giotto spacecraft) and comet Borrelly (right: acquired by the MICAS instrument on board NASA's Deep Space 1 spacecraft). The nuclei are shown to scale. Borrelly is clearly about half the size of Halley. The arrows indicate the direction to the Sun for the two images. Note that although Borrelly is quite active for a periodic comet, the activity is so weak that it is not easy to see. The image must be processed specifically to show jet structures. The dust emissions from Halley, on the other hand, are as bright as the nucleus itself near their sources.

Plate 5: This picture of comet P/Wirtanen (the target comet of the Rosetta mission) was obtained shortly after its recovery in 1996 when the comet was 2.526 AU from the Sun. Celestial north is up. The field of view in this small frame is 70 000 km^2 and the resolution is around 1500 km. By comparison with comets Halley and Borrelly, Wirtanen is a weakly active comet.

Plate 6: An image of comet C1999 T1 taken by Javier Licandro which shows the typical ground-based view of its coma and tail (courtesy of Luisa M. Lara).

Plate 7: This false-colour image shows the Mars Pathfinder lander seen with the cameras of the Sojourner rover. The imager for Mars Pathfinder can be seen sitting on its mast above the lander. The lander is surrounded by its deflated airbags on which it bounced onto the surface.

Plate 8: This picture shows the engineering model of the Mars Polar Lander (MPL) spacecraft. This model was used to train scientists and engineers on how to use the lander and its robotic arm. It was sited in a sandbox at the University of California at Los Angeles (UCLA).

Plate 9: One of the next major missions to a comet will be Rosetta, due for launch in January 2003. The spacecraft carries a complement of 11 experiments plus a lander with a surface science package. The main imaging system is called OSIRIS and is shown here at the Max-Planck-Institut fuer Aeronomie on a mock-up of the spacecraft just before delivery.

Plate 10: In this enhanced colour picture of Yogi and the rover, Sojourner, at the Pathfinder landing site, notice how the right side of Yogi looks less red than the rest of the rock. This effect was produced by the illumination conditions, not by a physical or chemical difference between the faces (photo: NASA).

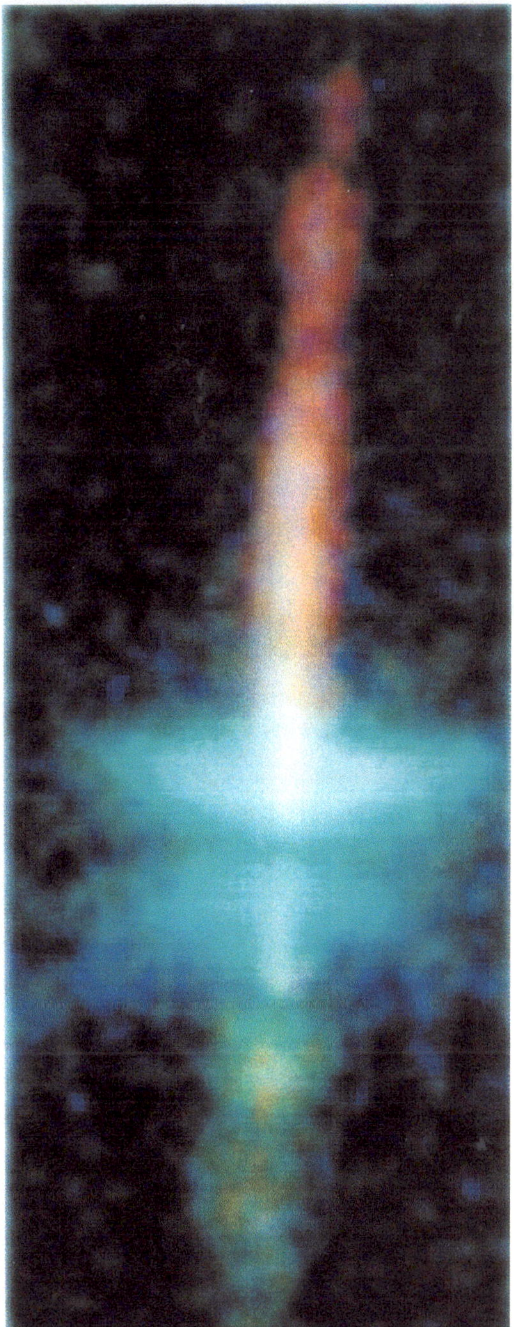

1

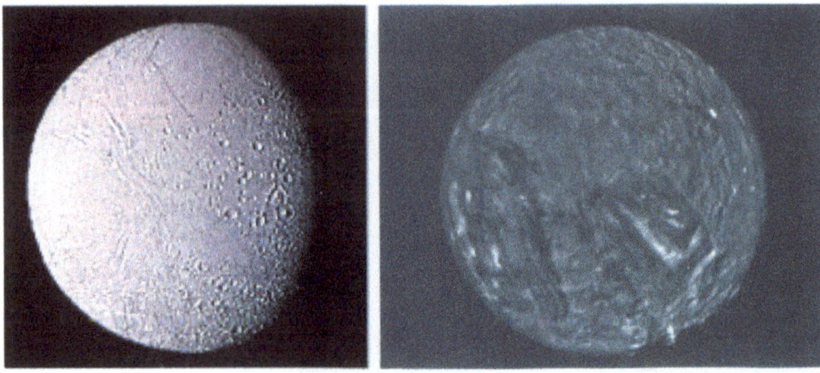

2

3

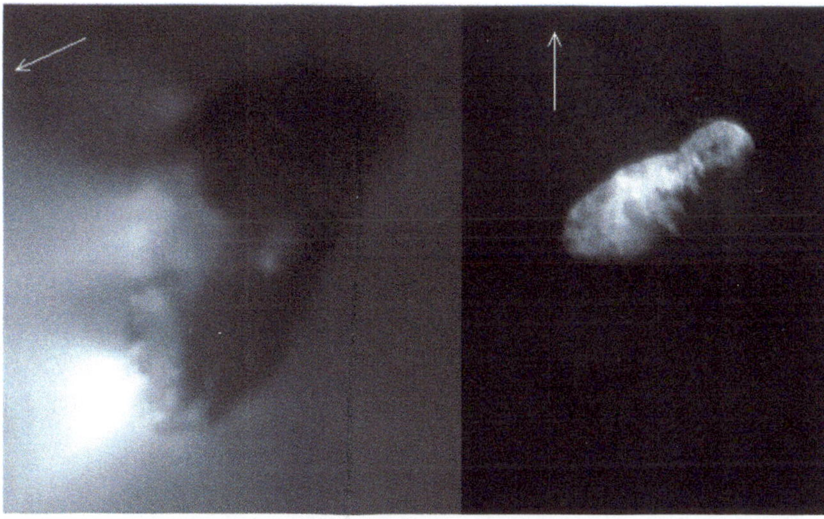

4

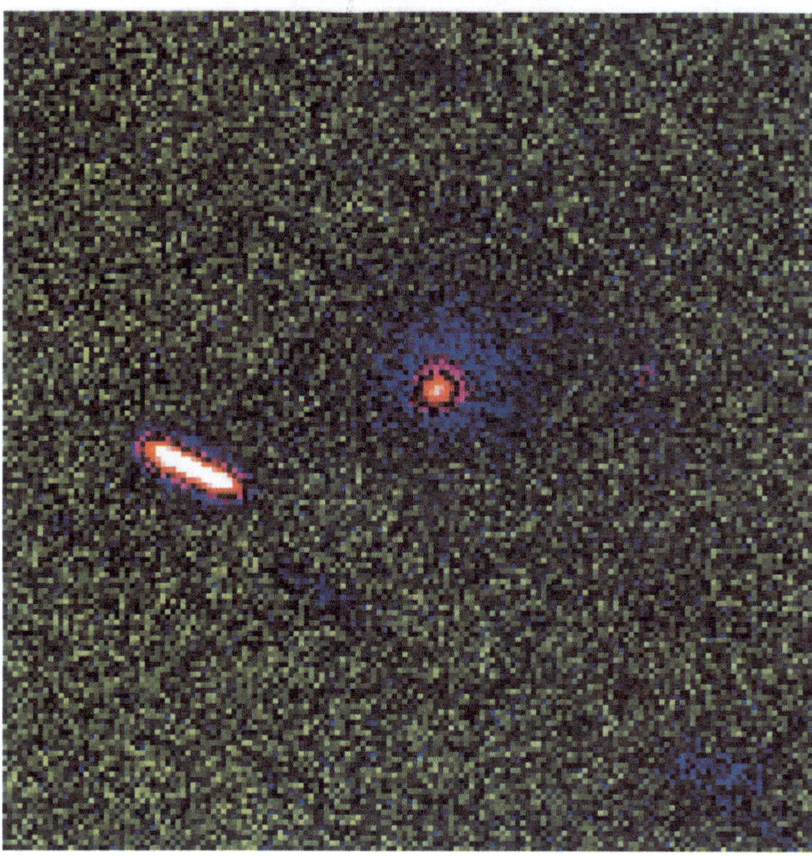

5

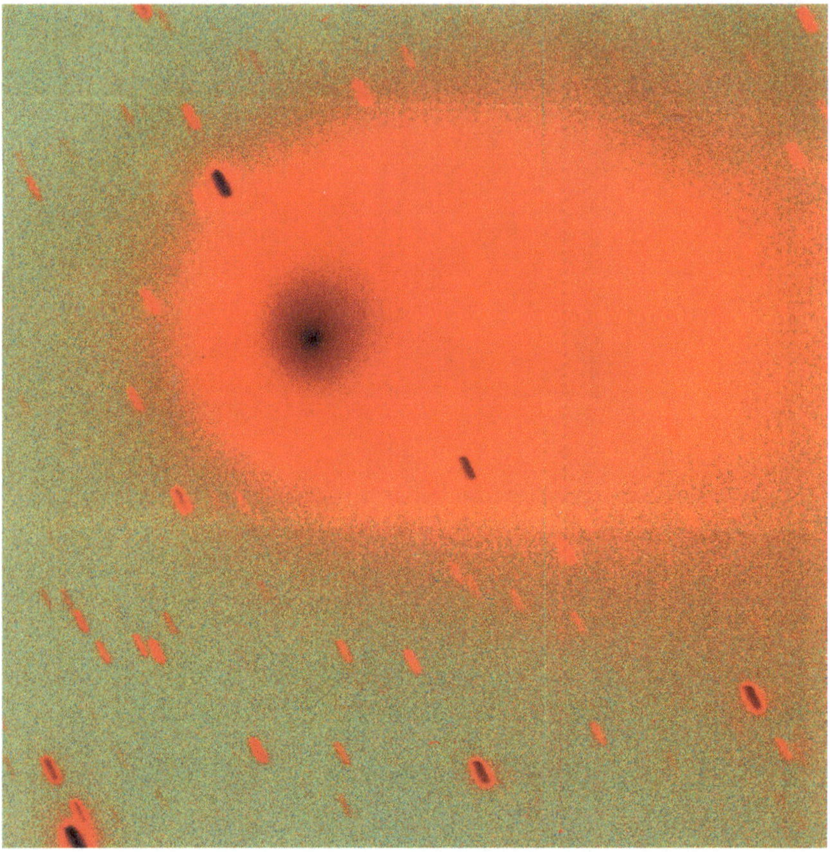

6

7

8

9

10

Introduction – Solar and Extra-Solar Planetary Systems

Iwan P. Williams

Over the last decade, there have been many exciting advances in all fields relating to our understanding of planetary systems. There has been a significant increase in our understanding of the general process of star formation, leading to an expectation that matter will be captured in a flattened envelope or nebula surrounding the young Sun. Theoretical models had predicted this for some time, but in the last decade, firm observational evidence of this has become fairly commonplace, with β Pictoris in particular displaying all the characteristics that were expected in systems where planets formed. The discovery of extra-solar planets has also confirmed the view that planetary formation is a normal phenomenon so that our system is no longer regarded as a 'one off' or special. Within the Solar System itself, both space exploration and improved facilities for ground-based observations have increased our knowedge of our own system dramatically. Pluto is now the only planet not to have been visited by a spacecraft, and spacecraft images also exist of asteroids and comets. In addition, it is now known that all the major planets have extensive satellite systems as well as complex ring structures. Finally, two new classes of objects have been discovered: the Centaurs orbiting between the major planets; and the Edgeworth–Kuiper objects beyond Neptune.

This book is based on the lectures given at a Pre-Doctoral Summer School held in Ballyvaughn, County Clare, Ireland during 7 – 18 September 1998, supported by the European Astrophysical Doctoral Network (EADN). The aim of the School was to give an authoritative account of these new developments so that a thorough general background in the state of our knowledge would be obtained by all participants. The scientific contents of the School can be divided into a number of broad fields. The areas are: **Formation of Planetary Systems; Planets and Satellites; and Small Bodies and Dust.** The chapter on Dynamics by Murray spans all of these areas.

Formation of Planets: Our current state of knowledge concerning the process of star formation was described and applied, in particular to Solar-type stars. This led to a discussion of the disk-like structures that are found around many young stars, particular attention being paid to those that have the rough dimensions of the Solar System. The general features of our own system were described, leading to a discussion of the similarities and differences between our system and systems containing planets around other stars. Building on this platform of knowledge, current theories for the origin of the Solar System were described and placed within their historical context. The chapters by T. R. Ray, R. P. Nelson, and T. Encrenaz describe the current state of knowledge in this area.

Planets and Satellites: The Solar System falls naturally into two groupings; objects that are essentially large, so that gravity is the dominant force and they are essentially spherical in shape; and smaller bodies. There are a few exceptions. Ceres, the largest asteroid by this definition should be with the large bodies rather than the asteroids, for example, but this definition forms a rough guide to the ordering of the chapters. Chapters by T. Encrenaz and N. Thomas cover this area.

Small Bodies and Dust: The small bodies are by far the most numerous, though they contribute only a small fraction to the total mass of the system. Each family of small bodies, Asteroids, Comets, Meteors and Trans-Neptunian Objects all have their own attractions. Asteroids present a danger to human civilization. Comets are in the news following Hale - Bopp. Meteors have become topical due to the expected Leonid display, and Trans–Neptunian Objects are new discoveries. The chapters by I. P. Williams, I. P. Williams and A. Fitzsimmons, and I. Mann deal with this area.

The Solar System: An Overview

Iwan P. Williams

Astronomy Unit, Queen Mary, University of London,
Mile End Road, London E1 4NS, UK

Abstract. The Solar System consists of a significant number of bodies, many of which are significantly different, both physically and chemically from both near and distant neighbours. The intention here is to give a brief overview, both as a single system and of the diversity of individual bodies.

1 Introduction

To understand fully the origin, evolution and structure of the Solar System, and indeed planetary systems around other stars, a knowledge of many different field is required. The system is a complex dynamical system within which many of the bodies evolve independently of each other, obeying the rules of chemistry and physics as they pertain to the particular environment of that body. The number of extra-solar planets discovered is still small, though increasing, but the difficulty of detection and observation of such planets means that in order to characterise a planetary system, we still have to rely very heavily on our own system. The five planets, Mercury, Venus, Mars, Jupiter and Saturn, together with the Sun and the Moon, have been recognised since antiquity as being different from the fixed stars because of their motion across the heavenly sphere. In all cases except the Sun, this recognition was correct, but it is now generally recognised that the Sun is in fact but one of the many stars that exist. In early days, there was also a difficulty in regarding the Earth as being essentially the same as some of the other planets. This difficulty persisted until the Copernican view (Copernicus 1543) of the Solar System became generally accepted.

With the invention of the telescope, both the number of members of the Solar System and the number of different types of bodies increased with the passage of time, starting with the discovery of the Galilean satellites of Jupiter by Galileo Galilei in January 1610. Later in the same year, curious appendages which appeared and disappeared were found about Saturn. The explanation that these were caused by a thin flat ring had to wait for about forty years until this was proposed by Huygens(1659). Nevertheless, by this date, the Solar System was recognised as consisting of a star, various planets, moons and a ring system. Comets had been known, like planets, since antiquity, but were generally regarded as omens or messengers of doom. However, in 1695, Edmund Halley suggested that they were also members of the Solar System that periodically return to the locality of the Sun, a suggestion that was proved true on Christmas Day 1758 by a farmer living near Dresden, when he recovered what we now know as comet 1P/Halley.

In 1781, the known Solar System roughly doubled in size with the discovery by Herschel of a new planet orbiting the Sun at 19 AU, compared to Saturn at 10 AU. However, this was important, not only because of the increased size, but also because a new planet had been discovered for the first time. In fact, Uranus had been observed, but not recognised as a planet by a number of other observers prior to Herschel, the list of potential discoverers includes Flamstead, Bradley, Le Monnier and Mayer. Twenty years later on the first day of the nineteenth century, Guiseppe Piazzi discovered what was first thought to be an other planet, Ceres, but was recognised as being too small to be a proper planet and became the first of a population of minor planets or asteroids that are located mostly between the orbits of Mars and Jupiter. For nearly two centuries, the Solar System settled down to be a system consisting of a star, various planets and their moons, comets and asteroids, until in 1992 Jewitt and Luu (1992) discovered an object called 1992 QB, orbiting beyond Neptune.

2 The Solar System

The Solar System consists of a star, various types of planets and assorted other minor bodies. To understand the system however we need to do more that to look at each member as an individual body, the totality is greater than the sum of its parts. On the other hand, there is a difficulty about generalising from the characteristics of the Solar System, namely that it is the only planetary system for which we have sufficient knowledge at present to be able to describe its main characteristics. Nearly twenty extra-Solar planets have been discovered at this time, but most represent the discovery of a single planet orbiting a star. It is thus not possible to describe properties of planetary systems from such a set. With only one data set, it is difficult to know what are characteristics of planetary systems and what are the result of chance. One obvious characteristic is that all the bodies, with the exception of long period comets tend to orbit close to the same plane. Since this planarity is also observed in circumstellar disks around young stars, there is a general belief that these are related and in consequence that most planetary systems would be co-planar. Such co-planarity has formed the backbone upon which most theories for the origin of the planetary system have been built. Though speculating on the origin of the Earth and the cosmos must be as old as the human race, theories only start to have relevance to our current thinking after the Copernican revolution. After all, forming the cosmos centred on the Earth is rather different from a formation scenario where the Earth is but one tiny component. Even within that much shorter time span, virtually every conceivable theory has been proposed. This is not surprising for there are not that many fundamentally different notions that can be proposed.

There are only three places material for the planets can have originated, the Sun, another passing body or the interstellar medium (including the cloud out of which the Sun formed). There are only two things that can be done to form planets, break up the cloud through some instability into planetary sized lumps, or build up planets through accretion. This gives only six possible variations, though sub-variants are much more numerous. Indeed a multitude of variants on all six themes have been proposed. Both star and planet formation are discussed later. I will now simple describe the main classes of objects in the system and their main characteristics. More details will be found in the following chapters.

3 The Sun

Most of the mass of the Solar System currently resides within the Sun, its mass being 1.999 x 10^{30} kg. The Sun is a normal main-sequence star of spectral type GIV with a luminosity of 3.86 x 10^{26} Js-1, and an effective temperature of 5785K. In terms of composition, the Sun is about 70% hydrogen with all other elements (mostly carbon, nitrogen and oxygen) other than helium forming only about 1.5%. At present the Solar energy output comes from nuclear burning, the conversion of hydrogen to helium. This was not always the case and prior to the ignition of hydrogen, the energy came from gravitation as the star contracted and the central temperature increased to nuclear burning temperatures. The Sun spent of the order of 10^7 years in its pre-main sequence phase and, near the end of this phase, it is likely that significant mass outflow from the Sun occurred through what is generally called the T-Tauri wind. The axis of rotation of the Sun is almost orthogonal to the plane of ecliptic, that is, the spin axis is nearly coincident with the angular momentum vector of the planets. Though 99.9% of the total mass of the system is within the Sun, the vast majority of the angular momentum resides within the planets. Over the last few decades, considerable advances have been made in the observations of young stars and proto-stars, starting from a very young stage within dense molecular clouds. An overview of the current state of our knowledge is given later in the book.

4 The Planets

The name "planets"comes from "wandering stars"and for much of history they were regarded as just that, though in the immediate post-Copernican era the only advance in understanding was in terms of why they wandered rather than what they were. They wandered because of their Keplerian motion, coupled with that of the Earth, about the Sun. Our planetary system consists of nine planets orbiting the sun on near co-planar and near circular orbits, the two planets that deviate the most from this definition being Mercury with a 7° inclination and Pluto with an inclination of 17°, while their eccentricities are respectively 0.2 and 0.25. In terms of distances from the Sun, they are spread from 0.4 AU to 40 AU, with corresponding orbital periods of 3 months and 250 years. Put another way, for every completed orbit of Pluto, Mercury has completed 1000 orbits. The spacing of the planets is also very uneven, with five planets within 5AU of the Sun and the remaining four roughly evenly spaced at 10, 20, 30 and 40 AU. In fact there are also many differences between individual planets, differences that make it impossible to discuss them all as a single entity. We shall thus discuss them in groups, starting with the most massive and working downwards. For convenience, we shall in this overview also group the satellites with their parent planet, though this distinction is not always followed within the individual chapters.

4.1 Jupiter and Saturn

These are truly the giants amongst the planets, having masses respectively of 19 and 5.7×10^{26} kg. (or 318 and 95 $\times$ the mass of Earth). They are located at 5.2 and 9.5 AU from the Sun. Both are primarily composed of hydrogen and helium, in fact a composition very similar to that of the Sun. Both planets have substantial and active atmospheres with significant meteorology and any solid core that they posses is fairly tiny compared to the general mass of the planet. Both have very significant satellite systems and in some ways resemble mini-solar systems. These satellites range from substantial bodies of the same general dimensions as our Moon, to small lumps no more than a few tens of kilometres across. Both also have an extensive ring system, though they differ significantly from each other with the Jovian one being far more tenuous and having much less detailed dynamical structures.

4.2 Uranus and Neptune

Uranus was the first "new"planet to be discovered in the Solar System, by Herschel in 1781. Neptune was discovered in 1846, following searches based on predictions by Adams (1847) and LeVerrier (1849), made in order to explain the anomalous residuals in the motion of Uranus. These two planets turned out to be very similar to each other, having masses of 8.7 and 10×10^{24} kg (15 and 17 $\times$ the mass of Earth). Thus, whilst being much bigger than the Earth, they are also considerably smaller than Jupiter and Saturn. They are often referred to as the "Icy Giants"because their composition is dominated by ices based on molecules of carbon nitrogen and oxygen, together with some hydrogen and helium, mostly in their atmospheres. They orbit the Sun at mean distances of 19 and 30 AU. Uranus has its rotation axis lying virtually in the plane of the ecliptic, rather than nearly orthogonal like the other planets. The satellite system of Uranus lies close to this rotation plane. Neptune also has a satellite system, dominated by Triton, which is the only large satellite in the Solar System that moves on a retrograde orbit.

4.3 Earth and Venus

These two planets have often rightly been called the twin planets. Their masses are respectively 6.0 and 4.9 $\times$ 10^{24} kg orbiting the Sun at 1 and 0.7 AU. The composition of both is dominated by iron-silicates (metal and rock) and both have substantial atmospheres. The Earth has a massive satellite (the Moon) while Venus has none.

4.4 Mercury, Mars and Pluto

These three planets have little in common with each other, other than the fact that they are all dissimilar to other planets. Mercury is the planet nearest the Sun at only 0.4 AU, while from February 1999 Pluto is the planet furthest from the Sun at a mean distance of 39.5 AU. Mercury and Pluto have highish inclinations and higher orbital eccentricities than the other planets. Because of the high eccentricity, the perihelion distance of Pluto's orbit is less than the mean distance of Neptune, so that for short periods, Pluto is not the furthest planet. Close encounters between Neptune and Pluto are not however possible because they are in mean-motion resonance, Pluto orbiting the Sun twice for every three completed orbits of Neptune. Mercury and Pluto are also by far the two smallest planets, more comparable in mass to the larger satellites. Mars is the fourth planet from the Sun and nearest to the Asteroid belt. It is more massive that either Mercury or Pluto but is still significantly smaller than Earth or Venus. Mars has two small satellites, more similar in mass to asteroids than to most of the satellites previously mentioned. Pluto on the other hand has a very large satellite, Charon and in terms of mass ratio, the Pluto-Charon system is more akin to the Earth-Moon system than any other satellite system.

5 The Minor Bodies

Within the Solar System there are also considerable numbers of smaller bodies than those already described. Indeed, if we include the smallest bodies, the interplanetary dust grains, then the number is so large that in normal discussion one would almost regard it as infinity. Hence, in discussion of the dust population, it is more usual to consider population characteristics rather than the properties of individual grains.

5.1 Comets

Like planets, the existence of comets had been recognised since time immemorial. However debates as to their nature and whether they were true astronomical objects or simply atmospheric phenomena, went on until fairly recent times and indeed, it is only with the fly-by of comet Halley by the spacecraft Giotto that the existence of a solid nucleus was finally confirmed. For most of history, comets have been regarded as prophets of doom or messengers foretelling of disasters. They have often been recorded and depicted in fine art (see for example Olson and Pasachoff, 1998). Halley was the first to demonstrate that comets could move on periodic orbits and of course predicted the return of the comet that now bears his name. This established their membership of the Solar System. By the middle of the nineteenth century, a connection between comets and meteor streams was generally accepted, following work by Kirkwood (1861) and Schiaparelli (1867). Confirmation of this came with the observed break up, and subsequent large meteor storms, of comet Biela (see Lovell, 1954 for an account). This association led to a widespread acceptance of the "flying sand bank"model for comets strongly defended for example by Lyttleton (1953) and it took decades before the "icy conglomerate"model, initially proposed by Whipple (1950) gained general acceptance. It is now believed that the cometary nucleus is a few kilometres in radius, composed of ices with some embedded grains, the ice being more akin to snow than blocks of ice. The well-known cometary tail is formed by sublimation of the ices, mostly occurring with astronomical units of the Sun when water ice sublimes. Earlier activity can occur through sublimation of other ices. Comets in the inner Solar System have a short life-time compared to the age of the planets and spend most of their lives far away from the Sun in the so called "Oort cloud"(Oort, 1950). Recent discoveries (see later) suggest that some comets also originate from the Edgeworth-Kuiper Belt.

5.2 Asteroids

The study of asteroids started on the first day of the 19th century when Piazzi, observing from Palermo, discovered a new object in the Solar System. Though the specific discovery by Piazzi was serendipitous, a campaign had been initiated by von Zach, whereby the zodiac was divided into 24 zones and a different astronomer was assigned to search each zone for a suspected planet. Piazzi was not one of these 24 astronomers. The expectation of finding a planet was based on the belief that the Titius-Bode law that predicted planetary distances was correct (see Nieto 1973 for a discussion of this law). We must remember that its correctness had only recently in 1781 been demonstrated through the discovery of the planet Uranus by Herschel. The only unexpected element in the discovery by Piazzi was that the new planet, named Ceres, was rather faint, much fainter than expected, indicating that the body was somewhat smaller than the "predicted"planet. Within the next four years, three further similar objects were discovered, Pallas, and Vesta, by Olbers and Juno by Harding. No further objects of this class was discovered for 40 years and it was during this period that the group were called, minor planets - for they clearly were not proper planets. It was also during this time interval that the hypothesis was first put forward that these minor planets were remnants of a proper planet that had been broken up by some mechanism.

By now, over 10 000 asteroids have well determined orbits, most of them lying in the so-called main belt between Mars and Jupiter. Several hundred are also known to be captured near the Lagrangian equilibrium points of Jupiter, thus librating in a 1:1 resonance. These are known as Trojan asteroids. Another significant sub-group, not least because they have the potential to collide with the Earth, is the Near Earth Asteroids. As the name suggests, their orbit passes close to that of the Earth.

Asteroids cover a fair size spread from Ceres at near 1000 km down to a host of mostly unknown bodies at under 1 km. Unlike comets, in general ice is not a major constituent.

5.3 Edgeworth-Kuiper Belt Objects

As already mentioned, Jewitt and Luu (1992) discovered an object a few hundred kilometres in diameters orbiting beyond Pluto. This was to be the first of many and at present, well over 100 are known. Two recent reviews are Jewitt (1999) and Williams (1999). In a cosmogonic sense these bodies are important since both Edgeworth (1943) and Kuiper (1951) had clearly stated the obvious fact that there is no reason for the Solar System to terminate at Neptune.

5.4 Meteors and Dust

Meteors are strictly defined as a streak of light in the upper atmosphere. However common usage has extended this term also to the small body responsible for the streak of light. Meteors have also been observed and recorded since pre-history, though like many other phenomena, their true nature only became apparent in recent times. Visible meteors are caused by small (millimetre to a few centimetre) dust grains burning in the upper atmosphere. These are generally regarded as the upper size range of dust grains ejected from comets, the smaller end being simply called interplanetary dust. If ejection from the comet was fairly recent, then these dust grains show a coherence of orbits, thus forming a meteor stream that produces a meteor shower on collision with the Earth.

6 Other Planetary System

Searching for planets around other stars has been an ongoing project for a considerable time (see for example Williams 1988). However successful detections are a fairly recent phenomena. At the present time about 50 such systems are known. In principle, they produce valuable constraints on theories for the formation and evolution of planetary systems. Many of the discovered systems vary markedly from the Solar System, in particular in having Jovian type planets orbiting close to the parent star. However, detection is still a difficult task so that there is a great observational bias towards finding big planets about small stars and planets with very short orbital periods.

7 Conclusions

Solar Systems are varied and fascinating, and within our own system, the composition is also varied and fascinating. The following chapters will deal in more detail with varies aspects touched upon in this short overview.

8 References

1. Adams, J.C., 1847, On The Perturbations of Uranus, Mem.R.astr.Soc., 16, 427-460
2. Copernicus, N., 1543, De revolutionibus Orbium Caelestium Libri Sex, Nuremberg
3. Edgeworth, K.E., 1943, The evolution of our planetary system, J. Brit. Astron. Assoc., 53, 181-188
4. Huygens, C., 1659, Systema Saturnum, in Oeuvres Completes de Christiaan Huygens vol XXI, The Hague
5. Jewitt, D.C., 1999, Kuiper Belt Objects, Annu. Rev. Earth Planet Sci., 27, 287-312
6. Jewitt, D. C. and Luu, J. X., 1992, IAU Circular No 5611
7. Kirkwood, D., 1861, Danville Quarterly Review, December 1861
8. Kuiper, G.P., 1951, On the Origin of the Solar System, in Astrophysics: a Topical Symposium, ed Hynek, J.A., 357-424, McGraw-Hill, New York
9. Le Verrier, U.J.J., 1849, Nouvelles recherchers sur les mouvementes des planetes, Comp. Rend. Acad. Sci. Paris, 29, 1-3
10. Lovell, A.C.B., 1954, Meteor Astronomy, Oxford University Press, Oxford
11. Lyttleton, R.A., 1953, The Comets and their Origin, Cambridge University Press, London
12. Nieto, M., 1973, Titius-Bode Law of Planetary Distances, Pergamon Press, London
13. Olson, R.J.M and Pasachoff, J.M., 1998, Fire in the Sky, Cambridge University Press, Cambridge.
14. Oort, P., 1950, The structure of the cloud of comets surrounding the Solar System and a hypothesis concerning its origin, Bull. Astron. Inst. Netherlands, 11, 91-110
15. Schiaparelli, G.V., 1867, Sur la relation qui existe entre les cometes et les etoiles filantes, Astronomische Nachrichten, 68, 331-332
16. Whipple, F.L., 1950, A comet model I, The acceleration of comet Encke, Ap.J., 51, 711-718

17. Williams, I.P., 1988, The evidence for other Planetary Systems, in The Physics of the Planets, Ed. Runcorn, S. K., J.Willey & Son, Chichester, 401-410.
18. Williams, I.P., 1999, The Solar System Beyond Neptune, in Dynamics of Small Bodies in the Solar System, Eds. Steves, B.A. and Roy, A.E., Kluwer.

Setting the Scene: A Star Formation Perspective

Tom P. Ray

School of Cosmic Physics, Dublin Institute for Advanced Studies,
5 Merrion Square, Dublin 2, Ireland

Abstract. We had the sky up there, all speckled with stars, and we used to lay on our backs and look up at them, and discuss about whether they was made or only just happened.
The Adventures of Huckleberry Finn, M. Twain

1 Introduction

This school is primarily about understanding the Solar System as we see it today but obviously an appreciation of its history and origins is fundamental to that understanding. I wish to set the scene so-to-speak by considering what observations of young stars have to tell us about conditions in the primitive Solar Nebula and the environment of the Sun soon after its birth. There are many important questions about the Solar System that star formation studies can help us address. For example, is it more likely that the Sun formed in a dense cluster of stars, that dispersed 5 billion years ago, or as a single isolated star? What was the probable mass of the primitive Solar Nebula and what role, if any, had an outflow from it in establishing the angular momentum distribution of the Solar System? Obviously observations of the circumstellar environments of young stars can help us address an even wider question: how likely is planetary formation in the first place?

The topic of star formation will, no doubt, be new to many of you, so I will try to be as comprehensive (although this is not necessarily the same thing as comprehensible!) as possible. That said, in a short introduction to the subject, such as this, one can only cover what, I hope, are the salient points. Thus I shall refer the reader where appropriate to more in-depth reviews where he or she can find further details if desired. Finally I will concentrate on the formation of low mass stars such as our sun, the birth of high mass stars (see, for example, [34] is a separate topic in itself!

2 Stellar Nurseries: Molecular Clouds

Since stars begin their lives in the dark dusty environments of molecular clouds, I should start by saying something about them. Although they come in a range of sizes, most of the gas and dust is in the form of giant molecular clouds (GMCs) with masses in the range $10^{6-7}M_{\odot}$, sizes $\approx 20 - 100$pc, and average molecular hydrogen densities $< n_{H_2} > \approx 10^2 cm^{-3}$ (Blitz 1993). There is a sharp cut-off in the number of GMCs with M $\gtrsim 5x10^6 M_{\odot}$([105]) although what limits their mass is not clear. Perhaps it is tidal effects from the Milky Way or the disruptive forces that are released when such clouds form massive stars ([105]).

In any event we know from CO surveys (e.g.[39]) that most of the mass in molecular clouds is confined to the spiral arms of our Galaxy. Since the spiral arms are moving with respect to the Galaxy, and there is only *atomic gas* between the arms, two firm conclusions are possible: first molecular clouds last at most one arm crossing time, i.e. about 10^7 years and second that they arise from the compression of atomic gas. The latter idea is reinforced by the discovery of HI "halos"around GMCs (e.g. [65]).

In addition to the GMCs, small molecular clouds (or SMCs) with M $\lesssim 10^2 M_{\odot}$ are known but, as already stated, the *total* mass of the ISM in this form is quite small (see, for example, [105]). Moreover, while some low mass stars do form in SMCs (although probably not very efficiently), they do not contain any massive stars: these are exclusively found in GMCs.

A cursory study of the individual structures seen in molecular clouds show they are far from uniform as they contain filaments, rings, clumps, and cores. It is from clumps that massive star clusters, such as the Trapezium Cluster in Orion (Fig. 1) are thought to form. Although individual clumps may be gravitationally bound, the resultant star clusters are not since a large fraction of the gas is lost in forming the cluster[1]

Molecular clumps in turn are found to contain finer sub-condensations known as "cores". It is thought that these cores ([10]), or at least some of them, are the fundamental building blocks for individual stars or multiple systems such as binaries. Cores have much higher densities than clumps ($\approx 10^5 cm^{-3}$ v. $3 \times 10^2 cm^{-3}$) and are hotter. Moreover while clumps are found to have a power law mass distribution with:

$$dN/dlnM \propto M^{-0.6 \rightarrow -0.8} \tag{1}$$

where N is the number of clumps with mass M (e.g. [52]), starless cores have recently been found to have a much steeper mass spectrum with a corresponding index of -1.1 ([102]). This is important since it implies the mass spectrum of cores is close to the initial (stellar or Salpeter) mass function (IMF) which has a slope of -1.35. In other words molecular cores are converted efficiently into stars, or at least the efficiency at which they are converted does not depend strongly on mass. If this is the case then the IMF is determined

[1] Cluster formation is thought ([107]) to be the dominant mode of star formation in the Galaxy, and so statistically it is more likely that the Sun formed in such a cluster. As we shall see, however, whether a star begins its life at the edges or the centre of such a cluster probably has a profound influence on whether or not it forms planets (see §Disks Around Young Stars).

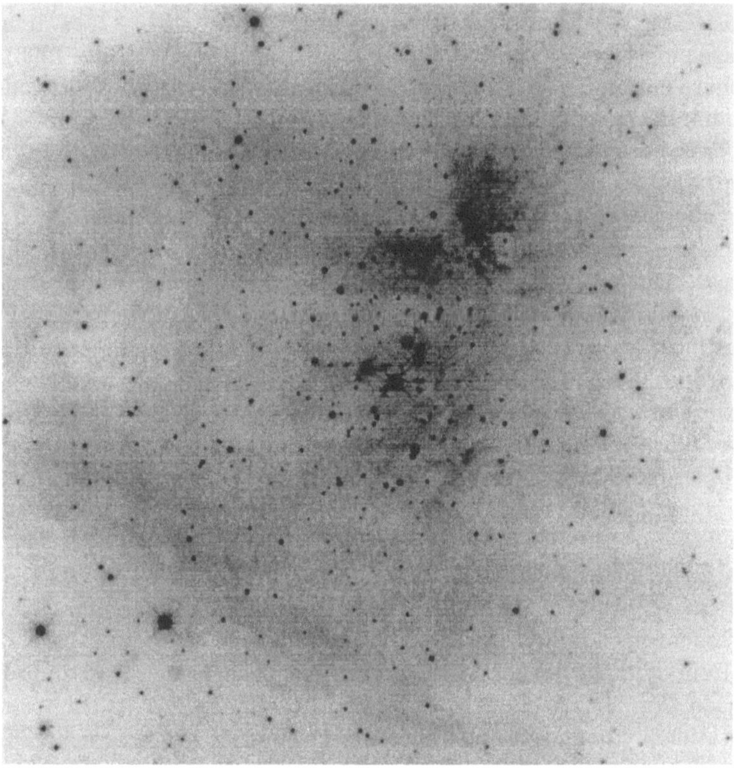

Fig. 1. The Trapezium Cluster in K as imaged using the infrared camera MAGIC on the Calar Alto 3.5m Telescope from [63]. North is up in this 5′x5′ image and the bright Trapezium stars can easily be seen towards the centre of the frame. The density at the centre of this cluster is an amazing 5.10^4 stars per cubic parsec

primarily by the cloud fragmentation process, it fixes the relative abundance of low and high mass stars rather than, for example, the outflow phenomenon which we will discuss later.

3 Classifying Young Stars

Before describing how a young star forms from a molecular core and evolves onto the main sequence, it is worth saying a little about the different types of young low mass stars observed, and to see if we can place them in some form of evolutionary sequence. The first young stars to be observed in detail where the so-called objects of [45], in the Taurus Auriga Cloud. These stars, which were named after their prototype T Tauri are now known as classical T Tauri stars (cTTSs). Although the spectroscopic characteristics of cTTSs were described by [45], it was [1] who first recognised their pre-main sequence nature. The spectrum of a typical cTTS (see, for example, the excellent review of T Tauri stars by [11]) shows not only a continuum with a number of absorption lines characteristic of its spectral class (G to M) but in addition strong emission lines, for example of the Balmer Series, and various permitted and forbidden lines (e.g. the H and K lines of CaII, [SII], [OI], etc.). An interesting feature of the absorption lines is that they tend to be "filled-in" to varying degrees compared to the corresponding main sequence star with the same rotational velocity. This effect is known as veiling (see, for example, [36]) and in some extreme cases the absorption lines are almost entirely obliterated.

The origin of the various features seen in the optical spectra of cTTSs is quite complex. The forbidden lines, for example, can, in the main, be attributed to one or more outflows (see §Outflows) and often both a low and high velocity component is present. The high velocity component is due to a highly collimated jet (see §Outflows) while the low velocity component probably arises from a poorly collimated disk wind ([54]). In contrast, the Balmer lines may contain signatures of both outflows and accretion ([17]) and finally the most likely cause of veiling of the photospheric lines in cTTSs is an additional hot continuum provided by accretion ([38]).

Along with the classical T Tauri stars, a second group of optically visible young stars, the so-called weak-line T Tauri stars (wTTSs) have been identified, in the first instance by their X-ray emission ([11]). As their name implies, they lack the strong emission lines seen in the cTTSs but they occupy a similar position in the HR diagram and are of comparable age (i.e. 10^{5-6}yrs). Unlike the cTTSs, there is little or no evidence for circumstellar material in the vicinity of wTTSs or that they have strong outflows. Thus it would seem that the presence of circumstellar material is the trigger for much of the activity observed in cTTSs. We do not know, with any degree of certainty, whether cTTSs evolve into wTTSs or whether the two groups have evolved separately from the early phases I am about to describe. Both scenarios are possible although the latter seems more probable (see [9]). In any event, both cTTSs and wTTSs are low mass young stars and, given that our Sun was surrounded by a proto-planetary disk, it must have passed through the cTTS phase.

With the launch of the Infrared Astronomical Satellite (IRAS) in January 1983, and improvements in the capabilities of millimetre astronomy, it was soon realised that molecular clouds like Orion and Taurus Auriga contain numerous highly embedded stars in addition to the optically visible ones. It turned out to be useful to classify young stellar objects (YSOs) according to their spectral energy distribution at infrared wavelengths i.e. their flux over the 1 to 100μm range. Plotting $\log\lambda F_\lambda$ against $\log\lambda$, those sources with a slope greater than 0 are referred to as Class I sources (see Fig. 2). Here F_λ is the flux at wavelength λ. Those with a slope between -1.5 and 0, are known as Class II objects, and those with a more negative slope as Class III. Since the stars themselves have surface temperatures of several thousand degrees, and radiate essentially as blackbodies, Class I and II sources have an infrared "excess"(see Fig. 2). The sequence from Class I to Class III almost certainly represents an evolutionary sequence in which a young star gradually emerges from the gas and dust that surrounds it at its birth.

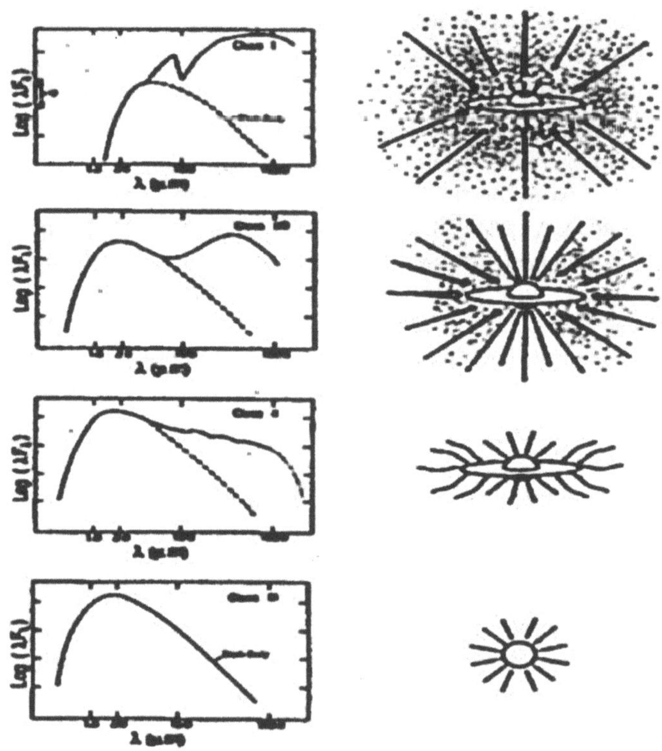

Fig. 2. A schematic from Wilking, [104] showing how the various spectral energy distributions of YSOs may arise. It is presumed that Class I YSOs eventually evolve to the Class III stage before joining the Zero Age Main Sequence (ZAMS). Class O objects, which were discovered later, have a peak in their spectral energy distribution, which are essentially blackbody, around 100 microns (see, [2]).

Soon a "cottage industry"developed in modelling IRAS spectral energy distributions. It was found that various combinations of pre-main sequence stars with spherical or quasi-spherical dust distributions could not account for the full range of observed spectral energy distributions and that the presence of an optically thick, but geometrically thin, dusty disk had to be invoked (e.g. [47]). Although this was indirect evidence for the presence of disks, their direct detection would have to wait a few years longer (see §Disks Around Young Stars).

Most cTTSs were found to have Class II IRAS spectral energy distributions, best modelled as a combination of a star+disk. Here the accretion rate has slowed down (to perhaps $10^{-7}M_{\odot}yr^{-1}$) and the star has driven much of its placental material away. In contrast the Class I sources are surrounded by a disk but in addition a tenuous envelope (e.g. [59]). Class I sources are typically younger (1-2x10^{5}yrs) in comparison to Class II sources ($\approx 10^{6}$yrs) and derive a large fraction of their luminosity from accretion. The earliest phases, what may be regarded as the true "protostar"phase, is represented by Class O objects ([2]). Class O objects have large submillimeter to bolometric luminosities suggesting that their envelope mass *exceeds* their central mass and that the young star is still in the process of accumulating much of its final mass. Class O objects can be distinguished from the pre-stellar cores (mentioned in the previous section) by the presence of a central source. Such a source may be indicated by radio continuum emission or an outflow for example.

4 How Does a Young Star Evolve onto the Main Sequence?

What are we to make of this zoo of low mass YSOs (remember we have said virtually nothing about their higher mass counterparts)? Let us start by imagining an idealised spherical core that is about to collapse inside a molecular cloud. At typical densities the core is transparent to thermal infrared/millimetre radiation and is efficiently cooled by molecular lines. It can therefore be characterised by a single temperature which, as I have already noted, is around 20 K. According to conventional theory (e.g. [95], [96]), the "inside-out"collapse of a self-gravitating isothermal sphere occurs at a rate:

$$\dot{M}_{infall} \approx c_{eff}^{3}/G \tag{2}$$

where c_{eff} is the effective sound speed in the sphere and G is the gravitational constant.[2] As matter builds up in the centre of the sphere it gradually becomes optically thick and starts to heat up. A quasi-equilibrium state is reached, in which the maximum mass of the "first"protostellar core is about $0.04M_{\odot}$ ([14]). This phase is reached relatively quickly in about 10^{4}yrs. The first protostellar core however continues to gain mass and once it reaches a temperature of about 2,000 K, the dissociation of molecular hydrogen acts as an energy sink leading to a second collapse phase. The core is now much more massive and and it is this stage which we can observationally identify with the Class O phase. When temperatures at the centre are high enough ($T \approx 10^{6}$K) deuterium burning commences ([99]). Since the protostar is fully convective, and the deuterium is burnt relatively quickly, deuterium burning

[2] Note that for such an isothermal sphere the density should decrease with radius like r^{-2}

will only last as long as copious amounts of matter continue to rain on the protostar. As with normal hydrogen burning, deuterium burning is very temperature sensitive, and so there is a line in the HR diagram (known as the stellar birthline) which is analogous to the main sequence.

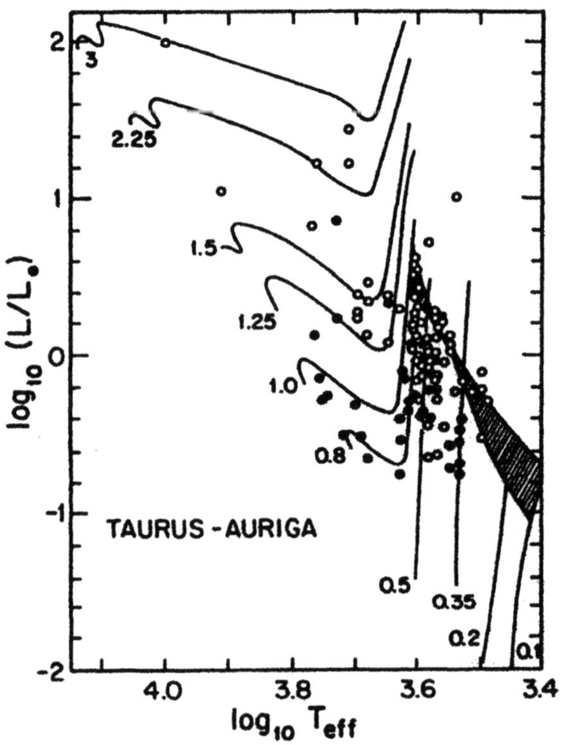

Fig. 3. Theoretical tracks for pre-main sequence stars and the birthline for low mass stars (from [98]). Each track is labelled by mass in solar units. Open and closed circles are observations of classical T Tauri and weak-line T Tauri stars respectively. The hatched region shows the expected small variation in the location of the birthline as a result of changing the mass accretion rate from $2\text{x}10^{-6}$ to $10^{-5} M_\odot$ yr^{-1}.

Once the main accretion phase is finished stars move down from the birth line and, since they are no longer surrounded by as much gas and dust as before, they become optically visible for the first time (Fig. 3). Note that they may still be surrounded by a disk at this stage.

The young star then enters the next phase, slow gravitational contraction taking approximately a Kelvin-Helmholtz timescale, i.e. of order a few million years, towards the main sequence. If its mass is low enough, it will move in the HR diagram almost vertically along a fully convective track (the so-called Hayashi track) with little variation in its surface temperature with decreasing radius. If, on the other hand, its mass is greater than about $0.6M_\odot$, then the star will at some point develop a radiative core and it will also move horizontally along a radiative (or Henyey) track (see Fig. 3).

Of course the simple theory outlined above, while reproducing the gross features of what we see, cannot be the whole story since spherical symmetry is presumed. In reality there are non-spherical phenomena such as disk accretion and outflows associated with star formation which may play an important role. For example, theory suggests (e.g. [48]) that outflows may be necessary to remove angular momentum from accreted matter although at the same time most of the matter is still accreted. To sum up, Class O sources would seem to correspond to the phase when the star is building up a significant fraction of its mass. By the time it becomes optically visible as a T Tauri star, the main accretion phase is finished even though the star may still be surrounded by a disk. Before discussing the properties of these disks, and the direct observational proof of their existence, I should say a little about outflows since they are perhaps the most dramatic signature of stellar birth. At the same time I appreciate that for most of the participants in this school, their primary interest is in "proto-planetary" disks.

5 Outflows

The existence of outflows from young stars (and, in particular, winds) has been known for many years (e.g. [53]) although their true extent and importance has only been realised in the last decade. For example the Balmer lines of many classical T Tauri stars (e.g. [17]) and their intermediate mass brethren, the Herbig Ae/Be stars (e.g. [32]), show a blueshifted absorption component indicating terminal wind velocities of up to several hundred kms^{-1}. On larger angular scales, [93] correctly surmised that the emission line nebulae known as Herbig-Haro (HH) objects were parts of outflows from young stars, their emission arising from radiative shocks.

With the development of large millimetre dishes and detectors in the early eighties, and observations, particularly using the J=1-0 rotational line of CO at 2.3mm, the true sizes, and power, of outflows from young stars became clear for the first time. Analysis of the CO lines showed that in addition to the bulk of the gas at the rest velocity of the cloud, high velocity emission was also present in the line wings both at blue and redshifted velocities. The red and blueshifted high velocity molecular gas was often found to be spatially offset so that it gave the appearance of two separate redshifted and blueshifted lobes centred on the YSO (e.g. [31];[75]; [89]).

Using ^{12}CO (or, if optically thick, ^{13}CO) line measurements it is possible to measure the mass of CO moving in the molecular outflow and, using standard abundances, to convert this to total masses of gas and dust. In some cases as much as a few solar masses are contained in the lobes emanating from a low mass young star. Since one can measure the extent, mass and typical velocity of a molecular outflow, both its dynamical timescale, τ_{D}, and mass flow rate, $\dot{\mathrm{M}}_{\mathrm{mol}}$, can be estimated. Although these vary considerably, typically $\tau_{\mathrm{D}} \approx 10^{4\rightarrow5}$yrs and $\dot{\mathrm{M}}_{\mathrm{mol}} \approx 10^{-4\leftrightarrow-7}\mathrm{M}_{\odot}\,\mathrm{yr}^{-1}$.

Generally speaking molecular outflows, unlike the Herbig-Haro jets which I will discuss shortly, are poorly collimated, at least at low (a few tens of kms^{-1}) velocities. With the development of millimetre interferometry however it has become possible to produce CO maps with high spatial resolution. Interestingly it is found that the highest velocity CO emission is often well collimated (see Fig. 4) although there is very little mass moving at high velocities.

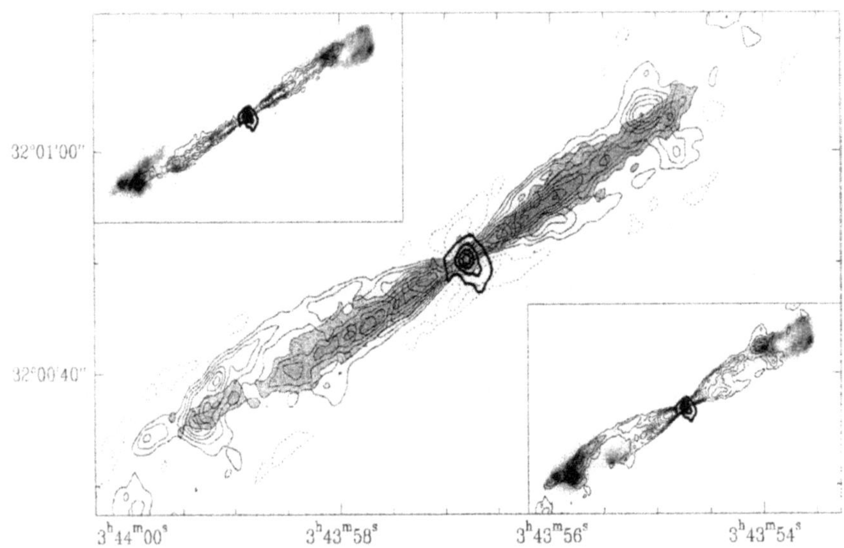

Fig. 4. A Plateau de Bure interferometer map of the CO emission from HH 211. The central panel shows a jet (high velocity gas) in grey-scale within an ovoid "cavity" of lower velocity CO emission. The upper left and lower right insets show the fast and slow CO emission (from [35]) overlaid on the shocked H_2 emission data supplied by M. McCaughrean and H. Zinnecker. Figure courtesy of John Richer.

What drives these molecular outflows? The currently favoured idea is that the ultimate prime mover (apologies to Thomas Aquinas!) is an underlying collimated, and partly ionized, Herbig-Haro like jet of the type discussed below ([75]; [25]). Certainly it appears that the low velocity molecular gas is entrained ambient material, rather than gas which was accelerated at the source. There are a number of reasons why this is thought likely although we cannot go into them here (see, for example, [85]) in this short review. Instead let us turn to discussing the ionized jet themselves.

When the emission line nebulae known as Herbig-Haro (HH) objects were first discovered ([41]; [38]), it was evident from the start that they were associated with star formation in some way although how they fitted into the overall picture was unclear. At one time it was even thought that stars might form inside HH objects. [93] was the first to notice that their spectra resembled those of very evolved slowly moving supernova remnants leading him to propose that they were in fact radiative shocks tracing an outflow from a young star. For a review of the early work on HH objects see [83]. With the availability of CCDs, it became possible to image regions containing HH objects with considerable depth and it was soon realised that many HH objects are either parts of jets or mark where a jet is ramming into its surroundings, the so-called "working-surface". The first HH jets to be recognised were from HL Tau, HH 30 and DG Tau B ([69]). After their initial discovery many more were then found ([68]; [101]; [88]; [84]). A very large number of such flows are now known and have been catalogued electronically ([88]). Many occur in groups, e.g. as in Serpens (see Fig. 5) which is not too surprising giving the propensity for stars to form in clusters.

As an example of a HH outflow, we will consider the case of HH 34. HH 34 was listed in an early catalogue of HH objects ([41]) but it only really attracted attention with the discovery of a jet pointing towards it ([67]; [88]; [13]). It then became clear that what was catalogued as HH 34 was, in fact, the bow shock of the newly discovered jet as it ploughed its way though its surroundings (see Fig. 6). [13] also found a counter-bow shock to the northeast of the HH 34 jet (see Fig. 6), implying that the flow is bipolar. No associated counter-jet has, however, been seen to date.

Early CCDs were small and thus only covered a small area on the sky at one time. Since it was possible to mosaic images, the area covered, however, was not just a function of the CCD size but also the observer's patience! With the development of larger format CCDs it became possible to image much bigger regions in one exposure. Images from these CCDs showed that HH flows, such as HH 34, were often much bigger than previously thought ([7] and [30]) although it has to be said that some large scale flows were known from early on (e.g. RNO 43, [84]). For example, the full extent of the HH 34 flow ([7]) is now known to be over 3 pc.

As stated earlier on HH objects, and their associated jets, are emission line objects. Their optical spectra are dominated by forbidden lines such as the [SII]$\lambda\lambda$6716,6731 doublet, [OI]λ6300, etc., as well as Balmer emission. These lines can be used not only to carry out diagnostic studies of the conditions in jets as, for example, obtaining their densities, temperatures, etc., e.g. [5] but also for kinematical studies i.e. obtaining radial velocities. For

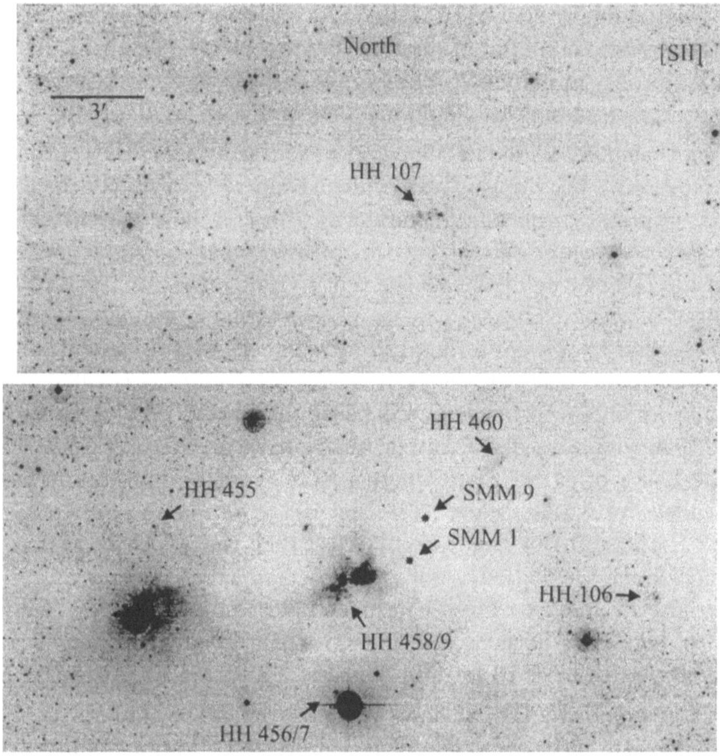

Fig. 5. A wide field [SII]λλ6716,6731 CCD mosaic showing a cluster of HH flows in the well known Serpens star forming region. Some are associated with molecular outflows and all arise from embedded sources (see [24]). This tendency for flows to occur in groups obviously follows from the fact that the clustering is common in star formation. Image from [24].

example, spectroscopic measurements of the different parts of the HH 34 flow have shown that its southern part, including the jet, are blueshifted while the northern counter-flow is redshifted. The average radial velocity in the jet and in the southern bow shock (sometimes known as HH 34S, see Fig. 6) is approximately -100 kms^{-1} ([13]). Spectroscopically determined radial velocities may be complemented by proper motion studies to determine tangential velocities and hence true spatial velocities. For the HH 34 flow, it has been shown that the velocities of both the northern and southern bow shocks are around 330 kms^{-1} along an axis at approximately 60–70° to our line of sight ([40]). Given the sound speed in the flow, such velocities are highly supersonic (see below).

From spectroscopy one finds typical temperatures and electron densities in HH jets to be around $5x10^3$–10^4 K and 10^2–10^4 cm^{-3} respectively. Their

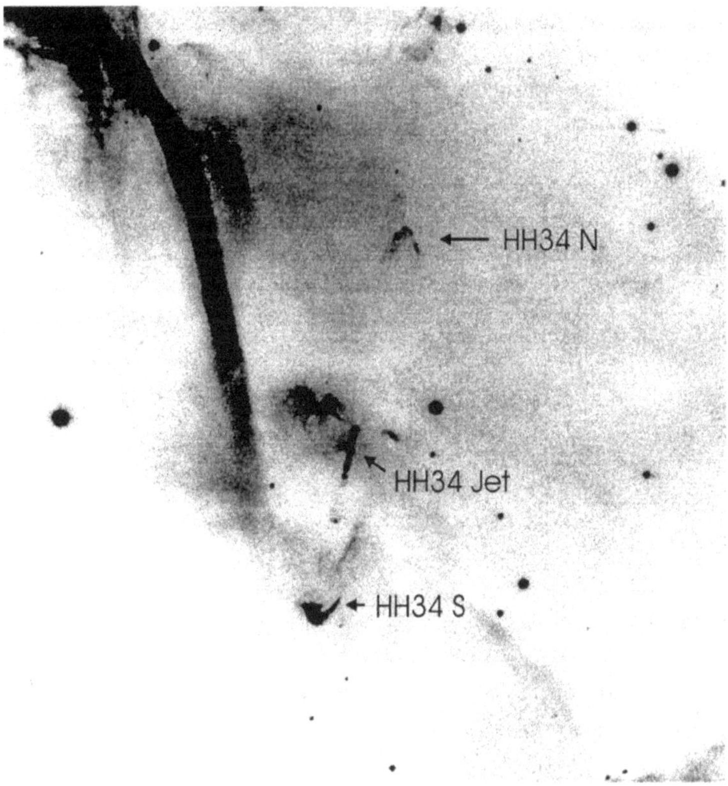

Fig. 6. The centre of the HH 34 flow as seen through a [SII]$\lambda\lambda$6716,6731 filter. The HH 34 jet emanates from an embedded infrared source, HH 34IRS, at its northern apex and points towards HH 34 S. Both bow shocks are approximately 100″ away from the infrared source. The HH 34 flow is itself much more extensive ([7]; [30]) as already hinted at by this frame since additional parts of the flow can be seen beyond HH 34 N and S. This image, courtesy of J. Eislöffel and R. Mundt, was taken with the 3.5-m telescope on Calar Alto, Spain. North is to the top and east is to the left.

ionization fraction is quite low and neutrals may constitute as much as 90% (e.g., [4]) or more of the flow. In the case of HH 34, I have already mentioned that the velocity of the flow is several hundred kms^{-1} and this is typical of jets from YSOs. The observed velocities do depend on, for example, the luminosity of the source with the most luminous sources, e.g. Z CMa ([79]), tending to have the higher velocity flows. Since the temperature, and hence the sound speed, for YSO jets are known, we can determine their Mach numbers. Typical values ($M_{jet} \approx 20$–100) imply these flows are hypersonic. Mass fluxes are estimated to be 10^{-8}–$10^{-6} M_{\odot} yr^{-1}$ again, to some degree, depending on the luminosity of the source ([28]) and, of course, its evolutionary status.

All HH jets, including the HH 34 jet, appear to be "knotty" at high enough resolution. Sometimes these knots are quasi-periodically spaced and may vary by large factors in brightness (see, for example, [86]). While it is not certain how these knots arise, there is no lack of explanations including the non-linear growth of instabilities ([33]; [64]) and that they are internal "working surfaces" ([81]) caused by variability in the flow from the young star. In the latter case the idea is that faster supersonic jet material catches up with previously ejected slower gas to form internal shocks i.e. working surfaces. Images from the HST (see for example [86]) seem to favour the internal working surfaces hypothesis but further studies are needed to conclusively prove this is the origin of the knots.

Highly collimated flows from young stars are also visible in the near-infrared. In fact there are clear advantages in observing at such wavelengths, particularly for the least evolved sources, since the flow can be hidden behind many magnitudes of visual extinction. The primary emission line used for imaging in the near-infrared is the ro-vibrational 2.12μm transition of hydrogen. Excitation of this line requires relatively high temperatures (a few thousand Kelvin) (e.g. [30]). As the cooling times for the molecule are quite short (of order a few years), the H_2 emission traces "current" regions of interaction between the jet and its ambient medium. Recently a number of highly collimated flows have been discovered in the infrared with little or no corresponding optical emission (see, for example, [108]).

Finally we should say something about the origins of the jets themselves. As this is primarily a review of the observations we will not say very much but instead refer the reader to some of the references mentioned below for further details. Briefly there is now strong evidence that jets and disks (see §Disks Around Young Stars) are intimately related phenomena in the sense that a disk has to be present for a jet to form. Current theories suggest that outflows are powered by disk accretion and that the inflow/outflow process is controlled by magnetic fields (see, for example, [48]). The basic idea is that if the disk is threaded by open magnetic field lines, then some of the material that is being accreted may get loaded onto the field lines and is then centrifugally ejected in the form of a wind. Such winds carry away angular momentum from the disk and may even be *necessary* for the accretion process to proceed. A primary advantage of centrifugally driven magneto-hydrodynamic winds is that, under a wide variety of conditions, they produce self-collimated narrow jets. This is because as the field lines are dragged back by the inertia of the wind, strong toroidal fields are generated which help to focus it. Of course the collimation process may also be assisted by pressure forces in an external medium. Magneto-centrifugal jet launching models, as they are known, have been studied in detail by many authors both analytically (e.g. [80]; [96]; [57]) and through numerical simulations (e.g. [73]; [91]). There is however considerable debate in the current literature as to whether the wind is launched from the surface of the disk (e.g.[73]) or from the boundary layer between the disk and the YSO's magnetosphere (e.g. [96]).

6 Disks Around Young Stars

Of particular interest to the participants of this school, is the discovery of disks around young stars since it is thought that a fraction of the disk material will, in at least some cases, form planets. I have already mentioned one indirect piece of evidence for disks: the need for flattened dust distributions around young stars to explain their spectral energy distributions. Other observations also indirectly point to the presence of disks around many YSOs. For example, the forbidden line emission in classical T Tauri stars, and in Herbig Ae/Be stars (their higher mass counterparts), is primarily blue-shifted (e.g. [17]; [23,?]). This asymmetry is naturally explained as a result of obscuration by an optically thick disk (e.g. [28]). The presence of "polarization disks" close to the star is also best understood in terms of multiple scattering of starlight in a dusty disk (e.g. [77]).

The above evidence is, however, indirect. Only in the past few years has it become possible to view disks directly; much of the problem in seeing them can be attributed to their small angular size. For example, in the case of the nearest star forming regions such as Taurus Auriga (150pc) and Orion (450pc), a putative disk as large as the orbit of Neptune (about 60AU) would subtend roughly $0''.4$ and $0''.1$ respectively. As this is just below the typical optical "seeing" at even the best astronomical sites, their detection is quite a challenge (!) requiring either the Hubble Space Telescope, as we will see, or adaptive optics techniques from the ground. In reality, the above estimate for the size of a typical YSO disk is somewhat conservative: for example in the case of the Solar System, we expect the primitive nebula from which it formed to have extended out at least into the Edgeworth-Kuiper Belt (see Fitzsimmons, this volume) i.e. 100–200 AU.

There are two main methods of imaging disks around young stars. The first is by mapping the thermal infrared/millimetre emission produced by the gas or dust in the disk. The second is by observing optical/near-infrared radiation from the star that is scattered or absorbed by the dust. Imaging in the thermal infrared from the ground is difficult so most observers have concentrated their efforts on the submillimeter/millimetre window. Because YSO disks are small, and the angular resolution of individual millimetre dishes is so poor, millimetre interferometers such as BIMA, OVRO, Plateau du Bure, etc. have to be used (e.g. [50]) to resolve them. At millimetre wavelengths, the gas can either be detected using a strong tracer emission line, such as the 2.3mm CO line, or we can observe the dust in the disk by its continuum emission ([56]). Note that at typical disk radii, i.e. about 50AU, the dust/gas is expected to have a temperature of about 30K assuming a $1L_\odot$ source. In principle, Doppler shifts in the CO line can be used to determine the velocity of the gas and thus test whether it is in quasi-Keplerian rotation around the central YSO. In practice, particularly when there is an envelope of molecular gas in addition to the disk, it is difficult to disentangle rotation from inflow and outflow (e.g. [50]; [69]). The clearest evidence for pure Keplerian disks comes from Class II sources, such as classical T Tauri stars, where the envelope is missing. For example, the motion of the disk around DM Tau is well fitted by a $V \propto r^{-0.5}$ curve ([27]).

The millimetre continuum emission from the dust in the disk is optically thin and so given the observed flux, the disk temperature, and the distance to the source, the mass of the disk can be determined assuming a standard dust/gas ratio. If an envelope is present, its mass can also be measured using the same technique. A general trend is seen ([69]) whereby the *envelope mass* is found to decrease in going from Class 0 to Class II as one might imagine. More surprising, however, is that there seems to be little variation in *disk masses* within the same class range ([58]) suggesting that the disk does not evolve significantly during the first 10^5 years in the life of a solar mass star. Although there are wide variations in disk masses, typical masses are around $0.01 - 0.1 M_\odot$ i.e. sufficient to form several giant planets. For further information on millimetre studies of YSO disks the reader is referred to the excellent reviews by [18] and [51].

Observing disks by their scattered/absorbed radiation is also a very effective means of detecting them although with some caveats as I will now describe. In principle very small amounts of dust, for example, a few Earth masses (!), can be detected providing it has similar size/composition properties as dust in the ISM and the central star is bright enough. On the other hand if the star is bright, there may be a significant contrast problem in trying to distinguish the faint circumstellar material from the star. For example, in the case of HST images (see, for example, [75] and Fig. 7), it is most important that the Point Spread Function of the telescope be well known to achieve optimum subtraction of the star: small errors can give rise to false results ([63]).

A disk is easiest to see in scattered or absorbed light when it is edge-on and occults its star. Obviously such a configuration is rare although perhaps not as rare as one might think: some disks are flared, i.e. thicken with increasing radius, and thus the chances of obscuring the central YSO is not insignificant. Such flaring of disks was predicted many years ago (see, for example, [47]). A nice example is HH 30 ([16]; [83]). This flow has a fine bipolar jet and two cusp-shaped nebulae that straddle the optically invisible source. Cutting across the nebulae is a dark lane, i.e. the disk, which grows in height with increasing longitudinal distance from the source. Based on the disk isophotes, the degree of flaring has been estimated to be about 15% (e.g. [106]). The mass in the HH 30 disk has been measured using both millimetre interferometry and by fitting the observed dust distribution ([100]). Both methods yield a surprisingly low mass, $10^{-3} M_\odot$, i.e. about 1 Jupiter mass.

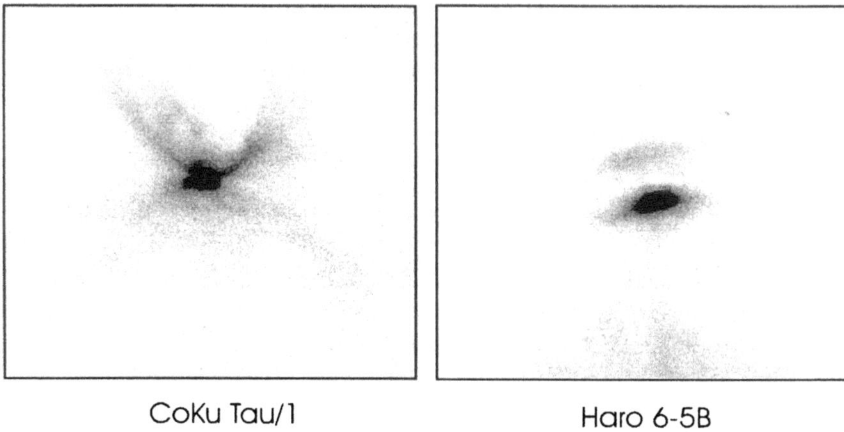

Fig. 7. This pair of near-infrared images, from the NICMOS camera on-board HST, illustrates the effect of what happens when a YSO disk axis changes angle with respect to our line of sight. On the left is the binary CoKu Tau/1, The "wings" of the reflection nebula outline the edges of a region in the stars' dusty surroundings, which have been cleared by an outflow. A thin lane (seen here in reverse video) extends to the left and to right of the binary, suggesting the presence of a circumbinary disk. The disk/outflow axes are at a moderate angle to our line of sight. On the right is Haro 6-5B the outflow of which is closer to the plane of the sky. Here we see two bright regions separated by a dark lane and reminiscent of HH 30. Note again that the dark lane appears white in this reverse video image. Optically the star is not seen but the infrared view reveals the young star just above the dust lane. Images courtesy of Deborah Padgett (see [75]).

Other examples of edge-on systems are seen near the Trapezium Cluster in Orion ([61]; [63]), where the visibility of the disk is improved by the enhanced (HII) background (these are the so-called silhouette disks). We shall discuss the disks seen near the Trapezium Cluster in more detail shortly.

Sometimes, especially in the case of the youngest sources, the central star is obscured even though its disk is not edge-on. An example is HL Tau, the disk of which has been imaged using the HST ([100]) and from the ground using adaptive optics techniques ([20]). Here the obscuration of the YSO is probably provided by the dusty envelope. The size of the disk in both scattered light, and as determined from millimetre interferometry ([56]), is approximately 150 AU. HL Tau is unusual, however, since the central source is not obscured for most classical T Tauri stars. As discussed by [63], imaging the disks around such YSOs is difficult although not impossible.

It is interesting, particularly from the perspective of planetary formation, that disks have also been observed around binary systems, such as GG Tau ([90]). Here an inner hole is seen in the disk that is approximately 3 times bigger than the semi-major axis of the binary. Such a hole is consistent with tidal stripping of the disk (see, for example, [3]) although it is likely that there are small disks a few AU in diameter around both components. UY Aur ([21]) and CoKu Tau/1 (see Fig. 7) are other examples of systems with circumbinary disks. UY Aur has a wider separation than that of GG Tau and the corresponding size of the disk hole is also larger.

We have already mentioned the cluster of low and intermediate mass stars surrounding the Trapezium in Orion (see Fig. 1). Infrared imaging ([61]) shows that this cluster consists of at least 700 stars, a large number of which are contained within the HII region ionized, primarily, by the most powerful of the Trapezium stars, θ^1C Orionis. Many of the low mass stars close to the Trapezium are associated with small ($\lesssim 1''$) compact emission line nebulae known as "proplyds"([71]). The word proplyd is an abbreviation for "protoplanetary disk"although this may be something of a misnomer as we shall see. The proplyds often show dusty cometary-shaped tails that point away from θ^1C Orionis and, in a number of cases, an associated ionization front on the side opposite the tail ([6]). Models for the proplyds generally assume that they have disks at their centre (e.g. [44]) and that the nebulae are photoionized gas from the disk surface which is in the process of being evaporated (photo-ablated) by UV radiation. In some cases, the disks at the centre of the proplyds are observed directly. [44] have questioned whether it is possible for planets to form in the harsh environment that the proplyds suffer: their disks may be ablating so rapidly that there is probably insufficient time for large bodies like Jupiter to form. Further from θ^1C Orionis, many disks are seen both edge-on and at various inclinations, their visibility increased by the presence of the bright background HII region (see Fig. 8). [63] have shown that there are young low mass stars at the centre of these "silhouette"disks. The silhouette disks near the Trapezium can be seen not only in the optical but in the near-infrared as well. In particular recent HST NICMOS observations ([19]) have shown that the surfaces of these disks emit copiously in the H_2 2.12μm line which is excited by FUV photons from the Trapezium.

How quickly do disks disperse and in what way does it happen? The timescales over which much of the disk mass is "lost", e.g. to planetary formation, is still poorly known. If we take the Solar System as being typical, then one might guess at a period of 10 Myr (e.g. [78]) for the dust grains to clump together to form large bodies. This means that the classical T Tauri phase, which lasts approximately 1 Myr, is the phase just before planetary formation. On the other hand disks have been found around main sequence Vega-type stars such as β Pic (e.g. [8]; [16]) the ages of which could be as large as 1 Gyr. One might then ask the question, is it right to speak of disk dispersal at all? There are however major differences between YSO and β Pic-type disks: first the amount of dust in β Pic-type disks is at most a few lunar masses ([109]) i.e. considerably less than that found in the circumstellar environment of a typical classical T Tauri star. Moreover, due to the Poynting

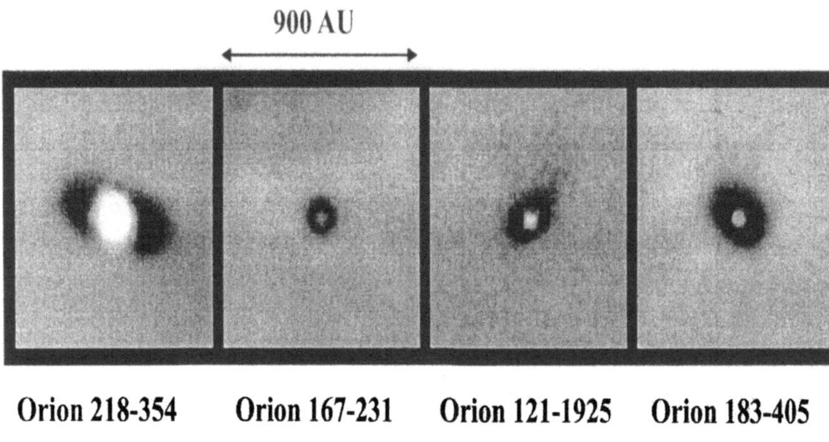

Fig. 8. HST images of four silhouette disks in the Orion Nebula from [62]. Each panel is 2" square which corresponds to 900 AU at the distance of Orion. These disks (seen using a Hα filter, stand out because of the relatively bright HII background. Image courtesy of Mark McCaughrean.

Robertson effect, the dust in these older systems has to be continually replenished, most likely as a result of collisions between larger bodies (e.g. [70]; [8]). Thus the β Pic-type systems have at least formed planetesimals, and perhaps even planets as suggested by the warping of the β Pic disk ([16]). Thus we are probably dealing with remnants of planetary formation rather than proto-planetary disks.

Once we can observe a YSO (i.e. it has at least reached the Class II stage), we can tell its age by fitting it in the HR diagram and matching its properties against theoretical isochrones. Assuming the disk and the star formed together we then have an age for the disk. Using this method [72] looked for evidence of an anti-correlation of disk mass with age amongst T Tauri stars. Despite some of the stars being as old as 10 Myr, they did not find any sign of disk evolution in their millimetre continuum survey. In contrast, [82] could not detect any millimetre emission from nearby post-T Tauri stars with ages of about 100 Myr. These two results in combination hint therefore that disk dispersal times should be measured in tens of millions of years. Note also that dispersal times may differ widely from system to system: for example it is theoretically expected that disks may disperse relatively quickly in binary systems with separations comparable to their size (e.g. [76]). An example in point is the nearly edge-on disk around HR 4796A ([43]; [51]) first imaged in the thermal infrared. HR 4796A is a main sequence A-type star with a M-type T Tauri companion some 500 AU away. The estimated age of the system is around 10 Myr but the mass of dust in the disk is very low indeed suggesting that much of the disk material may have already spiralled into the primary ([43]). It is also interesting that the disk around HR 4796A appears to have an inner "hole"that may have been caused by the action of a "sweeper"planet ([43]).

7 Conclusions

I hope that in this brief review I have given you a flavour of the contributions that star formation studies have made to unravelling the origin of our own and other planetary systems. In the past twenty years, enormous strides have been made in our understanding of how stars are born. For the most part this has been a result of improvements in technology at infrared and millimetre wavelengths although high resolution optical studies have also played a role. It is true to say that for the first time a reasonably coherent picture is beginning to emerge of how a star comes into being: it is perhaps a bit ironic that up to recently astrophysicists could speak with some authority on the first 3 minutes of the Universe but were somewhat hazy on what happened in the first 100 Myr in the life of our Sun and its attendant planets!

We now know that solar-like stars as well as accreting matter when they form also have outflows which regulate not only their angular momentum but probably that of any associated planets. These outflows can stretch for distances up to several parsecs and are perhaps the most dramatic signature of stellar birth. It would appear that a significant proportion (at least 50%) of low mass stars are surrounded by disks as soon as they form and it would seem highly likely that many of these disks develop into planetary systems like our own.

Acknowledgements

The author wishes to thank the Scientific Director of the EADN School, Iwan Williams for all his hard work in making the school the success that is was. I would also like to thank the students for their enthusiasm which did not falter despite the vagaries of the weather in the West of Ireland!

References

1. Ambartsumian, J.A., 1947, in Stellar Evolution and Astrophysics, Erevan: Acad. Sci. Armen. SSR
2. André, P., Ward-Thompson, D., Barsony, M., 2000, in Protostars and Planets IV, eds. V. Mannings, A.P. Boss and S.S. Russell (University of Arizona Press), in press
3. Artymowicz, P., Lubow, S.H., 1995, in Disks and Outflows around Young Stars, eds. S.V.W. Beckwith, J. Staude, A. Quetz and A. Natta, Lec. Notes in Phys., Vol. 465, (Springer Verlag), p115
4. Bacciotti, F., Chiuderi, C., Oliva, E., 1995, A&A, 296, 185

5. Bacciotti, F., Eislöffel, J., Ray, T.P., 1999, A&A, in press
6. Bally, J., Sutherland, R.S., Devine, D., Johnstone, D., 1998, AJ, 116, 293
7. Bally, J., Devine, D., 1994, ApJ, 428, L65
8. Backman, D.E., Paresce, F., 1993, in *Protostars and Planets III*, eds. E. Levy and J. Lunine (University of Arizona Press), p1253
9. Basri, G., Bertout, C., in *Protostars and Planets III,* eds. E. Levy and J. Lunine, (University of Arizona Press), p543
10. Benson, P.J., Myers, P.C., 1989, ApJS, 71, 89
11. Bertout, C., 1989, ARA&A, 27, 351
12. Blitz, L., 1993, in *Protostars and Planets III,* eds. E. Levy and J. Lunine, (University of Arizona Press), p125
13. Bührke, T., Mundt, R., Ray, T.P., 1988, A&A, 200, 99
14. Boss, A.P., Myhill, E.A., ApJS, 83, 311
15. Burrows, C.J., et al., 1995, Bull. Amer. Astron. Soc., 27,1329
16. Burrows, C.J., et al., 1996, ApJ, 473, 437
17. Calvet, N., 1997, in *Herbig-Haro Flows and the Birth of Low Mass Stars,* IAU Symposium No. 182, eds. B. Reipurth and C. Bertout, (Dordrecht: Kluwer Academic Publishers), p417
18. Chandler, C.J., 1998, in *Proceedings of the International Origins Conference,* eds. C.E. Woodward, J.M. Shull and H.A. Thronson, (ASP), p237
19. Chen, H., Bally, J., O'Dell, C.R., McCaughrean, M.J., Thompson, R.L., Rieke, M., Schneider, G., Young, E.T., 1998, ApJ, 492, L173
20. Close, L., Roddier, F., Northcott, M.J., Roddier, C., Graves, J,E., 1997, ApJ, 478, 766
21. Close, L. et al., 1998, ApJ, 499, 883
22. Corcoran, M., Ray, T.P., 1997, A&A, 321, 189
23. Corcoran, M., Ray, T.P., 1998, A&A, 331, 147
24. Davis, C.J., Matthews, H.E., Ray, T.P., Dent, W.R.F., Richer, J.S., 1999, MNRAS, in press
25. Downes, T.P., Ray, T.P., 1999, A&A, 345, 977
26. Dopita, M.A., Schwartz, R.D., Evans, I., 1982, ApJ, 263, L73
27. Dutrey, A., Guilloteau, S., Prato, L., Simon, M., Duvert, G., Schuster, K., Ménard, F., 1998, A&A, 338, L63
28. Edwards, S., Ray, T.P., Mundt, R., 1993, in *Protostars and Planets III,* eds. E. Levy and J. Lunine, (University of Arizona Press), p567
29. Eislöffel, J. 1997, in *Herbig-Haro Outflows and the Birth of Low Mass Stars,* IAU Symposium No. 182, eds. B. Reipurth and C. Bertout, (Dordrecht: Kluwer Academic Publishers), p93
30. Eislöffel, J., Mundt, R., 1997, AJ, 114, 280
31. Fich, M., Lada, C.J., 1997, ApJ, 484, L63
32. Finkenzeller, U., Mundt, R., 1984, A&AS, 55, 109
33. Ferrari, A., Massaglia, S., Bodo, G., & Rossi, P., 1996, in *Solar and Astrophysical Magnetohydrodynamic Flows*, Proceedings of the NATO ASI, Heraklion, Crete, ed. K. Tsinganos, (Dordrecht: Kluwer Academic Publishers), p607
34. Garay, G., Lizano, S., 1999, PASP, 111, 1049
35. Gueth, F., Guilloteau, S., 1999, A&A, 343, 571
36. Gullbring, E., Hartmann, L., Briceno, C,, Calvet, N., 1998, ApJ, 492, 323
37. Haro, G. 1952, ApJ, 115, 572
38. Hartigan, P., Edwards, S., Ghandour, L., 1995, ApJ, 452, 736

39. Heyer, M.H., Terebey, S., 1998, ApJ, 502, 265
40. Heathcote, S., Reipurth, B., 1992, AJ, 104, 2193
41. Herbig, G.H., 1951, ApJ, 113, 697
42. Herbig, G.H., 1974, Lick Obs. Bull. 658
43. Jayawardhana, R., Fisher, C., Hartmann, L., Telesco, C., Piña, R., Fazio, G., 1998, ApJ, 503, L79
44. Johnstone, D., Hollenbach, D., Bally, J., 1998, ApJ, 499, 758
45. Joy, A.H., 1945, ApJ, 102, 168
46. Kenyon, S.J., Calvet, N., Hartmann, L., 1993, ApJ, 414, 676
47. Kenyon, S.J., Hartmann, L., 1987, ApJ, 323, 714
48. Königl, A., Pudritz, R.E., 2000, in *Protostars and Planets IV*, eds. V. Mannings, A.P. Boss and S.S. Russell (University of Arizona Press), in press
49. Koerner, D.W., Sargent, A.I., 1995, AJ, 109, 2138
50. Koerner, D.W., 1997, Origins of Life and Evolution of the Biosphere, 27, 157
51. Koerner, D.W., Ressler, M.E., Werner, M.W., Backman, D.E., 1998, ApJ, 503, L83
52. Kramer, C., Stutzki, J., Rohrig, R., Corneliussen, U., 1998, A&A, 329, 249
53. Kuhi, L.V., 1964, ApJ, 140, 1409
54. Kwan, J., Tademaru, E., 1995, ApJ, 454, 382
55. Lada, E.A., Strom, K.M., Myers, P.C., 1993, in *Protostars and Planets III,* eds. E. Levy and J. Lunine, (University of Arizona Press), p245
56. Lay, O.P., Carlstrom, J.E., Hills, R.E., Philips, T.G., 1994, ApJ, 434, L75
57. Lery, T., Heyvaerts, J., Appl, S., Norman, C.A., 1999, A&A, in press
58. Looney, L.W., Mundy, L.G., Welch, W.J., 1999, ApJ, in press
59. Lucas, P.W, Roche, P.F., 1997, MNRAS, 286, 895
60. McCaughrean, M.J., Chen, H., Bally, J., Erickson, E., Thompson, R.I., Rieke, M., Schneider, G., Stolovy, S., Young, E.T., 1998, ApJ, 492, LL157
61. McCaughrean, M.J., O'Dell, C.R., 1996, AJ, 111, 1977
62. McCaughrean, M.J., Rayner, J.T., Zinnecker, H., Stauffer, J.R., 1996, in *Disks and Outflows around Young Stars*, eds. S.V.W. Beckwith, J. Staude, A. Quetz and A. Natta, Lec. Not. in Phys., Vol. 465, (Springer Verlag), p33
63. McCaughrean, M.J., Stapelfeldt, K., Close, L., 2000, in *Protostars and Planets IV*, eds. V. Mannings, A.P. Boss and S.S. Russell (University of Arizona Press), in press
64. Micono, M., Massaglia, S., Bodo, G., Rossi, P., Ferrari, A., 1998, A&A, 333, 1001
65. Moriarty-Schieven, G.H., Andersson, B.-G., Wannier, P.G., 1997, ApJ, 475, 642
66. Mundt, R., 1986, Can. J. Phys., 64, 407
67. Mundt, R., Bührke, T., Fried, J.W., Neckel, T., Sarcander, M., Stocke, J., 1984, A&A, 140, 17
68. Mundt, R., Fried, J.W., 1983, ApJ, 274, L83
69. Mundy, L.G., Looney, L.W., Welch, W.J., 2000, in *Protostars and Planets IV*, eds. V. Mannings, A.P. Boss and S.S. Russell (University of Arizona Press), in press
70. Nakano, T., 1988, MNRAS, 230, 551
71. O'Dell, C.R., Wong, S.K., 1996, ApJ, 111, 846
72. Osterloh, M., Beckwith, S.V.W., 1995, ApJ, 439, 288
73. Ouyed, R., Pudritz, R.E., 1997, ApJ, 482, 712

74. Padgett, D.L., Brandner, W., Stapelfeldt, K.R., Koerner, D., Tereby, S., 1999, AJ, 117, 1490
75. Padman, R., Bence, S., & Richer, J. 1997, in *Herbig-Haro Outflows and the Birth of Low Mass Stars*, IAU Symposium No. 182, eds. B. Reipurth and C. Bertout, (Dordrecht: Kluwer Academic Publishers), p123
76. Papaloizou, J.C.B., Lin, D.N.C. 1995, ARA&A, 33, 505
77. Piirola, V., Scaltriti, F., Coyne, G.V., 1992, Nature, 359, 399
78. Podosek, F.A., Cassen, P., 1994, Meteoritics, 29, 6
79. Poetzel, R., Mundt, R., Ray, T.P., 1989, A&A, 224, L13
80. Pudritz, R.E., 1991, in *The Physics of Star Formation and Early Stellar Evolution*, eds. C.J. Lada and N.D. Kylafis, NATO ASI Series, Kluwer, p365
81. Raga, A.C., Kofman, L., 1992, ApJ, 386, 222
82. Ray, T.P., 1996, in *Solar and Astrophysical MHD Flows*, NATO ASI, Heraklion Crete, ed. K. Tsinganos, (Kluwer Academic Publishers), p539
83. Ray, T.P., 1987, A&A, 171, 145
84. Ray, T.P., 2000, Astrophys. Space Sci., in press
85. Ray, T.P., Mundt, R., Dyson, J., Falle, S.A.E.G., Raga, A., 1996, ApJ, 468, L103
86. Ray, T.P., Sargent, A.I., Beckwith, S.V.W., Koresko, C., Kelly, P., 1995, ApJ, 440, L89
87. Reipurth, B., 1994, A General Catalogue of Herbig-Haro Objects, available at ftp.hq.eso.org, at /pub/Catalogs/Herbig-Haro, using anonymous ftp.
88. Reipurth, B., Bally, J., Graham, J.A., Lane, A.P., Zealey, W.J., 1986, A&A, 164, 51
89. Richer, J., Shepherd, D., Cabrit, S., Bachiller, R., Churchwell, E., 2000, in *Protostars and Planets IV*, eds. V. Mannings, A.P. Boss and S.S. Russell (University of Arizona Press), in press
90. Roddier, C., Roddier, R., Northcott, M.J., Graves, J,E., Jim, K., 1996, ApJ, 463, 326
91. Romanov, M.M., Ustyugova, G.V., Koldoba, A.V., Chechetkin, V.M., Lovelace, R.V.E., 1998, ApJ, 500, 703
92. Sargent, A. I., Beckwith, S.V.W., 1991, ApJ, 328, L31
93. Schwartz, R.D., 1975, ApJ, 195, 631
94. Shu, F.H., 1977, ApJ, 214, 488
95. Shu, F.H., Adams, F.C., Lizano, S., 1987, ARA&A, 25, 23
96. Shu, F.H., Najita, J., Ostriker, E., Wilkin, F., Ruden, S., Lizano, S., 1994, ApJ, 429, 781
97. Stahler, S.W., 1988, ApJ, 332, 804
98. Stahler, S.W., Walter, F.M., 1993, in *Protostars and Planets III,* eds. E. Levy and J. Lunine, (University of Arizona Press), p405
99. Stapelfeldt, et al., 1995, ApJ, 449, 888
100. Stapelfeldt, K.R., Padgett, D.L., 1999, in preparation
101. Strom, S.E., Strom, K.M., Grasdalen, G.L., Sellgren, K., Wolff, S., 1985, AJ, 90, 2281
102. Testi, L., Sargent, A.I., 1998, ApJ, 508, L91
103. Wilking, B.A., Lada, C.J., Young, E.T., 1989, ApJ, 340, 823
104. Williams, J.P., Blitz, L., McKee, C.F., 2000, in *Protostars and Planets IV*, eds. V. Mannings, A.P. Boss and S.S. Russell (University of Arizona Press), in press

105. Williams, J.P., McKee, C.F., 1997, ApJ, 476, 166
106. Wood, K., Kenyon, S.J., Whitney, B., Turnbull, M., 1998, ApJ, 497, 404
107. Zinnecker, H., 1993, in *Protostars and Planets III,* eds. E. Levy and J. Lunine, (University of Arizona Press), p429
108. Zinnecker, H., McCaughrean, M.J., Rayner, J., 1996, Nature, 394, 862
109. Zuckermann, B., Becklin, E.E., 1993, ApJ, 414, 793

Extrasolar Planets: A Review of Current Observations and Theory

Richard P. Nelson

Astronomy Unit, Queen Mary, University of London,
Mile End Road, London E1 4NS, UK

Abstract. Since the discovery of a planet orbiting the star 51 Peg, there has been a tremendous increase in both theoretical and observational work aimed at examining the occurrence and nature of extrasolar planetary systems. In addition to the radial velocity searches that have led to the discovery of close to 50 extrasolar planets to date, we are now beginning to witness the first results from alternative searching methods such as microlensing and transit observations. Recent observations of a planet transiting the star HD 209458 have yielded detailed information on the nature of extrasolar planets. Proposed future ground and space based observing programmes promise to provide us with a detailed view of planetary systems in the Galaxy, including terrestrial as well as giant planets, on a time scale of a decade.

At the current juncture, the major challenge facing planet formation theorists is to provide an explanation of the current data on extrasolar planets. At present there is no unified picture that provides a tidy explanation for the diversity of systems observed, but progress in our understanding of planet formation is continuing to develop.

In this article we will review the current state of the observations of extrasolar planets, as well as the current theoretical models of their formation and structure. Future directions for both observational and theoretical work will also be indicated.

1 Introduction

Philosophers down through the millennia have speculated on the origin, diversity and number of planets in the universe. Historically, the broad acceptance of a geocentric view of the universe advocated by Aristotle put scientific enquiry into the question of 'other worlds'on hold, and ensured that discussion on the subject was confined primarily to the realm of religious speculation until the sixteenth century. The introduction of the heliocentric view by Copernicus in 1543, and the telescopic observation of solar system bodies by Galileo in 1609 finally revealed the true nature of planetary objects, and reintroduced the role of science into the debate. The recent discovery of extrasolar planets orbiting nearby stars represents the next chapter in the history of human inquiry into this subject.

The observation that the planets in the solar system orbit the sun in the same sense, and essentially lie in the same plane, led [16] and [19] to theorise that the planets had formed by the accumulation of material from a flattened disc of material in orbit about the sun. This basic paradigm still remains the favoured one amongst most researchers to the present day, and is strengthened by observations of gaseous protostellar discs in orbit around young stars (e.g. [3]). Definitive proof, however, that planets do indeed form within protostellar discs, remains elusive. Recent work indicates that the stellar mass function may in fact extend all the way down into the mass regime usually reserved for giant planets ([42]), so that the precise definition of what constitutes a planet rather than a brown dwarf is unclear. The current resurgence of interest in extrasolar planets and planet formation scenarios has arisen as a result of the discovery by [27] of a planet orbiting the nearby solar–type star 51 Peg. This discovery was confirmed shortly afterwards by [24]. Both of these groups use the radial velocity technique for finding extrasolar planets, which measures the reflex motion of the host star as it orbits the centre of mass of the star–planet system. This technique, as employed by a number of international collaborations, has now led to the discovery of about 50 extrasolar planets (see for example [25]; see also the extrasolar planets encyclopedia website http:||www.obspm.fr/encycl/encycl.html). The technique is most sensitive to the detection of closely orbiting, high mass planets, and as a consequence of this the planetary masses so far detected lie in the range $0.17 \leq M_p \sin i \leq 11.02$ $M_{Jupiter}$ (*i.e.* in the giant planet range). Note that the estimated planetary mass is modulated by a factor of $\sin i$, where $i = 0^o$ for a system whose orbital plane is seen face-on and $i = 90^o$ for a system that is edge-on. The range of estimated semimajor axes is $0.038 \leq a \leq 3.3$ AU, and the range of orbital eccentricities is found to be $0 \leq e \leq 0.71$. Obviously the planetary systems discovered so far are very different to the Solar System, leading to questions about their formation and evolutionary histories that remain unresolved. The radial velocity technique has been extraordinarily successful at finding planetary systems, and characterising their orbital parameters. However, two weaknesses lie in its inability to detect planets in systems where the reflex velocity of the star is ≤ 12 m s^{-1} [25], and in the lack of information that it provides about the structure of the planet. It is apparent that alternative planet search techniques are required in order to provide a more representative census of the diversity of planetary systems, and also to provide detailed information on the nature of the planets themselves. The method of detecting planetary transits has recently been used with great success. This method relies on the fact that a planetary system whose orbit plane is seen edge-on will lead to the planet passing across the disc of the host star once per orbit. The partial occultation of the host star will lead to a diminution of its luminosity which should be detectable using sensitive photometric monitoring. Recent ground based and HST observations have been used to measure the transit of an orbiting planet across the star HD 209458 ([13]; [9]; [15]; [10]). Observations of this type provide information about the size and structure of the planet, as well as determining its mass through the knowledge that $\sin i \simeq 1$ in systems that show transits.

The detection of microlensing events induced by stars with orbiting planets provides an alternative technique for indirectly observing planets around stars. This technique relies on the careful photometric monitoring of a selection of background field stars, such as in the Galactic bulge or Large Magellanic Cloud. Such surveys are currently underway to examine the composition of the galactic halo. In the event that a star passes directly in front of one of these background field stars, the observed luminosity of the star will be enhanced for a period of time due to the effect of the intervening star acting as a gravitational lens. The light curve generated by a binary system acting as the lense, or star plus planet system, has characteristics that can distinguish it from a single star. This information can be used to place statistical constraints on the ubiquity of planetary systems in the Galaxy for a wide range of planetary masses and semimajor axes that are not accessible to the radial velocity technique (e.g. [37]).

As a result of the discovery of extrasolar planets, and the apparent diversity in these systems, there has been a resurgence in theoretical work aimed at understanding planet formation. Most (but not all) of this work examines the issue of planet formation from within the frame work of the planet–forming–in–disc scenario, though there is still lively discussion on the precise mode(s) by which giant planets form. The data on planetary orbital parameters contains a number of surprises that require explanation. For example, why are there a number of giant planets orbiting close in to their host stars (given that the giant planets in our solar system orbit much further out)? Did they form there *in situ*, or did they form further out and migrate inwards? Why are so many of the planets on eccentric orbits ? Whilst no definitive answers to these questions have been provided as yet, recent work is helping to clarify the overall picture.

This article is organised as follows. Section 2 will contain a review of the current observations that provide information on extrasolar planets. Section 3 will contain a brief review of the current state of theoretical work on giant planet formation, and post–formation evolution. Section 4 will examine the unresolved questions, and will describe proposals for future observations. Section 5 will contain concluding remarks.

2 Observational Data

The discovery of extrasolar planets orbiting about nearby solar–type stars represents one of the most exciting astronomical discoveries of recent times. The number of known extrasolar planetary objects now numbers about 50, with there being some systems containing (at least) three giant planets, (see the extrasolar planets encyclopedia website http:||www.obspm.fr/encycl/encycl.html). The large number of planets being discovered now means that meaningful statistical and scientific analysis of the data can be performed, whilst bearing in mind the observational biases.

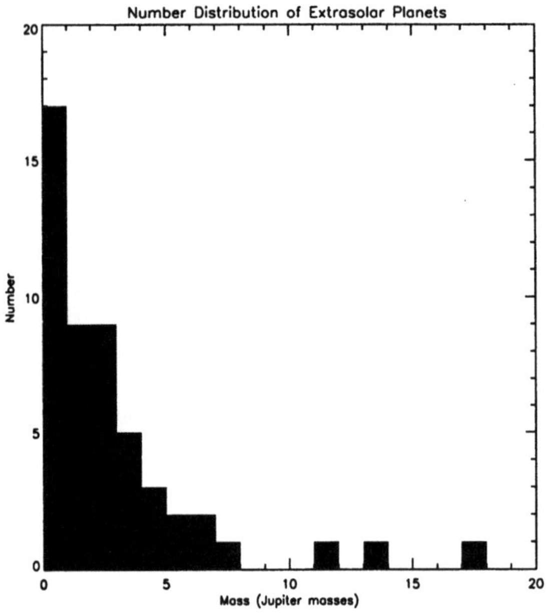

Fig. 1. This figure shows the number distribution of the currently known extrasolar planets as a function of $M_p \sin i$, where the data bin width $= 1\ \mathrm{M}_{Jupiter}$. The data for this figure (and figures 2,3, and 4) were taken from the extrasolar planets encyclopedia website.

2.1 Radial Velocity Measurements

All of the known planets orbiting nearby solar–type stars were first detected using the radial velocity technique, with the first discovery being the planet in orbit about 51 Peg ([27]; [24]). This technique relies on the detection of the doppler shift induced in the light signal from the host star as it orbits about its common centre of mass with the planet. The detection of this doppler shift requires the detection of a specific spectral line in the star's spectrum, whose position in frequency space is seen to shift back and forth periodically with respect to a fixed reference line. The intrinsic turbulent photospheric motion of solar–type stars has an amplitude of ~ 3 m s^{-1}, limiting the effectiveness of this technique to systems in which the observed orbital motion of the target star is ≥ 12 m s^{-1} (see [25]). We note that the sun orbits about the sun–Jupiter barycentre with a velocity of $\simeq 13$ m s^{-1}.

The radial velocity technique is able to provide direct information on the mass of the planet (modulated by $\sin i$), the period of its orbit, and its orbital eccentricity. This latter property is obtained from the fact that the amplitude of the velocity shift as a function of time (or orbital phase) is a non-sinusoidal function for a non-circular orbit. Given an estimate of the planetary mass and knowledge of the orbital period, the semimajor axis can

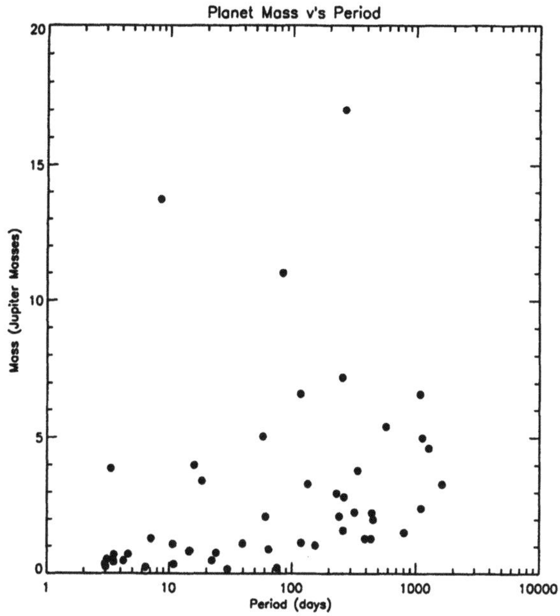

Fig. 2. This figure shows the estimated masses of the extrasolar planets plotted as a function of their orbital period (in units of days). It is clear that no strong correlation between these planetary characteristics exists in the data.

be obtained. Figure 1 shows a histogram of the number of planetary objects as a function of their mass, with the data bin width being 1 $M_{Jupiter}$. It is immediately obvious that there is a strong tendency for planets to be more numerous as their mass decreases, with there being very few objects with masses $M_p > 5$ $M_{Jupiter}$. This lack of higher mass objects has been dubbed as the 'brown dwarf desert'. It turns out that there are very few objects with masses below the stellar limit (0.08 $M_{\odot}$) known to orbit solar-type stars. These data raise very interesting questions about why lower mass objects are formed preferentially, and indicate that some process operates during planet formation that limits planetary growth to be ≤ 5 $M_{Jupiter}$.

Figure 2 shows a plot of planetary mass versus orbital period. It must be remembered that larger mass objects are easier to detect than lower mass objects, so that the lack of low mass objects with periods > 200 days is probably due to observational bias. In the region of the plot below this limit, it is apparent that there are no strong correlations in the data.

Figure 3 shows the orbital eccentricity plotted as a function of planetary mass. For objects with $M_p \sin i \leq 5$ $M_{Jupiter}$, it is clear that no correlation exists between orbital eccentricity and planetary mass. This fact is of great importance when considering the origin of the observed eccentricity, since certain eccentricity driving mechanisms predict a positive correlation between

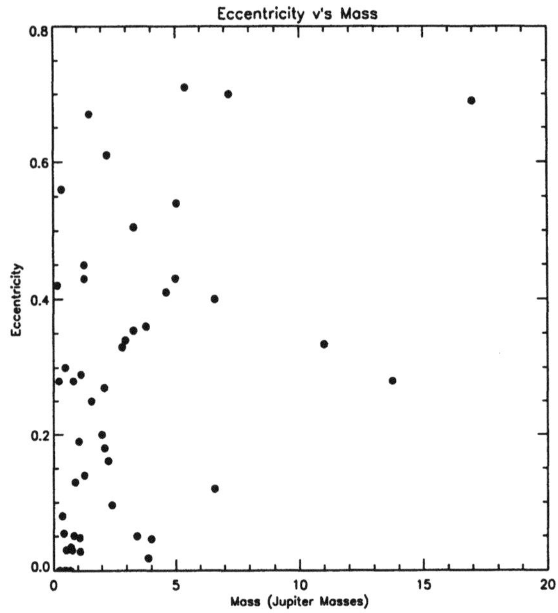

Fig. 3. This figure shows the orbital eccentricity of the extrasolar planets plotted as a function of their mass (in units of $M_{Jupiter}$). It is clear that again no strong correlation between these planetary characteristics exists in the data. This has relevance to understanding the physical mechanism responsible for the eccentric orbits.

eccentricity and mass (see Section 3). The data beyond $M_p \sin i = 5\ M_{Jupiter}$ is too sparse for meaningful statements to be made. Figure 4 shows a plot of eccentricity versus orbital period. Short period planets undergo significant tidal interactions with their host stars, leading to the circularisation of their orbits ([36]; [39]). For periods beyond about 10 days, tidal circularisation has not had time to operate efficiently, and the data show an essentially random distribution of eccentricities as a function of orbital period.

2.2 Planetary Transits of HD 209458

It has recently been established that the short period planet orbiting the star HD 209458 actually transits the star once per orbit by virtue of its orbital plane being almost exactly edge on to the line of sight ([13]; [9,10]; [15]). This exciting discovery has now opened up a whole new branch of planetary astronomy, concerned with the detailed structure and evolution of planets outside our solar system. As the planet passes in front of its host star, it blocks out a small amount of light, causing a measurable dimming of its apparent luminosity, as illustrated by figure 5. The fractional change in the

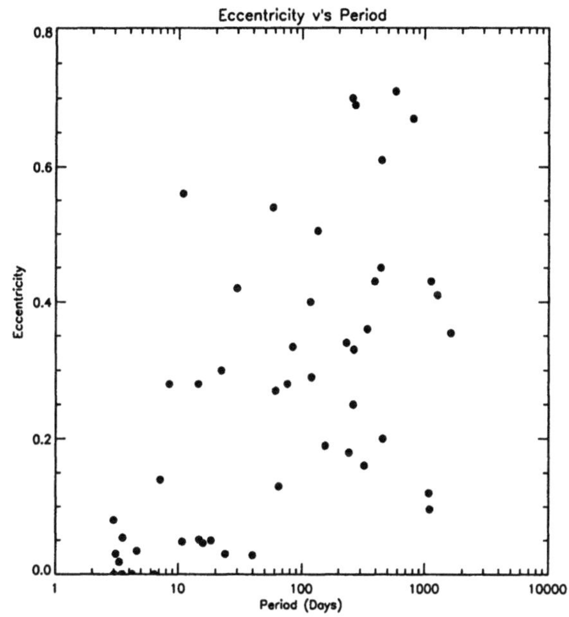

Fig. 4. This figure shows the orbital eccentricity of the extrasolar planets plotted as a function of their orbital period (in units of days). It is clear that again no strong correlation between these planetary characteristics exists in the data.

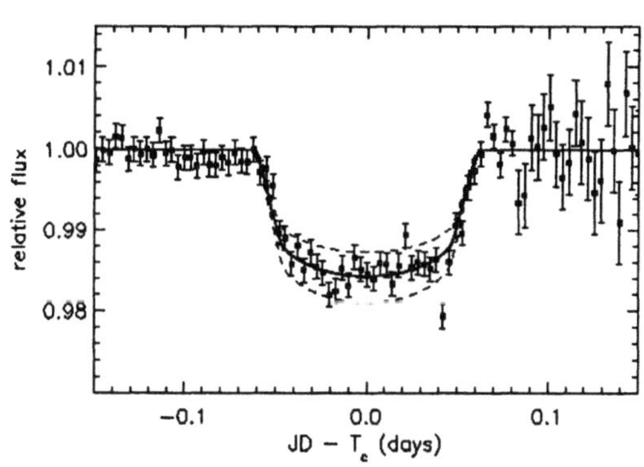

Fig. 5. This figure (taken with kind permission from [10]) shows the dimming of the starlight from HD 209458 as the planet passes in front of the star dust during transit.

apparent brightness is primarily a function of the planet radius, so that by measuring the dimming of the starlight one can estimate this radius.

In addition to being a function of the planetary radius (R_p), the precise form of the light curve generated by a transit depends on the stellar mass (M_s), the stellar radius (R_s), the degree of stellar limb–darkening, and the inclination i between the normal to the line-of-sight and the orbit plane of the star–planet system ([9]). Estimates of the radius of the planet around HD 209458 lie in the range $1.27 \leq R_p \leq 1.55$ $R_{Jupiter}$, depending on the precise values adopted for M_s, R_s, and i. The most recently reported values, obtained from HST photometry ([10]) indicate $i = 86.67 \pm 0.25$ degrees, $R_p = 1.35 \pm 0.05$ $R_{Jupiter}$, $R_s = 1.15 \pm 0.04$ $R_{\odot}$. The mass of the planet M_p can be obtained by combining this value of i with the radial velocity measurements, giving $M_p = 0.63$ $M_{Jupiter}$. The obtained value of R_p agrees extremely well with the theoretical radius–versus–age trajectories computed by [7] for a gas–giant planet forming close to its host star. Interestingly, these calculations suggest that the planet around HD 209458 must either have formed *in situ* close to its star, or else must have migrated inwards to its current location within 10^7 yr of formation if it formed further out, as suggested by currently favoured planet formation models.

More accurate estimates of the stellar and planetary properties in this system will lead to the refinement of the evolutionary models, and provide detailed information on the structure and atmospheric properties of the planet. The current HST photometry is sufficiently sensitive to suggest the non existence of planetary rings, as well as the absence of a planetary satellite of radius ≥ 1.5 R_{earth} or mass ≥ 3 M_{earth} ([10]). More sensitive photometric observations will enable the indirect detection of reflected starlight from the planet. During phases when neither the planet is transiting the star, or the star is occulting the planet, the luminosity of the system comes from a combination of direct starlight and starlight reflected by the planet. As the planet orbits around the far side of the star, and is occulted by it, the total luminosity will decrease slightly.

Sensitive spectroscopy will enable the detection of absorption lines induced by the passage of the starlight through the upper atmosphere of the planet. These observations will provide valuable information on the planetary albedo and atmospheric chemistry for comparison with theoretical models currently being computed (e.g. [7]; [38]).

2.3 Microlensing Searches

The microlensing searches currently underway to detect baryonic dark matter in the Galactic halo are sensitive to the detection of Jupiter–mass extrasolar planets orbiting their stars with semimajor axes of several AU. This is because the usual symmetric light curves generated by single lenses become more complicated when the lense is a binary or star-plus-planet system, giving a tell-tale signature of a planet or binary companion ([37]).

To date there have been no firm detections of planetary companions among the confirmed microlensing events, but this non-detection may be used to put statistical limits on the frequency of planetary companions around the most common type of stars in the Galaxy. Recent estimates, based on a sample of 43 microlensing events, suggest that less than 1/3 of ~ 0.3 $M_\odot$ stars have Jovian mass companions in the range $1.5 < a < 4$ AU ([37]).

3 Theoretical Models

In this section we will review the basic ideas of how planetary systems form, and how current theories are attempting to confront and explain the observational data on extrasolar planets. We will begin by describing currently favoured planetary formation scenarios, and then go on to discuss how protoplanets may continue to evolve due to interaction with their protoplanetary discs or with other protoplanets that may be forming within the same system.

3.1 The Basic Picture

The idea that planets form within the thin gaseous discs observed around young T Tauri stars is accepted by the majority of researchers working in the field of planetary science and planet formation. The precise modes by which planets form, however, are still a matter of intense debate. Crudely speaking there are two schools of thought: one holds that giant planets form through gravitational instability during the early lifetime of a protostellar disc, leading to the creation of massive fragments in the Jovian planet mass range; the other believes that giant planet formation occurs *via* a multistage process that involves the growth of solid bodies through binary collisions, starting with the coagulation of dust grains, and ending with the formation of solid protoplanetary cores of mass $\sim 10 - 15$ M_{earth}. At this stage the gas in the protostellar disc is unstable to rapid accretion onto the solid core, leading to the formation of a gas–giant planet. This model of giant planet formation is known as the 'core instability' model. The formation of lower mass terrestrial planets is debated less intensely, and is generally believed to occur through the slow process of coagulation of solid material though various stages until planetary mass objects are formed. This is expected to require about 10^8 yr in the terrestrial planet region (e.g. [8]).

Disc Fragmentation Recent work by [5] suggests that giant planets may form due to gravitational instability in the early stages of protostellar disc evolution, when the disc is relatively massive and gravitationally active. Numerical simulations of the evolution of massive protostellar discs performed by Boss show the formation of nonlinear spiral density waves, and in some cases the formation of dense clumps, with masses ~ 1 $M_{Jupiter}$, that appear to be relatively long lived. Due to the computationally intensive nature of this work, it is unclear whether these dense clumps are transient features that will eventually be sheared out, or are indeed stable objects that are able to survive, since the simulations were not run for long enough. Only longer time scale calculations will be able to determine this. It is worth noting that similar simulations were performed by [20], who found that the generation of spiral density waves in massive discs lead to the rapid redistribution of mass and angular momentum, without the formation of dense clumps. Recent work performed by [33] also suggests that a more complete treatment of the thermodynamics of protostellar disc gas, combined with different simulation boundary conditions, leads to a disc evolution in which gravitationally unstable dense clumps do NOT form. Obviously this idea remains contentious, and only time will tell what the final outcome will be.

Core Instability Model When considering the formation of planets in the core instability model, researchers usually use a disc model known as the Minimum Mass Solar Nebula (MMSN). This model is obtained by considering the radial distribution of mass contained in the planets of the solar system. The disc model is obtained by smearing out this mass to form a continuous distribution of matter, which is augmented with the required amount of hydrogen and helium to obtain a gas with solar abundance. The resulting disc has a surface density distribution given by

$$\Sigma(R) = 1700(R/1AU)^{-3/2}\mathrm{g\,cm}^{-2} \tag{1}$$

(see [14]). Giant planet formation theories are constrained by the observation that protostellar discs have observed lifetimes $\leq 10^7$ yr. An interesting fact is that the temperature distribution within this model falls below 170 K beyond a radius of $R \simeq 4$ AU, which is the point at which ices, such as water ice, are able to form. This radius is known as the 'ice condensation radius' or 'snow-line', beyond which the icy material increases the amount of solid material available to form solid planetary cores by about a factor of 5. Calculations of the rate of formation of planetary cores suggest that giant planets must form beyond the snow-line if giant planets are to form within the gaseous disc lifetime of $\leq 10^7$ yr.

In the core–instability scenario, the protostellar disc is envisaged to consist of a mixture of micron sized dust grains and gas that are initially well mixed. Collisions between the grains due to turbulence and Brownian motion cause them to grow *via* sticking, leading to the grains sinking towards the disc midplane. Estimates suggest that this sedimentation requires $< 10^5$ yr, and results in the formation of a dense dust layer consisting of cm sized particles. Early investigation into the properties of this dust layer (Goldreich & Ward 1973) suggested that it may become gravitationally unstable if the density in the dust layer approaches

$$\rho = 3M/4\pi a^3 \tag{2}$$

where M = mass of central star, a = orbital radius in disc. leading to the formation of km sized planetesimals. More recent analysis, however, suggests that turbulence in the disc may prevent the dust layer settling to such high densities ([41]; [12]). In this case the formation of planetesimals occurs through the continued process of binary collisions and coagulation of solid objects. Calculations suggest that this requires an additional 10^5 yr after the formation of the dust layer ([2]). Once objects are formed of size $\sim$ 10 km, the effect of gravitational focusing becomes important in the further growth of planetary cores, and can lead to the runaway growth of objects up to masses of $\sim 15\ M_{earth}$ within the required time scale ([34]). In the case of terrestrial planets forming interior to the snow-line, this runaway growth stalls when objects reach masses of a few lunar masses, because they exhaust the supply of the material within their local feeding zone. Here the objects reach their isolation masses and must undergo orbit crossing before further planetary growth can occur. Numerical simulations suggest that this requires around 3×10^8 yr.

Having formed a protoplanetary core of $\sim 15\ M_{earth}$ within the protoplanetary disc, the nebula gas becomes unstable to accretion onto the core, leading to the rapid growth of a gas–giant planet ([28]; [4]; [34]; [32]). The calculations of [34] suggest that this is just possible within 10^7 yr, providing the protostellar disc is a few times more massive than the MMSN model.

3.2 The Role of Disc–Planet Tidal Interaction

Having formed a protoplanet within a disc, there is a gravitational interaction between the disc and planet leading to an angular momentum exchange. The effects of this may be two-fold, depending on the mass of the planet. If the mass of the planet is such that

$$q < 3\left(\frac{H}{R}\right)^3 \tag{3}$$

where q = the mass ratio M_p/M_s, H = disc scale height, and R = radius, then the response of the nebula to the presence of a planet is linear, in the sense that nonlinear structures such as shocks do not form in the vicinity of the protoplanet. In this case the structure of the disc is relatively unaffected by the presence of the planet, and the interaction between disc and protoplanet leads to orbital migration of the planet. Angular momentum exchange occurs because the presence of the planet in the disc leads to periodic forcing of the disc material as it shears past the planet, leading to the excitation of spiral density waves at Lindblad resonances in the disc. These spiral density waves propagate away from the resonant locations, carrying with them an associated angular momentum flux. This angular momentum is deposited locally in the disc at the position where the waves are damped. The interaction is such that the planet exerts a positive torque on the more slowly moving material that lies beyond its orbital radius, and exerts a negative torque on the faster moving material that orbits interior to it. By Newton's third law each side of the disc exerts an equal but opposite torque on the protoplanet. Asymmetry in the Lindblad resonant locations results in the torque exerted by the outer disc being dominant, causing the protoplanet to migrate inwards. The migration time is given by (see [40]):

$$\tau_{mig} \simeq \frac{2}{c_1} \left(\frac{M_s}{M_p} \right)^2 \left(\frac{M_p}{\Sigma R^2} \right) \left(\frac{H}{R} \right)^3 \Omega^{-1} \tag{4}$$

where c_1 accounts for the torque asymmetry between the inner and outer disc, and should be proportional to H/R, M_s is the mass of the central star, M_p is the planet mass, Σ is the disc surface density, R is the orbital radius of the planet, Ω is the orbital angular velocity of the planet, and H/R is the disc aspect ratio. Evaluating this expression for typical disc parameters gives a migration time of a $\sim 10^5$ yr for a 1 M_{earth} planet initially at $r = 5$ AU, and $\sim 10^4$ yr for a 10 M_{earth} planet. This rapid migration during the earlier stages of planet formation represents a serious theoretical problem, given that the growth of a gas–giant planet requires about 10^7 yr ([34]). It seems that theory currently predicts that gas–giant planets should not form at all. It is worth noting two points at this stage. First, the calculations of gas–giant formation ([4]; [34]) do not incorporate migration into their 1-D spherically symmetric models. Migration may actually increase the rate of planet formation by allowing planetary cores to accrete material from a wider area in the disc. Second, the migration of planets in the intermediate mass range that generate a weakly nonlinear disc response has yet to be examined. It may turn out that the migration rates described above in equation (4) differ substantially in this regime.

When equation (3) no longer holds, then the disc response becomes non-linear, such that spiral shocks form in the close vicinity of the planet ([40]; [18]). In this case the angular momentum exchange between protoplanet and disc occurs locally, and an annular gap may be formed around the orbital location of the planet ([30]). Gap formation also requires the condition

$$q > \frac{40}{\mathcal{R}} \tag{5}$$

where $\mathcal{R}$ is the Reynolds number in the disc. This expression derives from the fact that a gap may only be maintained if the tidal torques exerted by the planet on the disc exceed the internal viscous torques that try to close the gap.

The formation of a gap can have important consequences for the rate at which the protoplanet accretes gas from the disc, and on the migration rate. The accretion of gas by a Jovian mass protoplanet embedded in a MMSN disc model was examined using numerical simulations by [6]. It was observed that the accretion rate was a sensitive function of planetary mass and disc viscosity, and the calculations suggested that giant planets should have masses

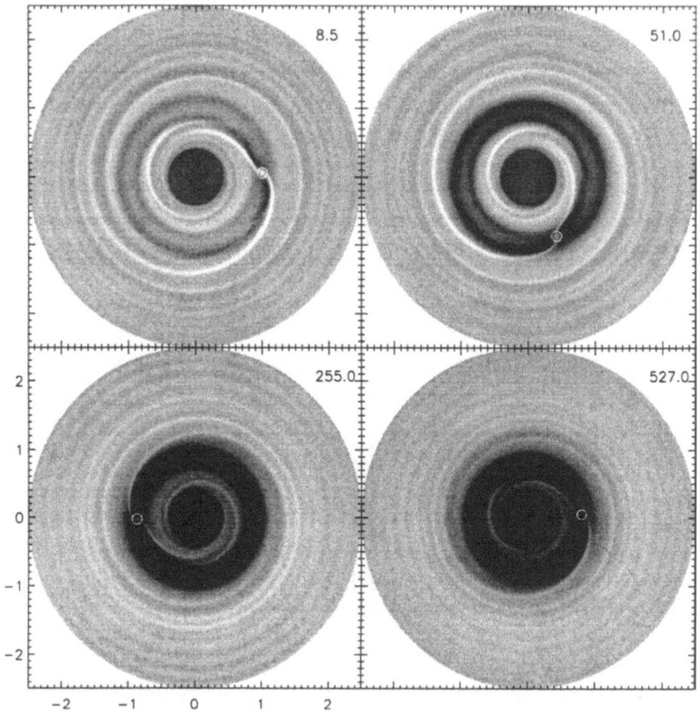

Fig. 6. This figure shows the evolution of a protoplanet embedded in a protostellar disc. The relative surface density of the disc material is represented by the grey scale. The white circle represents the position of the protoplanet. The disc is initially unperturbed at time $t = 0$, and the times corresponding to each panel are shown in the top right hand corners, in units of the initial orbital period.

in the range 1 − 10 $M_{Jupiter}$. Similar results were obtained in calculations by [17] and by [23]. More recent simulations performed by [29] examined the combined effects of gas accretion and orbital migration of Jovian mass protoplanets embedded in protostellar discs. These calculations showed that protoplanets initially located at 5 AU would migrate into the vicinity of their host stars within a few $\times 10^5$ yr, and would accrete an additional ~ 2 $M_{Jupiter}$ whilst doing so. In this situation, migration of the protoplanet occurs at a rate controlled by the viscous evolution of the disc (e.g. [22]), so that the migration time scale is similar to the viscous evolution time scale at the initial location of the planet. Figure 6 shows a time sequence for a Jupiter mass planet embedded in a disc, undergoing both migration and accretion (taken from [29]). The formation of a deep gap is apparent in this figure, as is the fact that the inner disc accretes onto the central star, leaving the protoplanet

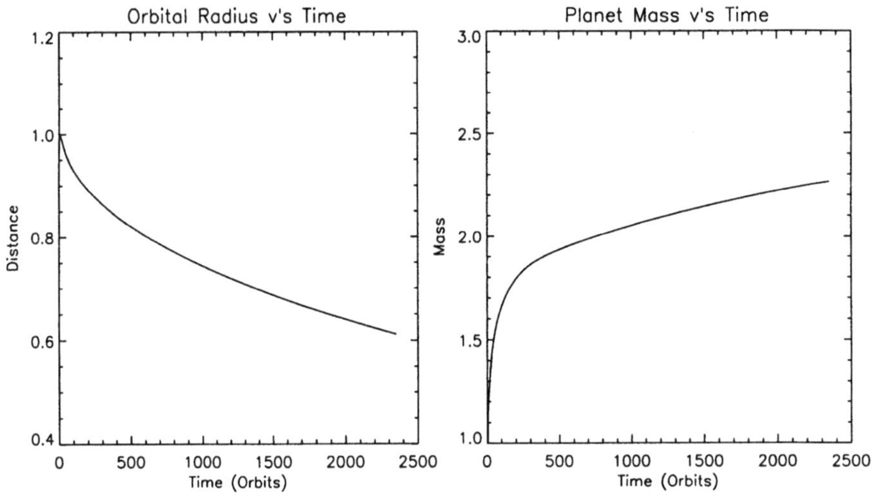

Fig. 7. This first panel of this figure shows the evolution of the orbital radius versus time for a migrating protoplanet emedded in a protostellar disc (see also fig. 6). The second panel shows the evolution of the mass of the protoplanet versus time, due to the protoplanet accreting gas from the disc.

sitting in an inner cavity. The viscous evolution of the outer disc pushes the planet inwards, as shown in the first panel of figure 7, which shows the orbital radius of the planet as a function of time. Extrapolation of this figure indicates that the planet will reach the central star after about 2×10^5 yr. The mass evolution of the protoplanet is shown in the second panel of figure 7. Extrapolation of this figure for 2×10^5 yr indicates that the planet will grow to a final mass of 3 $M_{Jupiter}$ before reaching the star. This is within the correct range for the observed extrasolar planets.

It is obvious from the first panel of figure 6 that a Jovian mass planet remains on a circular orbit when interacting with a standard disc. The origin of the eccentric orbits observed in the extrasolar planet data cannot be explained by interaction with a disc. Recent work performed by [32] indicates that eccentricity growth through disc-companion interaction only occurs for objects with masses ≥ 20 $M_{Jupiter}$ – *i.e.* brown dwarfs, not planets. A possible explanation for the origin of the observed eccentricities is that gravitational scattering with additional planets that have since been ejected, or which exist at orbital radii not yet accessible to the radial velocity searches, has pumped up the eccentricity of the observed planets.

3.3 N–body Effects

Recent work suggests that disc–planet interaction is probably not responsible for the observed eccentricities of the extrasolar planets, though it may be relevant to the semimajor axis and mass distribution. One possibility is that the observed eccentricities are generated by mutual gravitational interactions between systems of one or more planets.

The interaction between two Jovian mass planets (in the absence of a gaseous disc) was considered by [35]. These authors presented the results of computer simulations which showed that in some cases one of the planetary companions was ejected from the system, leaving behind a planet on an eccentric orbit. Under extreme cases, the eccentricity was sufficiently large that tidal interaction with the central star is likely to circularise the orbit, giving rise to a short period system on a circular orbit. In general, however, one expects that such interaction will lead to systems in which both objects survive in eccentric orbits. Further work is required to determine whether the statistical distribution of semimajor axes and eccentricities of this scenario match the observations, and whether a large number of the known planetary systems show signs of addition companions on more distant orbits.

A scenario has been suggested by [21] in which a system of massive planets forms within a disc, where the number planets may be as great as eight. In this situation, gravitational interaction between the planets will perturb them onto crossing orbits, such that they will collide and develop eccentric orbits. Computer simulations showed that massive planets (i.e. $M_p > 5$ M_{Jup} on eccentric orbits could be formed, with eccentricities in the range $0.2 \leq e \leq 0.9$. This scenario, however, does not explain the observed eccentricity of lower mass planets, and may in fact lead to a prediction that the eccentricity should scale with planetary mass. In addition, uncertainties remain about the plausibility of forming a system containing a large number of Jovian mass planets within a disc.

An analysis of the interaction between systems of three Jovian mass planets was undertaken by [26]. It was reported that typically one planet was ejected from the system, leaving two remaining objects on eccentric orbits in the range $0.1 \leq e \leq 0.9$. The implication is that interactions of this type should lead to the formation of systems with eccentricities of $e \sim 0.9$, which have not been observed in the data. This suggests that if the observed eccentricities are induced by scattering events between systems of planets, then additional processes must be acting to reduce the eccentricities. This may be tidal circularisation for the closer systems, or on–going interaction between the planets leading to a longer term modulation of their eccentricities. Alternatively, differing initial conditions with a different planet mass spectrum may lead to different final eccentricities.

4 Future Directions

The next decade promises to be a very exciting time for researchers working in the fields of extrasolar planets and planet formation. Future proposals (too numerous to mention) include both ground-based and space-based observing missions that will characterise the number and type of nearby extrasolar planetary systems (see http:||www.obspm.fr/encycl/encycl.html).

Ground-based radial velocity searches will continue to search for, and presumably detect, systems of giant planets around nearby stars. Microlensing searches will place increasing severe constraints on the occurrence and nature of planetary systems. In addition there are a number of ground-based projects searching for planetary transits, such as the STARE and VULCAIN projects. Infrared interferometers such as Keck and the VLT may reveal the presence of planets orbiting in discs by virtue of the gaps and spiral density waves that they generate, and similar structures may be observed by ALMA (Atacama Large Millimetre Array).

Future space-based projects include COROT, EDDINGTON, and KEPLER which will search for the transits of planets, SIM (Space Interferometry Mission) which will use interferometry to perform high resolution astrometry to detect the gravitational pull of planets on their host stars, and DARWIN/TPF (Terrestrial Planet Finder) which will use nulling interferometry to obtain direct images of planets and may detect spectroscopic signatures of planetary atmospheres.

Future observing programmes should also reveal the link between the low mass end of the stellar initial mass function and the upper mass limit of planetary objects, which has become topical due to the recent discovery of free–floating planetary mass objects in young star clusters [42].

On the theoretical front, future research will lead to an understanding of how planets interact with discs in which MHD turbulence is generated by the Balbus–Hawley instability ([1]). The examination of planets embedded in realistic disc models will allow a greater understanding of gap formation, migration, and mass accretion. In addition research will examine the role of disc fragmentation as a means of forming planets and brown dwarfs. The core instability model will be re–examined with the effects of planetary migration being included in order to understand how the time scales for planet formation may change, and the migration of intermediate mass planets (i.e. $M_p \simeq 10$ M_{earth}) will be examined. The role of multi–planet interactions in the presence and absence of discs will be examined to determine the role of these interactions in determining the orbital parameters of planetary systems for comparison with observations. A number of questions concerning the build–up of solid objects through sticking will also be addressed to examine how well large protoplanetary objects may be assembled by this process. The issue of planet formation in binary systems will also be addressed, as well as a close examination of the theoretical lower mass limit for the formation of planets and brown dwarfs by the fragmentation of molecular clouds.

In summary, future observational and theoretical research over the next decade promises to greatly enhance our understanding of planetary systems in the universe, ultimately providing information on the likely existence of habitable planets around other stars.

5 Conclusions

We have presented a (incomplete !) review of current observational data on extrasolar planets, and of current theoretical attempts to explain these data. Whilst the current data probably present an incomplete picture of the diversity of planetary systems in the Galaxy, they have nonetheless yielded some interesting surprises concerning the masses and orbital elements of extrasolar giant planets. Theoretical calculations indicate that disc–protoplanet tidal interactions may explain the distribution of mass and semimajor axes, and gravitational scattering within systems of protoplanets may account for their eccentric orbits. Further work is required to examine the detailed predictions of these models for comparison with the observational data.

The transit observations of HD 209458 confirm the gas–giant nature of the extrasolar planets, and suggest either formation *in situ* or a period of rapid inward migration post–formation for the short period systems. Future space-based and ground-based observations over the next decade promise to reveal a more complete picture of the nature of planetary systems. Theoretical work over the same time scale will address and clarify some of the outstanding theoretical questions posed by the observational data, hopefully providing us with a detailed picture of the formation mechanisms for planets in both our own and in extrasolar systems.

References

1. Balbus, S.A., Hawley, J.F., 1991, A Powerful Local Shear Instability in Weakly Magnetized Disks, Ap. J., **376**, 214
2. Beckwith, S.V.W., Henning, T., Nakagawa, Y., 2000, Dust Properties and Assembly of Large Particles in Protoplanetary Disks, Protostars and Planets IV (Tucson: University of Arizona Press; eds. Mannings, V., Boss, A.P., Russell, S.S.), p. 533
3. Beckwith, S.V.W., Sargent, A.I., 1996, Circumstellar disks and the search for neighbouring planetary systems, Mon. Not. R. astr. Soc., **383**, 139
4. Bodenheimer, P., Pollack, J.B., 1986, Calculations of the Accretion and Evolution of Giant Planets: The Effects of Solid Cores, Icarus, **67**, 391
5. Boss, A.P., 2000, Possible Rapid Gas Giant Planet Formation in Solar Nebula and other Protoplanetary Disks, Ap. J., **536**, L101
6. Bryden, G., Chen, X., Lin, D.N.C., Nelson, R.P., Papaloizou, J.C.B., 1999, Tidally Induced Gap Formation in Protostellar Disks: Gap Clearing and Suppression of Protoplanetary Growth, Ap. J., **514**, 344
7. Burrows, A., Guillot, T., Hubbard, W.B., Marley, M.S., Saumon, D., Lunine, J.I., Sudarsky, D., 2000, On the Radii of Close-in Giant Planets, Ap. J., **534**, L97-L100
8. Chambers, J.E., Wetherill, G.W., 1998, Making the Terrestrial Planets: N-body Integrations of Planetary Embryos in Three Dimensions, Icarus, **136**, 304
9. Charbonneau, D., Brown, T.M., Latham, D.W., Mayor, M., 2000a, Detection of Planetary Transits Across a Sun-Like Star, Ap. J., **529**, L45-L48

10. Charbonneau, D., Brown, T.M., Gilliland, R.L., Noyes, R.W., Burrows, A., 2000b, Probing the Outskirts of an Extrasolar Planet: HST Photometry of HD 209458, Poster at IAU Symposium 202 'Planetary Systems in the Universe', Manchester, U.K.
11. Goldreich, P., Ward, W.R., 1973, The Formation of Planetesimals, Ap. J., **183**, 1051
12. Cuzzi, J.N., Dobrovolskis, A.R., Champney, J.M., 1993, Particle - Gas Dynamics in the Midplane of a Protoplanetary Nebula, Icarus, **106**, 102
13. Henry, G.W., Marcy, G.W., Butler, R.P., Vogt, S.S., 2000, A Transiting '51 Peg-like' Planet, **529**, L41-L44
14. Hyashi, C., Nakazawa, K., Nakagawa, Y., 1985, Formation of the Solar System, (Tucson: University of Arizona Press), Protostars and Planets II, P.1100-1153.
15. Jha, S., Charbonneau, D., Garnavich, P.M., Sullivan, D.J., Sullivan, T., Brown, T., Tonry, J.L., 2000, Multicolor Observations of a Planetary Transit of HD 209458, **540**, L45-L48
16. Kant, I, 1755, Allegmeine Naturgeschichte und Theorie des Himmels (Leipzig)
17. Kley, W., 1999, Mass Flow and Accretion Through Gaps in Accretion Discs, Mon. Not. R. Astron. Soc., **303**, 696
18. Korykansky, D.G., Papaloizou, J.C.B., 1996, A Method for Calculations of Nonlinear Shear Flow: Application to Formation of Giant Planets in the Solar Nebula, Ap. JS., **105**, 181
19. Laplace, P.S., 1796, Exposition du Systѐme du Monde (Paris)
20. Laughlin, G., Bodenheimer, P., 2000, Nonaxisymmetric Evolution in Protostellar Discs, Ap. J., **436**, 335
21. Lin, D.N.C., Ida, S., 1997, On the Origin of Massive Eccentric Planets, Ap. J., **477**, 781
22. Lin, D.N.C., Papaloizou, J.C.B., 1986, On the Tidal Interaction between Protoplanets and the Protoplanetary Disk III - Orbital Migration of Protoplanets, Ap. J., **309**, 846
23. Lubow, S.H., Seibert, M., Artymowicz, P., 1999, Disk Accretion onto High Mass Planets, Ap. J., **526**, 1001
24. Marcy, G.W., Butler, R.P., 1995, The Planet around 51 Pegasi, AAS, 187.7004M
25. Marcy, G.W., Cochran, W.D., Mayor, M., 2000, Extrasolar Planets around Main-sequence Stars, Protostars and Planets IV, (Tucson: University of Arizona Press, eds: Mannings, V., Boss, A.P., Russell, S.S.), P.1285
26. Marzari, F., Weidenschilling, S.J., 1999, On the Eccentricities of Extrasolar Planets, American Astronomical Society DPS Meeting 31,
27. Mayor, M., Queloz, D., 1995, A Jupiter-mass companion to a solar-type star, Nature, **378**, 355
28. Mizuno, H, 1980, Prog. Theor. Phys., **64**, 544
29. Nelson, R.P., Papaloizou, J.C.B., Masset, F., Kley, W., 2000, The Migration and Growth of Protoplanets in Protostellar Discs, Mon. Not, R. Astron, Soc., **318**, 18
30. Papaloizou, J.C.B., Lin, D.N.C., 1984, On the Tidal Interaction between Protoplanets and the Primordial Solar Nebula, **285**, 818
31. Papaloizou, J.C.B., Nelson, R.P., Masset, F., 2001, Orbital Eccentricity Growth Through Disc-Companion Tidal Interaction, Astron. & Astrophys., In Press.

32. Papaloizou J.C.B, Terquem, C., 1999, Critical Protoplanetary Core Masses in Protoplanetary Disks and the Formation of Short-Period Giant Planets, Ap. J., **521**, 823
33. Pickett, B.K., Durisen, R.H., Cassen, P., Mejia, A.C., 2000, Protostellar Disc Instabilities and the Formation of Substellar Companions, Ap. J., **540**, L95
34. Pollack, J.B., Hubickyj, O., Bodenheimer, P., Lissauer, J.J., Podolak, M., Greenzweig, Y., 1996, Formation of Giant Planets by Concurrent Accretion of Solids and Gas, Icarus, **124**, 62
35. Rasio, F.A., Ford, E.B., 1996, Dynamical Instabilities and the Formation of Extrasolar Planetary Systems, Science, **274**, 954
36. Rasio, F.A., Tout, C.A., Lubow, S.H., Livio, M., 1996, Tidal Decay of Close Planetary Orbits, Ap. J., **470**, 1187
37. Sackett, P., 2000, Results from Microlensing Searches for Extrasolar Planets, to appear in the proceedings of the IAU Symposium 202 'Planetary Systems in the Universe', eds. A.J. Penny, P. Artymowicz, A.-M. Lagrange, S.S. Russell, A.S.P. Conference Series, Manchester, U.K.
38. Seager, S., Sasselov, D.D., 2000, Theoretical Transmission Spectra during Extrasolar Giant Planet Transits, Ap. J., **537**, 916
39. Terquem, C., Papaloizou, J.C.B., Nelson, R.P., Lin, D.N.C., 1998, On the Tidal Interaction of a Solar-Type Star with an Orbiting Companion: Excitation of g-mode Oscillations and Orbital Evolution, **502**, 788
40. Ward, W.R., 1997, Protoplanet Migration by Nebula Tides, Icarus, **126**, 261
41. Weidenschilling, S.J., 1980, Dust to Planetesimals - Settling and Coagulation in the Solar nebula, Icarus, **44**, 172
42. Zapatero-Osorio, M., Bejar, V., Martin, E., Rebolo, R., Barrado Y Navascues, D., Bailer-Jones, C., Mundt, R., 2000, Discovery of Young Isolated Planetary Mass Objects n the Orionis Star Cluster, Science, **290**, 103

The Giant Planets

Therése Encrenaz

Observatoire de Paris, 92195 Meudon, France

Abstract. The four giant planets of the solar system fall into two classes: the gaseous giants, Jupiter and Saturn, mostly composed of hydrogen and helium, and the icy giants, Uranus and neptune, which are less massive and mostly made of ices. This chapter reviews their thermal structures, their chemical compositions, and their cloud structures. The abundance ratios of the giant planets, inferred from their spectral properties, are reviewed, and the implications of these parameters in terms of formation and evolution models are discussed.

1 Introduction

The four giant planets - Jupiter, Saturn, Uranus and Neptune - are characterized by a large mass and volume, a low density, a large number of satellites and a ring system (Table 1). Their atmospheric composition is dominated by hydrogen and helium, with traces of other minor elements, in particular methane, in a reduced form. As explained in an accompaying chapter this classification is a direct consequence of their formation scenario. Because they were far from the Sun in a cold environment, most of the mass of the nebula (apart from hydrogen of helium) was in the form of ices and could be incorporated into planetesimals. The cores built by the giant planets were thus far bigger than those of the terrestrial planets, which were formed at

Table 1. Physical properties of the planets

Name	Symbol	Equatorial diameter relative to the Earth	Equatorial diameter (km)	Flattening	Mass relative to the Earth[1]	Mean density	Surface gravity (in ms^{-2})	Escape velocity (in $km\,s^{-1}$)	Sideral rotation (in days, or in hours, minutes and seconds)	Inclination of the equator to the orbital plane
Mercury	☿	0.382	4878	0	0.055	5.44	3.78	4.25	58 646 d	0°
Venus	♀	0.949	12 104	0	0.815	5.25	8.60	10.36	243 d (r)[2]	2°07′
Earth	⊕	1	12 756	0.003 353	1	5.52	9.78	11.18	23 h 56 min 04 s	23°26′
Mars	♂	0.533	6794	0.005	0.107	3.94	3.72	5.02	24 h 37 min 23 s	23°59′
Jupiter	♃	11.19	142 800	0.062	317.80	1.24	24.8	59.64	9 h 50 min to 9 h 56 min	3°04′
Saturn	♄	9.41	120 000	0.0912	95.1	0.63	10.5	35.41	10 h 14 min to 10 h 39 min	26°44′
Uranus	♅	3.98	50 800	0.06	14.6	1.21	8.5	21.41	17 h 06 min[3]	98°
Neptune	♆	3.81	48 600	0.02	17.2	1.67	10.8	23.52	15 h 48 min	29°
Pluto	♇	~0.2	~3000	?	0.002	1?		?	6 d 9 h 18 min	?

[1] $m_E = 5.976 \times 10^{24}$ kg
[2] (r) indicates that the rotation is retrograde
[3] rotation period of the magnetic field

higher temperature and could thus incorporate only metals and silicates. It is now believed that all four giant planets formed from an initial icy core of about 10 to 15 terrestrial masses, the gravity fields were sufficient to accrete the surrounding primordial nebula, mostly formed of hydrogen and helium.

The giant planets fall into two classes: the gaseous giants, Jupiter and Saturn, and the icy giants, Uranus and Neptune. The relative mass fraction of the initial core is about 4% for Jupiter, 13% for Saturn, and more than 60% for Uranus and Neptune. The lower quantity of primordial gas accreted by Uranus and Neptune, as compared to the two other giants, may be the consequence of their later formation, as their collapsing phase may have occurred after most of the primordial gas was dissipated in the T-Tauri phase of the early Sun.

The presence of a ring system and a large number of satellites is also the consequence of the giant planets' formation. The collapse of the surrounding subnebula around the icy cores probably generated the formation of a disk in which satellites and rings formed. This scenario explains why many satellites of the outer planets are located in the vicinity of the planet's equatorial plane, on quasi-spherical orbits, rather like the planets around the Sun. Note, however, that some satellites do not follow this rule, in particular several outer Jovian satellites; these satellites, of different origin, are probably captured asteroids.

Jupiter and Saturn have been observed by astronomers from the ground since the appearance of the refracting telescopes, over three centuries ago. Galileo discovered the four Galilean satellites in 1610, and the Jovian Great Red Spot (GRS) was identified by Cassini in 1665. In 1659, Huygens discovered the nature of Saturn's rings, responsible for the long-term variations of the planet's disk. As a result of the use of large telescopes by astronomers, Uranus was discovered by Herschel in 1781. In 1846, Neptune was finally discovered by Galle, on the basis of calculations of celestial mechanics, simultaneously performed by Adams in England and Le Verrier in France. Information about the chemical composition of the giant planets was obtained with the development of ground-based spectroscopy, especially in the infrared range. A major improvement in our knowledge of the giant planets has come from space exploration, by Pioneer 10 and 11 in the 1970s, Voyager 1 and 2 between 1979 and 1989, and finally Galileo since 1995. The next step of this exploration will come when the Cassini-Huygens mission, launched in 1997, will approach the Salimian system in 2004.

2 Thermal Structure

The thermal structure of the giant planets is characterized by a convective troposphere where the gradient is adiabatic (or very close to it), an inversion level (the tropopause) at a pressure level of about 100 mbar, and a radiative region, the stratosphere, where the temperature increases with altitude

(Fig. 1). This temperature increase is caused by the absorption of the solar flux by methane and aerosols. The fact that the tropopause occurs at about the same pressure level for all the giant planets reflects, to first order, their similar chemical composition. The temperature minima decrease as the heliocentric distance increases, from 100 K in the case of Jupiter to 50 K for Neptune. This cold trap determines the physical state of several condensable compounds, such NH_3 which condenses in the upper tropospheres of Jupiter and Saturn, and CH_4 which saturates in the upper tropospheres of Uranus and Neptune.

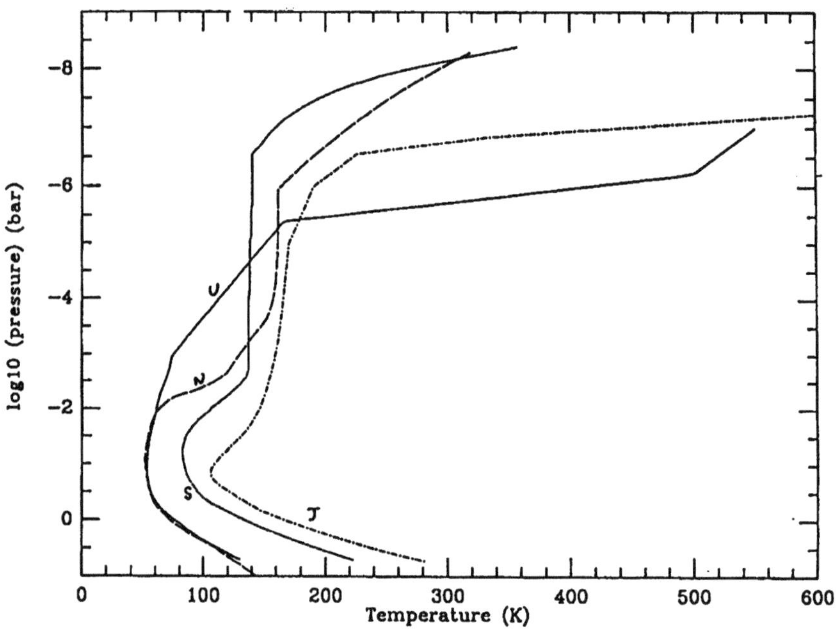

Fig. 1. The thermal profiles of the giant planets (from Encrenaz, 1999).

The thermal structure of the giant planets has been deruved from radio-occultation and UV-occultation measurements from Voyager 1 and 2. In addition, the thermal profile of Jupiter has been directly measured by the Galileo probe, down to a pressure level of 22 bars.

In the upper stratospheres of the giant planets, at pressure levels lower than 1 mbar, the temperature increases rapidly, and the temperature profiles strongly differ from one planet to another. The Jovian profile is the only one to have been measured accurately, thanks to the in-situ measurements of the Galileo probe. The heating mechanisms could be due to gravity waves propagating from below, and/or to precipitating high-energy particles coming from the magnetosphere.

3 Chemical Composition

The atmospheres of the giant planets are dominated by molecular hydrogen, helium (expected with a mixing ratio of about 10% by volume), and traces of CH_4, NH_3, H_2O, etc. Many minor species have been detected by infrared spectroscopy, either from the ground at selected wavelengths (especially in the 4.5-5.2 mm and 7-13 mm regions where the terrestrial atmosphere is transparent) or, more recently, with the Infrared Space Observatory which operated from Earth's orbit between 1995 and 1998.

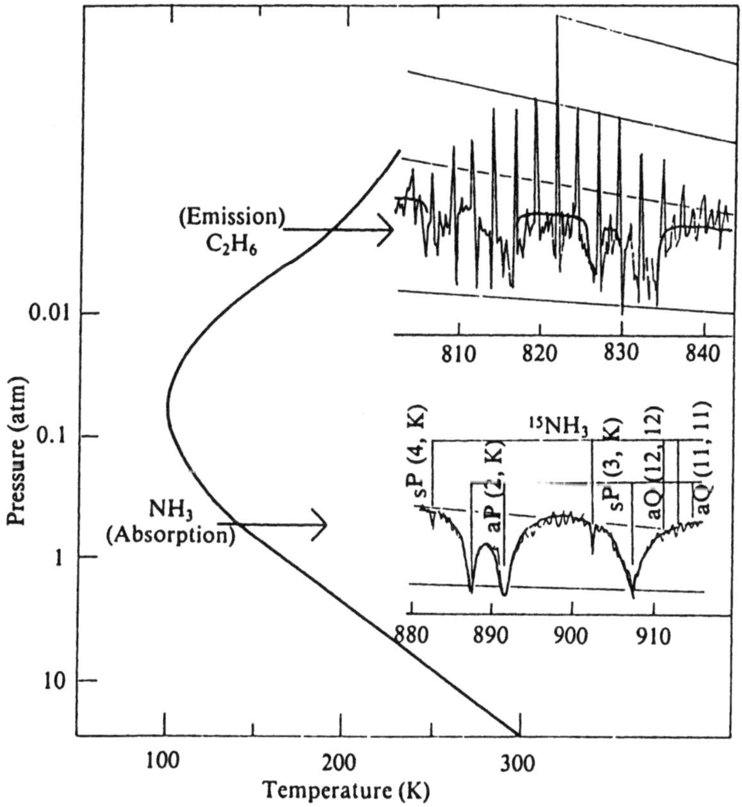

Fig. 2. Formation of molecular lines in the atmospheres of the giant planets. NH_3, present in the jovian troposphere, appears in absorption. C_2H_6, present in the stratosphere, is superimposed in emission. The data are from Tokunaga et al. (1979).

The infrared spectrum of a giant planet is (as in the case of any solar system object) the combination of a reflected solar component which dominates below about 4 mm, and a thermal component, which prevails at longer wavelengths. In the reflected solar component, atmospheric species are observed as absorption bands or lines in front of the solar continuum (typically a 5770 K blackbody curve, peaking at 0.5 μm). In the thermal regime, the observed flux strongly depends upon the thermal profile of the planet, as it probes an atmospheric level where the opacity is close to 1. The opacity depends upon the strength of the molecular band or line, and upon the abundance of the atmospheric molecule. As a result, stratospheric species are observed in emission because the temperature increases with height in this region, while tropospheric species appear in absorption, because the tropospheric temperature gradient is negative (Fig. 2). In the near-infrared range, where the reflected solar component dominates, the spectra of the giant planets are dominated by strong methane absorption (Fig. 3). In the thermal regime, the infrared spectra of Jupiter and Saturn are a combination of emission bands (by CH_4 and hydrocarbons) and absorption bands (by NH_3 and PH_3). In the case of Uranus and Neptune, the thermal spectra only show hydrocarbon emissions (Fig. 4).

Table 2 lists the atmospheric compounds presently identified in the giant planets. They fall in different categories.

The first category includes H_2, HD, CH_4, NH_3 and their isotopes, and other minor species associated with hydrogen, mostly found in the deep tropospheres of Jupiter and Saturn (PH_3, GeH_4, AsH_3, H_2O). In the case of Jupiter, new information has come from the in-situ measurements of the Galileo probe. These data are used to determine abundance ratios in the giant planets.

The second class of species, found in the stratospheres of the giant planets, includes hydrocarbons produced by the photodissociation of CH_4 (C_2H_2, C_2H_6,...). New species have been recently discovered by ISO: CH_3C_2H and C_4H_2 in Saturn, CH_3 in Saturn and Neptune. These species constrain photochemical models of the giant planets.

The third category includes stratospheric species of external origin, recently discovered by ISO (H_2O, CO_2). Water vapor was discovered in the stratospheres of all giant planets (Fig. 5) as well as Titan, above the condensation level (about 140 K), with a mean mixing ratio of about 10^{-9} in all cases. Because of the presence of the tropopause which acts as a cold trap, this water cannot come from the interior (where H_2O is also present) and must have an external origin. The inferred incoming water flux is remarkably equal for all giant planets and Titan, in the range of 10^6 cm^2s^{-1}. Two sources of oxygen have been proposed: an interplanetary meteoritic flux or a local source originating from rings and/or icy satellites. In addition, CO_2 (previously known to be present in Titan's stratosphere) was also detected

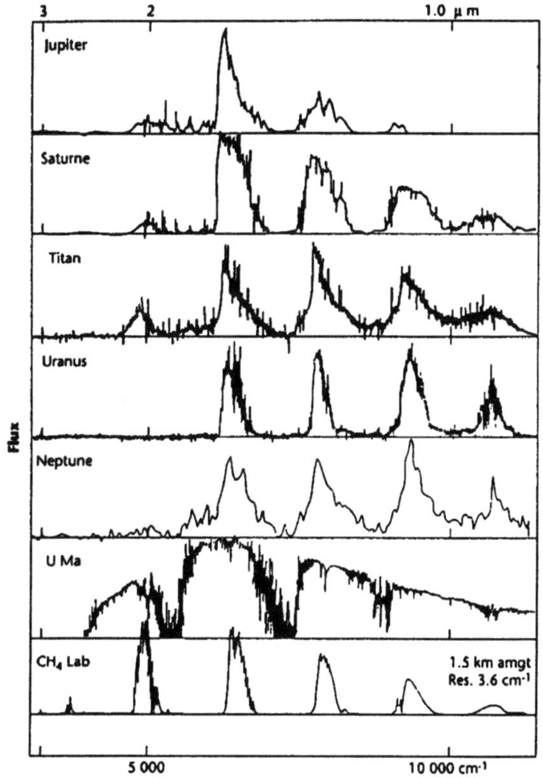

Fig. 3. The near-infrared spectrum of the giant planets and Titan between 1 and 3 μm, compared to the spectrum of a reference star (showing the spectral windows of the terrestrial atmosphere) and a laboratory spectrum of CH_4 showing the methane absorption bands (after Larson, Ann. Rev. Astron. Astrophys. 18, 43, 1980).

in Jupiter, Saturn and Neptune by ISO. It could have an external source, as H_2O, but could also be the result of chemical reactions in the stratosphere.

The last category includes HCN and CO the origins of which are still under debate. CO has been detected in Jupiter and Saturn by 5 μm spectroscopy, but its origin, stratospheric or tropospheric, is unclear. More recently, CO and HCN have been detected in Neptune's stratosphere, by heterodyne millimetre spectroscopy, in surprisingly high amounts. The high abundance of stratospheric CO in Neptune, while it is not detected in Uranus, might explain the detection of stratospheric CO_2 in Neptune but not on Uranus, as CO_2 could be produced from a chemical reaction between OH and CO. The question, however, is still under debate. The origin of stratospheric HCN in Neptune is attributed to a chemical reaction between CH_3 (coming from methane photodissociation) and nitrogen atoms, coming either from outside (i.e. Triton's tenuous nitrogen-dominated atmosphere) or from N_2 coming

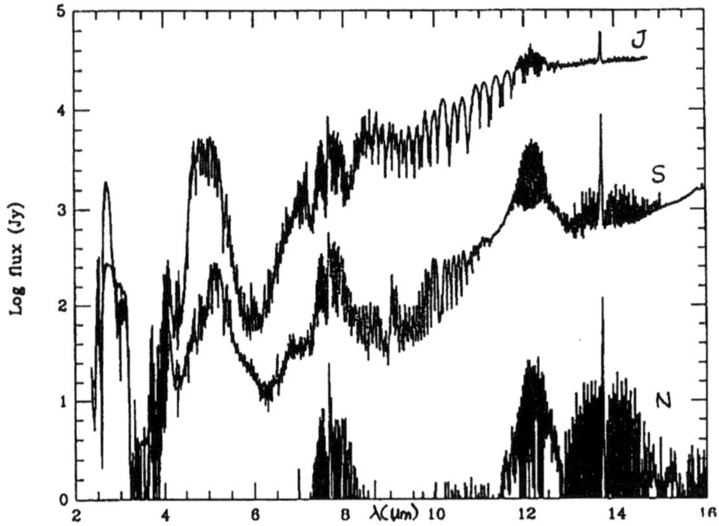

Fig. 4. The spectrum of the giant planets (Jupiter, Saturn and Neptune) between 2 and 16 μm, recorded by the Short-Wavelength Spectrometer (SWS) of ISO. The Uranus spectrum is only visible in the C_2H_2 band at 13.7 μm, where its flux in 3 times weaker than Neptune's.

from Neptune's interior. It has to be noted that, at the time of the collision of comet Shoemaker-Levy 9 with Jupiter, in July 1994, CO and HCN were detected in Jupiter's stratosphere, together with other transient species formed after the explosions (Table 3). Some of these species, including HCN, were observed for more than a year, after the impact.

4 Cloud Structure

4.1 The Nature of Clouds

On the basis of the chemical composition and the thermal profiles of the giant planets, it is possible to determine, by use of thermodynamical equilibrium models, the expected cloud structure of the giant planets. This method predicts, in the case of Jupiter, a NH_3 cloud near 140 K (0.6 bar), a NH_4SH cloud at 210 K (2 bar) and an H_2O cloud at about 250 K (4 bar). A comparable cloud structure is found for Saturn. In the case of Uranus and Neptune, a CH_4 cloud is expected around 80 K, at about 1 bar.

It is not easy to identify the nature of condensables by spectroscopy, because the spectroscopic signatures of solid phases are generally broad and difficult to interpret. For this reason, our knowledge of the nature of clouds is much less advanced that in the case of the gaseous composition. In the case of Jupiter, however, the NH_3 cloud has recently been identified, at 3 μm, by

Table 2. Chemical composition of the giant planets

GASEOUS SPECIES	JUPITER	SATURN	URANUS	NEPTUNE
H_2	1	1	1	1
HD	$1.8\ 10^{-5}$	$2.3\ 10^{-5}$	$5.5\ 10^{-5}$	$6.5\ 10^{-5}$
He	0.157	0.03	0.18	0.23
CH_4 (trop.)	$2.1\ 10^{-3}$	$4.4\ 10^{-3}$	$2\ 10^{-2}$	$4\ 10^{-2}$
CH_3D (trop.)	$2.5\ 10^{-7}$	$3.2\ 10^{-7}$	10^{-5}	$2\ 10^{-5}$
C_2H_2		$3.5\ 10^{-6}$ (0.1 mbar)	$2\text{-}4\ 10^{-7}$ (0.1-0.3 mbar)	$1.1\ 10^{-7}$ (0.1 mbar)
C_2H_6	$4.0\ 10^{-6}$ (0.3-50 mbar)	$4.0\ 10^{-6}$ (<10mbar)		$1.3\ 10^{-6}$ (0.03-1.5 mbar)
CH_3C_2H	detected	$6.0\ 10^{-10}$ (<10mbar)		
C_4H_2		$9.0\ 10^{-11}$ (<10mbar)		
C_2H_4	$7\ 10^{-9}$			
C_3H_8	$6\ 10^{-7}$			
C_6H_6	$2\ 10^{-9}$	detected		
CH_3		$0.2\text{-}1\ 10^{-7}$ (0.3 μbar)		$2\text{-}9\ 10^{-8}$ (0.2 μbar)
NH_3 (trop.)	$2\ 10^{-4}$ (3-4 bar)	$2\text{-}4\ 10^{-4}$ (3-4 bar)		
PH_3 (trop.)	$6\ 10^{-7}$	$1.7\ 10^{-6}$		
H_2S (trop.)	$7\ 10^{-6}$ (8 bars)			
GeH_4	$7\ 10^{-10}$	$2\ 10^{-9}$		
AsH_3	$3\ 10^{-10}$	$2\ 10^{-9}$		
CO (trop.)	$1.5\ 10^{-9}$	$2\ 10^{-9}$		
CO	$1.5\ 10^{-9}$	$2\ 10^{-9}$		10^{-6}
CO_2 (strat.)	$3\ 10^{-10}$ (<10 mbar)	$3\ 10\ ^{-10}$ (<10 mbar)		$5\ 10^{-10}$ (<5 mbar)
H_2O (trop.)	$1.4\ 10^{-5}$ (3-5 bar)	$2\ 10^{-7}$ (>3 bar)		
H_2O (strat.)	$1.5\ 10^{-9}$ (<10 mbar)	$2\text{-}20\ ^{-9}$ (<0.3 mbar)	$5\text{-}12\ 10^{-9}$ (<0.03 mbar)	$1.5\text{-}3.5\ 10^{-9}$ (<0.6 mbar)
HCN				$3\ 10^{-10}$)
H_3^+ (high strat.)	detected	detected	detected	

ISO and by the Near Infrared Mapping Spectrometer (NIMS) of the Galileo orbiter. However, we still have no precise assignment for the nature of the so-called "chromophores" responsible for the orange and brown colours of the clouds of Jupiter, and for the red colour of the GRS. Possible candidates include phosphorus and sulfur compounds or hydrocarbon condensates. In the case of Saturn, the white-yellow colour could be due to a larger abundance of NH_3 ice. The greenish colour of Uranus could be due to gaseous methane, which, in larger amounts, would be responsible for the deep blue colour of Neptune, while the white spots of Neptune could be duc to high-altitude cirrus clouds.

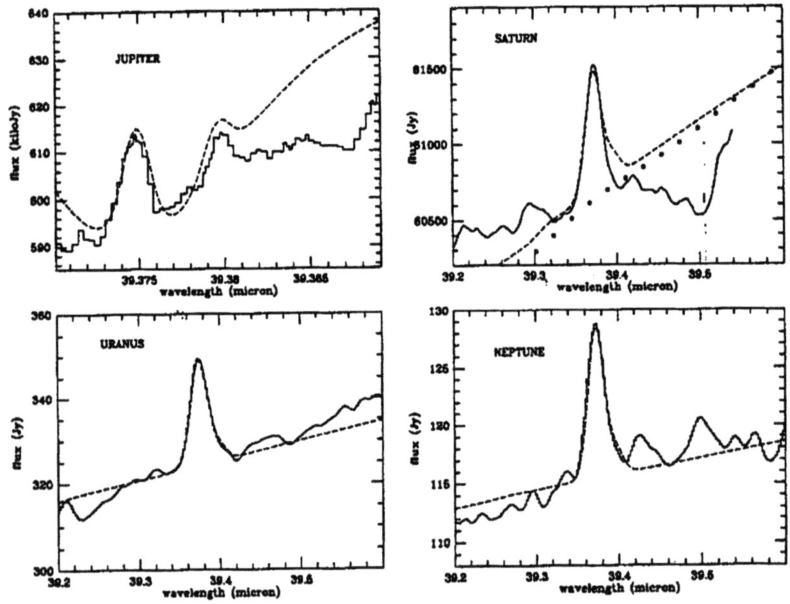

Fig. 5. H_2O emission lines detected in the four giant planets by ISO-SWS at 39.4 μm. The Jupiter data were recorded in the Fabry-Perot mode (R=3100) to avoid saturation. The other data were recorded in the grating mode (R=2000). The figure is taken from Lellouch et al., ESA-SP419, 131, 1997).

Table 3. New molecules formed in Jupiter at the time of the collision of comet Shoemaker-Levy 9.

Species	Total Mass (g)
CO	$2.5\ 10^{14}$
H_2O	$2\ 10^{12}$
S_2	$7\ 10^{11}$
CS_2	$1.5\ 10^{11}$
CS	$5\ 10^{11}$
OCS	$3\ 10^{12}$
HCN	$6\ 10^{11}$
C_2H_4	$3\ 10^{12}$
NH_3	$1\ 10^{13}$
PH_3	?
H_2S	?

after Lellouch, in "The collision of comet Shoemaker-Levy 9 and Jupiter", [12]

4.2 Jupiter Cloud Structure: The Galileo Results

Surprisingly, the in-situ measurements of the Galileo probe did not confirm the expected 3-cloud structure of the Jovian atmosphere. A weak cloud level was found at 0.5 bar, presumably due to NH_3 ice. Another weak layer was found around 1.25 bar, possibly due to NH_4SH, but there was no trace of any H_2O cloud at lower levels. Consistently with this result, the Galileo probe measured a very low abundance of water in the deep troposphere. The current explanation is that the Galileo probe entered an atypical region of Jupiter, "a hot spot". These regions are known to be strong infrared emitters at 5 μm, where the radiation comes from deep atmospheric levels (5-7 bars). It is now generally believed that these hot spots, and in particular the Probe entry site, are dry, cloud-free regions of downdraft motion; thus, they are not representative of the whole Jovian disk.

4.3 Jupiter and Saturn: The Belt-Zone System

The visible aspect of the jovian cloud structure, with the white zones, the dark belts and the Great Red Spot, has been known for centuries. The occurence and the stability of the belts and zones can be explained as an effect of the fast rotation of the planet. Observations of the jovian disk at different wavelengths, including the thermal radiation, show that the zones are colder and probably higher than the belts. Prior to Galileo, the interpretation was that zones were moist regions of upward motion, with cloud condensation at the top, whereas the belts were drier regions of downward motion. This interpretation was based on the fact that belts were regions of strong thermal infrared radiation, implying that the infrared radiation comes from depth. Within the belts, the so-called "hot spots", identified from their high radiation at 5 mm, were special examples of very dry and cloud-free regions. The Galileo probe entered one of these regions.

This general scheme of the jovian circulation is still valid after the Galileo mission. However, the observations have shown that the circulation pattern is much more complicated than a regular series of upward motion in the zones and downdraft in the belts. Such a regular pattern may exist, but many more complex small-scale motions appear to be superimposed upon it. As indicated by their temperature, the plumes and white ovals appear to be isolated areas of ascending motions while, as discussed above, hot spots are downdraft regions, in some cases at very small scale.

Belts and zones are also visible in Saturn, but they show much less contrast, possibly because of the thicker NH_3 cloud. The hot spots are also show less contrast than in the case of Jupiter. The reason for this difference is not clearly understood. The disk of Uranus shows very little contrast. The Neptune disk shows distinct features, such as the dark spot identified by Voyager 2 in 1989. Like the GRS in Jupiter, this feature, also in the southern hemisphere, was interpreted as a giant anticyclonic phenomenon of ascending motion. However, it did not have the stability of the GRS, since it was no longer visible in the HST images taken 15 years later.

5 Abundance Ratios

The aim of determining abundance ratios in the giant planets is to infer, as well as possible, the bulk composition of these objects. It is thus necessary to measure the mixing ratios of the minor species as deep as possible in the troposphere, in order to avoid changes due to circulation, condensation or photodissociation. D/H, He/H and C/H, which can be reliably determined, have important implications for the early conditions of the primordial nebula and for the formation processes of the giant planets. N/H, P/H, S/H and O/H are more difficult to determine for the reasons mentioned above.

5.1 D/H

According to the Standard Big Bang Model (SBBM), deuterium, initially produced by nuclear nucleosynthesis, is destroyed in stars and converted into ^{3}He. A determination of D/H in the primordial nebula therefore provides a determination of this ratio.

5.2 Gy Ago

The solar system formation scenario described in the accompanying chapter suggests that the D/H ratio in Jupiter (and, to a lesser extent, in Saturn) might be representative of the protosolar value. In contrast, D/H in Uranus and Neptune is expected to be enriched, due to deuterium enrichment in the ices of their initial cores. This enrichment comes from deuterium fractionation in icy planetesimals, from isotopic exchange occurring in ion-molecule and/or molecule-molecule reactions at low temperature. This deuterium enrichment is observed in several species in the interstellar medium.

The D/H ratio in giant planets was poorly determined prior to the ISO and Galileo data. Two methods had been used: the analysis of HD lines in the visible range, which suffered from uncertainties due to scattering effects, and the study of CH_3D signatures, the interpretation of which was limited by the uncertainty in the CH_3D fractionation factor. The observation of infrared rotational lines of HD with the ISO satellite provided both a direct measurement of D/H (HD being the main deuterated species) and an homogeneous determination of this ratio in the four giant planets. In addition, in the case of Jupiter, a direct measurement of D/H has been obtained by the mass spectrometer of the Galileo probe.

HD has been detected in the 4 giant planets by the ISO satellite (Fig. 6). Results are shown in Fig. 7, together with the Galileo results and D/H determinations in other solar-system objects. It can be seen that, in the case of

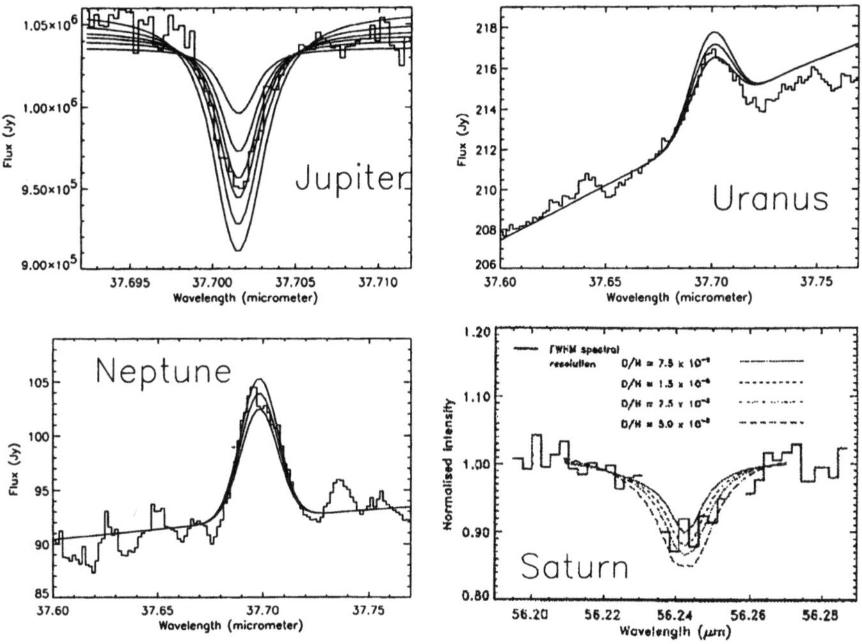

Fig. 6. Detection of HD in the four giant planets with ISO. The R(2) line at 37.7 μm was detected in Jupiter, Uranus and Neptune by SWS in the FP mode (R=3100). The R(1) line at 56.2 μm was detected in Saturn by the Long-Wavelength Spectrometer (LWS) in the FP mode (R=8000). The figure is taken from Lellouch, ESA-SP 427, in press, 1999).

Jupiter, the ISO measurement of D/H is in good agreement with the Galileo result which shows smaller error bars. A second interesting conclusion is that the ISO determinations of D/H in the giant planets show an increase with heliocentric distances, ranging from about 2 10^{-5} for Jupiter to 6 10^{-5} for Neptune, as expected from the nucleation model. Using models of Uranus' and Neptune's interiors, it is possible to derive an estimate of the D/H ratio in the ices of their initial cores. The result is about 10^{-4}, which is about 3 times less than the D/H ratio measured in comets. This difference has important implications about where and when comets formed in the early stages of solar system history.

Taking into account the error bars, the D/H ratio in Jupiter inferred by the Galileo probe is in good agreement with the estimated protosolar value as derived from ^{3}He measurements. This supports the idea that D/H in Jupiter is indeed representative of the protosolar value, as suggested by the nucleation model. On the other hand, the D/H value in Jupiter is slightly above the current estimates of D/H in the local interstellar medium. This result suggests a moderate deuterium depletion over the past 4.6 billion years. Using an

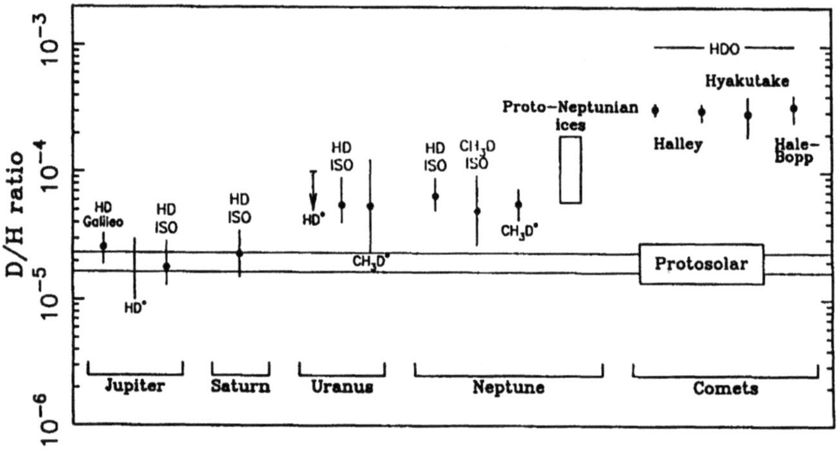

Fig. 7. The D/H ratio is several solar system objects. It can be seen that D/H increases from Jupiter to Neptune. The Jupiter value appears to be representative of the D/H value in the primordial nebula (protosolar value). The figure is taken from Encrenaz (1999).

evolutionary model with infall of primordial composition, it is possible to retrieve a primordial D/H ratio in the order of about 4 10^{-5}. This value corresponds, in the SBBM, to a baryon density of about 5 10^{-31}g/cm^3, a density which would be too small to close the Universe.

5.3 He/H

Prior to the Voyager observations, the expectation was that, since hydrogen is the main constituent and since helium is chemically inactive, no modification of the He/H would be expected in the giant planets' atmospheres. As a result, the He/H ratio should be the same in the four giant planets, and would be representative of its protosolar value. The unexpected conclusion which came from Voyager, however, was that both Jupiter and Saturn exhibit a helium depletion (small for Jupiter, very significant for Saturn) with respect to the Uranus and Neptune values.

Two remote sensing methods have been used to determine the He/H ratio in the giant planets. The first one is the inversion of the infrared spectrum of the planets between 20 and 50 μm. Because this spectrum is dominated by the pressure-induced spectrum of hydrogen, sensitive to H_2-H_2 and H_2-He collisions, a double iteration allows the retrieval of both the temperature profile and the He/H ratio. The second method is the inversion of occultation curves (either from ground-based observations of stellar occultations by the planets, or from radio-occultation experiments on space missions). As the occultation curve is sensitive to the refractive index of the atmosphere, which

Table 4. Helium mass fractions, Y, in the giant planets (from von Zahn et al., J. Geophys. Res. 103, 22815, 1998)

Source	Y
Jupiter (Galileo)	0.234 +/- 0.005
Jupiter (Voyager)	0.18 +/- 0.04
Saturn (Voyager)	0.06 +/- 0.05
Uranus (Voyager)	0.262 +/- 0.048
Neptune (Voyager)	0.32 +/- 0.05
Sun (helioseismology)	0.24 +/- 0.01
Protosolar	0.275 +/- 0.01
Primordial	0.232 +/- 0.005

is proportional to T/m (T being the temperature and m the mean molecular weight), a double iteration of these data also leads to a determination of T and He/H.

The two methods were usefully combined on the infrared and radio-occultation data of the Voyager spacecrafts for a double retrieval of these atmospheric parameters. Infrared data were obtained by the Infra-Red Interferometric Spectrometer (IRIS). In addition, a direct measurement of the helium abundance in Jupiter has been recently obtained from two experiments onboard the Galileo probe. The Galileo determination is higher than the Voyager one, although in marginal agreement taking into account the error bars. The helium mass fraction Y in the four giant planets is given in Table 4, together with the protosolar value, estimated from evolutionary models, and the primordial value, inferred from the observation of extragalactic H II regions. It should be noted that the protosolar value of is significantly larger than the current estimate of its primordial value. Table 4 shows that, taking into account the error bars, the values of Y in Uranus and Neptune are compatible with the protosolar estimate while the Jupiter value (which, by coincidence, agrees with the Y value in the convective regions of the Sun) is slightly lower, and the Saturn one is very significantly lower.

What could be the explanation for such a depletion? It has been argued that helium condensation could take place in the liquid hydrogen phase which is expected to exist in the interiors of Jupiter and Saturn. As Jupiter's, the interior of Saturn has been slowly cooling since its formation. Theoretical models predict that, a few billion years after the planets' formation, Saturn's adiabat crossed the saturation curve of helium in liquid hydrogen, leading to the condensation of helium droplets falling towards the planet's center (Fig. 8). This differentiation mechanism has had the effect of depleting the outer envelope of Saturn in gaseous helium, (as observed by Voyager) but has also released internal energy. Such an effect could contributing to the internal energy source required by the Voyager IRIS experiment (0.8 times the absorbed solar energy). In the case of Jupiter, the helium differentiation mechanism could also apply, but it should have started at a later stage and its effect should be lower, since the planet initiated its cooling from a higher

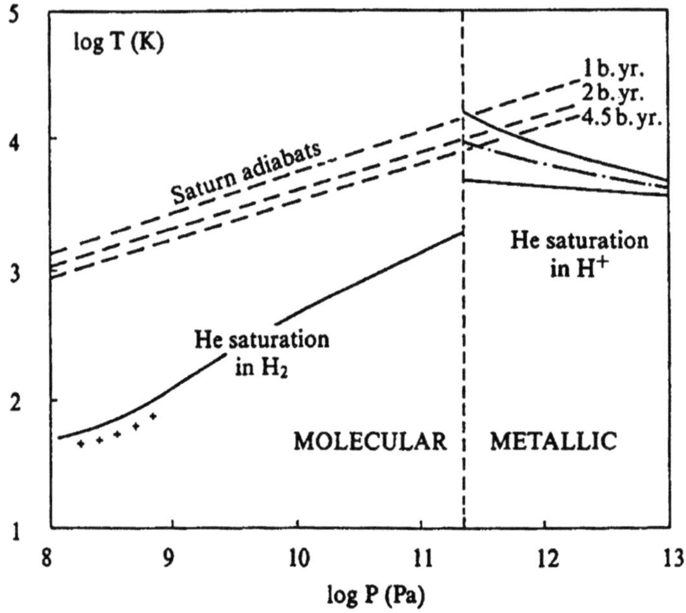

Fig. 8. Helium condensation in Saturn's interior. As Saturn cools down, its T(P) adiabat crosses the saturation curve of helium in liquid hydrogen and helium differentiation takes place. As a result, helium droplets sink toward the centre and helium is depleted in the outer envelope. The mechanism is less efficient in Jupiter because Jupiter started from a higher temperature. It does not take place in Uranus and Neptune because their internal pressure is not sufficient for hydrogen to be in liquid phase (after D. Stevenson, Ann.Rev.Earth Plan.Science, 10, 257, 1982).

temperature. This is consistent with the observed jovian value of Y. Finally, on Uranus and Neptune, helium differentiation is not expected to have taken place, because the pressures in their interiors are not high enough for hydrogen to be present in the liquid phase. Their He abundance could thus be representative of its protosolar value.

Noble gases other than helium have been identified for the first time in Jupiter by the Neutral Mass Spectrometer (NMS) of the Galileo probe. Ne in Jupiter is strongly depleted with respect to its solar value, with a Ne/H ratio of ≤ 0.13 times the solar value. The Ar/H jovian ratio is ≤ 1.7 times the solar value. Kr/H and Xe/H in Jupiter have upper limits of 5 times their solar values. With the on-going improvement of calibration procedures to the NMS, there is hope that these upper limits will ultimately be converted into measured mixing ratios. Isotopic ratios of Ne, Ar, Kr and Xe were found to have the solar (or terrestrial) values.

5.4 C/H

The C/H ratio in the four giant planets was first estimated from the strong methane absorption shown by the near-infrared spectra of the planets (Fig. 3). In all cases, these measurements refer to the troposphere, below the condensation level which appears in Uranus' and Neptune's upper tropospheres. Since no condensation takes place in Jupiter and Saturn, stratospheric measurements of CH_4 can also be used for these two planets. In these two cases, C/H has been inferred from the inversion of the ν_4 band of methane at 7.7μm, using the infrared Voyager data. More recently, the C/H value in Jupiter has been derived by the Galileo probe, confirming earlier results by Voyager.

All observations show the the C/H ratio in the giant planets is enriched with respect to its value in the Sun, by a factor which increases from 3 in the case of Jupiter to about 60 in the case of Neptune (Table 2).

5.5 P/H, S/H, N/H, O/H in Jupiter and Saturn

PH_3, NH_3 and H_2O have only been detected in the tropospheres of Jupiter and Saturn. Their condensation in Uranus and Neptune prevents them being detected in the tropospheres of these planets. In addition, H_2S has been detected in Jupiter by the Neutral Mass Spectrometer (NMS) of the Galileo probe.

The P/H ratio is difficult to establish because the PH_3 mixing ratio is depleted in the upper troposphere and above by photodissociation. According to thermochemical models, PH_3 is not expected to be present in the upper troposphere of Jupiter as, between 800 and 300 K, it should be oxidised by H_2O to form P_4O_6. Still, it has been detected at various infrared and at millimetre wavelengths. The currently accepted explanation is that strong vertical mixing transports PH_3 to the upper atmosphere at a rate faster than the conversion of PH_3 into P_4O_6 (a similar explanation holds for other disequilibrium species found in the troposphere, GeH_4, AsH_3 and CO). The 5 μm observations provide the deepest value of the mixing ratio. Both the Voyager analysis and the NIMS-Galileo data indicate a PH_3 mixing ratio close to the solar value. No significant additional information has so far been provided by the Galileo probe measurements. In the case of Saturn, a P/H enrichment by a factor of about 3 is found. In any case, the interpretation of P/H in terms of elemental ratio is probably meaningless, as the PH_3 mixing ratio is probably more diagnostic of vertical mixing.

As with PH_3, NH_3 shows a strongly depleted vertical profile in the upper troposphere, first because of the condensation of ammonia in a cloud at about 0.5 bar, and second because of solar photolysis. Information about the NH_3 mixing ratio in the deep atmosphere has come from the analysis of the radio spectrum of Jupiter and Saturn which is strongly influenced by the NH_3 inversion band centred at 1.35 cm. This study indicates a possible N/H enrichment by a factor 2 in Jupiter and by a factor 2 - 4 in Saturn. Additional information about the deep NH_3 mixing ratio has been provided by the radio signal of the Galileo probe, which indicated a possible N/H enrichment by a factor about 4 at pressures higher than 7 bars. This result, however, requires further confirmation.

H_2S was detected for the first time by the Galileo NMS experiment. Hydrogen sulfide is strongly depleted at about 2 bars as it reacts with NH_3 to form NH_4SH, which is expected to condense around this level. Consequently, it could not be identified by remote-sensing spectroscopy. NMS has measured a S/H enrichment by a factor 2.5 at pressures higher than 16 bars.

The case of H_2O raises an important still unresolved question. Prior to Galileo, several different 5 μm analyses of hot spots regions, using IRIS-Voyager data, indicated a H_2O tropospheric abundance strongly depleted with respect to the solar value of O/H. The measurements of H_2O made by the Galileo probe have confirmed a very low value of the water vapor abundance at the levels probed by the 5 μm radiation (3 - 4 bars), but have shown that the H_2O mixing ratio does increase with pressure to reach a value corresponding to an O/H ratio about 0.35 times solar at a pressure of 19 bars.

The explanation which is becoming generally accepted is that the hot spot which the probe entered is, as the other hot spots, not representative of the whole jovian disk. Other evidence is provided, as mentioned above, by the absence of clouds the Galileo probe nephelometer data. The fact that the H_2O mixing ratio increases with pressure in the hot spot, as measured by the probe, provides additional indication that hot spots (mostly located in the belts) would be dry, cloud-free regions of subsidence while the zones would be cloudy regions of ascending motions. However, the circulation pattern appears to be much more complicated than a simple zone/belt convective system and is still far from being understood. The conclusion regarding the deep O/H ratio is that the Galileo probe may not have explored deep enough in the jovian troposphere for the H_2O mixing ratio to be representative of its value in the jovian deep interior.

In the case of Saturn, recent measurements by ISO in the 5 μm region also indicate that water is strongly depleted with respect to the solar O/H value. As in the case of Jupiter, ISO spectra of the whole Saturn disk are mostly sensitive to the hot spots, which emit most of the 5 μm radiation. This result tends to indicate a similar behaviour in both planets, where hot spots would be dry, cloud-free subsidence regions. However, hot spots in Saturn are less localized and of lower contrast than in Jupiter, and the circulation systems may not be similar on both planets.

The conclusion of the above study is that, in the deep troposphere, below the levels where condensation may take place, there is an enrichment in carbon (factor ranging from 3 in Jupiter up to 60 in Neptune), sulfur (2.5 in Jupiter), and nitrogen (from 2 to 4 in Jupiter and Saturn) with respect to the solar value, while results about a possible oxygen enrichment are not conclusive. The C, N and S enrichments are most likely due to the ices contained in the initial cores, which were probably mixed in the atmospheres of the giant planets in the heating phase generated by the collapse of the surrounding nebula around their cores. It can be seen that the scenario of giant planets formation mentioned above, called the "nucleation scenario", receives strong support from the measurements of elemental abundances.

6 Titan

Titan, Saturn's biggest satellite, was discovered by Huygens in 1655. In 1908, Comas Sola concluded, from limb-darkening observations of its disk, found that the satellite was surrounded by a thick atmosphere. The nature and the thickness of this atmosphere, however, remained a puzzle until the exploration of the satellite by the Voyager 1 spacecraft in 1981. From its radio-occultation and UV spectrometry experiments, Voyager was able to determine the thermal structure and the main composition of Titan's atmosphere.

6.1 Thermal Structure and Atmospheric Composition

Titan's atmospheric composition is dominated by molecular nitrogen. The methane mixing ratio is about 2%. Argon might contribute up to a few percent, but this gas, chemically inactive, has not been firmly identified yet. The surface pressure is 1.5 bar and the surface temperature is 94 K. As most planetary atmospheres, Titan's thermal profile (Fig. 9) is characterized by a troposphere where the temperature decreases as the altitude increases, a minimum temperature level (the tropopause, at about 100 mbar, where T = 71 K), and a stratosphere where the temperature increases again with height. Methane probably saturates in the upper troposphere, and is photodissociated in the stratosphere. In the upper stratosphere, nitrogen is also dissociated by high-energy particles coming from Saturn's magnetosphere. As a result, a complex chemistry takes place in Titan's stratosphere, leading to the formation of a large number of hydrocarbons and nitriles (Table 5), most of them were first identified by the infrared spectrometer (IRIS) of Voyager 1. Some of the hydrocarbons are expected to condense, in particular C_2H_6, and may contribute to the thick cloud layer which hides Titan's surface at visible wavelengths.

In addition, an oxygen source is also present in the stratosphere, leading to the presence of CO_2 (detected by Voyager), CO (identified from ground-based observations) and H_2O (recently detected by ISO; Fig. 10). It is interesting to note that the incoming water flux needed to account for the H_2O observations is similar to that required for Saturn. This analogy is not fully understood presently, as, no matter what the origin of the source may be (interplanetary or local), one would expect a larger flux on Saturn because of the focussing effect due to its larger gravity field. The question of the origin of the oxygen source is thus still open.

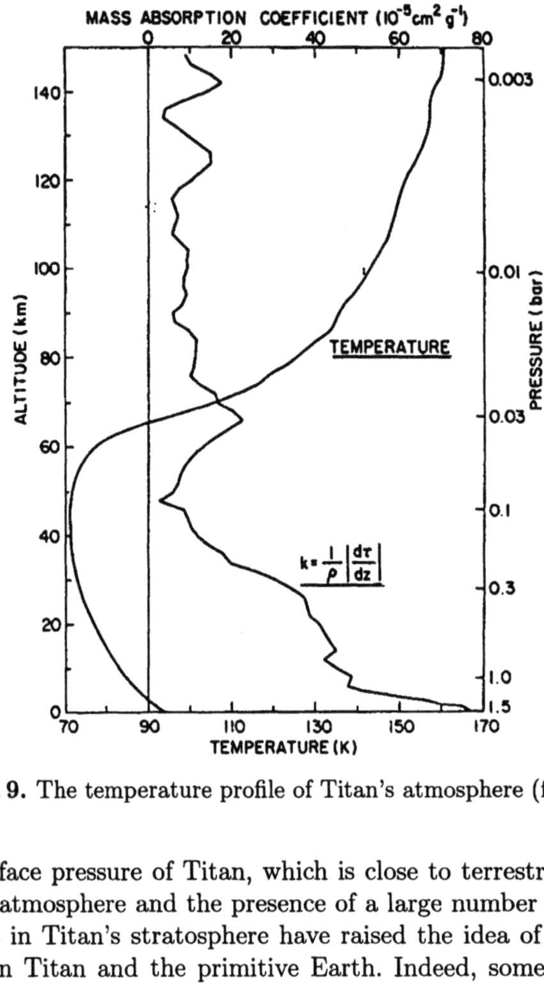

Fig. 9. The temperature profile of Titan's atmosphere (from [19])

The surface pressure of Titan, which is close to terrestrial, its nitrogen-dominated atmosphere and the presence of a large number of hydrocarbons and nitriles in Titan's stratosphere have raised the idea of a possible analogy between Titan and the primitive Earth. Indeed, some of the complex molecules found in Titan's atmosphere (HCN, HC_3N, CH_3CN) are among the ones which are formed in laboratory simulation experiments of prebiotic chemistry, aimed at the formation of amino-acids from a mixture of simple molecules (H_2, CH_4, NH_3...) submitted to electrical discharges or strong UV radiation. However, the very low temperature of Titan's atmosphere must have strongly reduced the speed of all chemical reactions.

6.2 The Surface of Titan

What is the nature of Titan's surface? This is still an open question, which might stay so until the arrival of the Cassini-Huygens mission in 2004. On the basis of the atmospheric composition and the physical conditions at the surface, it has been suggested that Titan's surface could be at least partly covered with an ocean containing CH_4, C_2H_6 and N_2. Radar images of the satellite, however, have failed to confirm this. Recent high-resolution images

Table 5. Physical properties and atmospheric composition of Titan (from [18]

Surface Radius	2575 km	
Mass	1.35 x 10^{23}kg (=0.022 x Earth)	
Mean Density	1880 kg m^{-3}	
Distance from Saturn	1.23 x 10^9m (=20 Saturn radii)	
from Sun	9.546 AU	
Orbital Period	15.95 days	
around Sun	29.5 years	
Obliquity	26.7°	
Surface Temperature	94 K	
Surface Pressure	1.44 bar	
Composition (mole fractions):		
Nitrogen	N_2	90-97%
Argon	Ar	0-6%
Methane	CH_4	0.5-4%
Hydrogen	H_2	0.2%
Ethane	C_2H_6	1 x 10^{-5}
Acetylene	C_2H_2	2 x 10^{-6}
Propane	C_3H_8	5 x 10^{-7}
Ethylene	C_2H_4	1 x 10^{-7}
Diacetylene	C_4H_2	1 x 10^{-9}
Hydrogen Cyanide	HCN	1 x 10^{-7}
Carbon Monoxide	CO	$\sim 10^{-5}$
Carbon Dioxide	CO_2	1 x 10^{-8}
Water	H_2O	8 x 10^{-9}

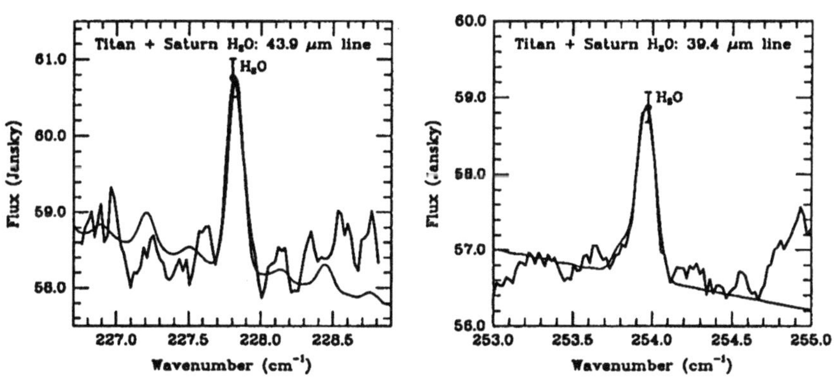

Fig. 10. Detection of H_2O in the stratosphere of Titan by ISO-SWS in the grating mode (R=2000). The figure is taken from Coustenis at al., Astron. Astrophys. 336, L85, 1998.

of Titan's surface in the near-infrared range, obtained with the HST and from the ground, using an adaptive optics system, have revealed some heterogeneity, which might possibly be due to a high ground. Near-infrared ground-based spectroscopy also suggests the possible presence of water ice.

6.3 The Origin of Titan

As illustrated by its quasi-circular orbit, located in the vicinity of Saturn's equatorial plane, Titan was most likely formed within Saturn's subnebula. Its atmosphere was probably outgassed from the globe, with the possible addition of meteoritic infall. An illustration of the "secondary" nature of Titan's atmosphere is given by the measurement of D/H in Titan, which was found to be close to 10^{-4}, i.e. significantly higher than its value in the primordial nebula. Because of the low temperature of Saturn's subnebula ($<$ 100 K), it has been suggested that Titan's bulk may have been formed from clathrates, i.e. water ice lattices trapping various molecules like CH_4, Ar, NH_3, CO or N_2. As CH_4 and NH_3 were probably preferentially formed in Saturn's subnebula, NH_3 may have been present first in Titan's atmosphere, and later converted into N_2 by photolysis.

7 The Future of the Exploration of the Giant Planets

Following the success of the Galileo mission, the next step in the space exploration of the giant planets will be the Cassini mision. Launched in 1997, Cassini observed Jupiter in December 2000, extending the database achieved by the Galileo mission. The spacecraft will then approach the system of Saturn in 2004 for an in-depth exploration of Titan's atmosphere and surface, and a 4-year monitoring of Saturn, its rings and its satellites. The Cassini mission consists of an orbiter, built by NASA, equipped with a dozen remote-sensing experiments, and a probe, built by ESA and called Huygens. The probe is equipped with 6 instruments and will enter Titan's atmosphere and hit its surface (or ocean). The Cassini mission should perform a complete investigation of Saturn's atmosphere and magnetosphere, its ring system and its satellites. There is presently no planned space mission beyond Cassini devoted to a giant planet.

The exploration of giant planets, however, should greatly benefit from future satellites in Earth orbit. After the recent success of HST and ISO, the New Generation Space Telescope (NGST), developed by NASA and planned for a launch in 2010, should allow the infrared observation of giant planets and Titan with a major increase in sensitivity. FIRST (Far Infrared Submillimeter Telescope recently renamed Herschel), an ESA cornerstone scheduled for launch in 2007, will allow the study of the giant planets' stratospheres with increased sensitivity and spectral resolution.

Finally, Uranus, Neptune and Titan, which are faint objects in the near infrared, will be privileged sources for planetary programs using 8-m class telescopes. The observations of Titan's surface will greatly benefit from the use of adaptive optics associated wih large telescopes. Infrared observations of Jupiter and Saturn will also be possible with improved sensitivity and spectral resolution, while the continuous monitoring of the planetary disks at visible wavelengths will provide the database required for long-term variability studies and for complementing the space programs.

References

1. S.K. Atreya: Atmospheres and ionospheres of the outer planets and their satellites,(Springer-Verlag 1986)
2. S.K. Atreya, et al., edts.: Origin and evolution of planetary and satellite atmospheres, (University of Arizona Press 1989)
3. S.K. Atreya, et al.: Comparative atmospheres of Jupiter and Saturn: Deep atmospheric composition, cloud structure, vertical mixing and origin, Plan. Space Science, in press, 1999
4. Th. Encrenaz: ISO observations of solar-system objects, Proceedings of the Les Houches Summer School "Infrared Astronomy, today and to-morrow", F. Casoli and J. Lequeux, edts., 1999
5. Th. Encrenaz, et al.: The solar system, (Springer-Verlag 1995)
6. Th. Encrenaz, et al.: The atmospheric composition and structure of Jupiter and Saturn from ISO observations: A preliminary review. Plan. Space Science, in press, 1999
7. J.S. Lewis: Physics and chemistry of the solar system, (Academic Press 1995)
8. M.S. Marley: Interiors of the Giant Planets, in "Encyclopedia of the solar system", P. R. Weissman et al., edts., (Academic Press 1999)
9. D. Morrison, T. Owen: The planetary system, (Addison-Wesley 1996)
10. T.R. Taylor: Solar system evolution, (Cambridge 1992)
11. R.A. West: Atmospheres of the Giant Planets, in "Encyclopedia of the solar system", P. R. Weissman et al., edts., (Academic Press 1999)
12. K.S. Noll, et al., edts.: The collision of comet Shoemaker-Levy 9 and Jupiter, (Cambridge 1996)
13. Galileo Probe: Reports, Science, 272, 837-860, 1996
14. Galileo Orbiter: Reports, Science, 274, 377-413, 1999
15. Articles, J. Geophys. Res. Special Issue, 103, 22775, 23069, 1998
16. S.K. Atreya, et al.: Chemistry and clouds of Jupiter's atmosphere: A Galileo perspective, in "The three Galileos: the Man, the Spacecraft, the Telescope", C. Barbieri et al., edts., (Kluwer 1997)
17. Th. Encrenaz: The planet Jupiter, Astron. Astrophys. Rev., in press, 1999
18. A. Coustenis and R. Lorenz: Titan, in "Encyclopedia of the Solar System", P. R. Weissman et al., edts., (Academic Press 1999)
19. T. Gehrels, M.S. Matthews, edts.: Saturn, (University of Arizona Press 1984)
20. J.T. Bergstrahl, et al., edts.: Uranus, (University of Arizona Press 1989)
21. D.P. Cruikshank, et al., edts.: Neptune and Triton, (University of Arizona Press 1996)

The Formation of Planets

Therése Encrenaz

Observatoire de Paris, Meudon/France

Abstract. This chapter presents the main characteristics of the formation scenario of the solar system. The fist sections summarize the observational constaints and the input of stellar physics, and review previous formation models. The next section describes the currently accepted model, based on the collapse of a protoplanetary nebula into a disk and the subsequent accretion of planets from planetesimals within this disk. The last section presents the formation and the evolution of planetary atmospheres within this formation scenatio.

1 The Observational Constraints

The planets of the solar system mainly fall into two distinct categories (Table 1). The terrestrial planets (Mercury, Venus, the Earth and Mars), at heliocentric distances closer than 2 astronomical units (AU), have a relatively large density (3.9-5.5 g/cm^3) and a small number of satellites (between 0 and 2). Apart from Mercury which has no stable atmosphere, their atmospheres exhibit a wide variety of physical conditions (pressure, temperature) but have a common chemical composition based on CO_2 and N_2, with variable amounts of O_2, H_2O and CO. At larger distances from the Sun (above 5 AU), the giant planets (Jupiter, Saturn, Uranus and Neptune) have low densities (0.7-1.7 g/cm^3), a large number of satellites (between 8 and 21 discovered so far), and thick atmospheres mostly composed of H_2 and He, with traces of CH_4 and other minor constituents in reduced form. Pluto, the most distant planet, does not belong to any of these categories and rather resembles the satellites of the outer planets; it is now recognized as the largest example of the newly-discovered population of trans-neptunian objects. Between the terrestrial and the giant planets, the asteroids (or minor planets) are mostly found at about 3 AU from the Sun, in the main asteroidal belt.

The orbital properties of the planets and their satellites show several important characteristics

- The orbits of all planets (except Pluto's) are close to the ecliptic plane (this plane being defined as the plane of the Earth's orbit).
- All these orbits (except Pluto's) are quasi- circular around the Sun, and the planets all rotate in the same direction.
- All planets (except Venus) are in direct rotation around their axis (in addition, the rotation axis of Uranus is very close to the ecliptic plane).
- Many satellites of the giant planets are also on quasi-circular orbits close to the equatorial plane of the planet; these planetary systems thus behave like "mini" solar systems.

These basic facts have to be accounted for in any scenario of solar system formation.

Table 1. Physical and atmospheric parameters of the planets

Planet	Helioc. Distance (A.U.)	Radius (R⊕)	ρ (g/cm^3)	Number of satellites	Atmosph. Composition
Terrestr. Planets					
Mercury	0.39	0.38	5.44	0	–
Venus	0.72	0.95	5.25	0	CO_2(96%) N_2(4%)
Earth	1.0	1.0	5.52	1	N_2(77%) O_2(21%)
Mars	1.52	0.53	3.91	2	CO_2(95%) N_2(3%)
Giant Planets					
Jupiter	5.20	11.19	1.31	16	H_2(91%) HE(9%)
Saturn	9.55	9.41	0.69	21	H_2(96%) He(4%)
Uranus	19.22	3.98	1.21	15	H_2(~90%) He(~10%)
Neptune	30.11	3.81	1.67	8	H_2(~90%) He(~10%)
Pluto	39.44	0.18	2	1	CH_4,N_2?

2 The Input of Stellar Physics

The information we can obtain from the study of stars, from their birth to their death, is of great interest for understanding the early stages of solar system formation. The study of the interstellar medium and star formation regions, the observation of protoplanetary disks, of young stellar objects, the nucleosynthesis, are especially relevant to our study.

2.1 Gravitational Collapse and Protoplanetary Disks

Star formation is found to take place in high density rotating regions of interstellar clouds. The density increase leads to a gravitational instability

(Jeans instability), resulting in a gravitational collapse. The collapse can lead to the formation of a single star surrounded by a disk, or a multiple system. Such multiple systems, frequently observed, seem to be associated with rapidly rotating cloud cores, while slowly rotating clouds preferentially produce single stars and disks.

The first protoplanetary disk to be discovered was found around Vega, in 1983, by the IRAS satellite which was able to measure the infrared excess of the star. IRAS also identified several other possible candidates. In 1984, a protoplanetary disk was directly imaged around the star β Pic from a ground-based coronographic measurement. Many protoplanetary disks have been recently found around young stars, especially with the HST.

2.2 T-Tauri/FU-Orionis Phases

The observation of young stellar objects gives information on the early stages of violent stellar activity, called the T-Tauri or FU-Orionis phases. T- Tauri stars have masses ranging between 0.2 and 3 solar masses. They have a high mass loss (10^{-8} $M_{S/y^{-1}}$) and a short lifetime (2 10^5 to 2 10^7 y). The T-Tauri stars are surrounded by dusty disks, show strong bipolar flows and an intense magnetic activity. FU-Orionis stars exhibit even more extreme conditions, with even more violent outbursts and a very high mass loss (10−5 $M_{S/}^{y-1}$). It is now believed that T-Tauri/FU-Orionis activity may represent stellar activity in the early stages of stellar evolution (10^5 - 10^7 y), and could have occurred in the beginning of the solar system history (see chapter by Ray).

Table 2. Abundances of elements in the Sun

Element	Relative abundance (by number)	Fraction of total mass
H	3.18 x 10^{10}	0.980 0
He	2.21 x 10^9	
C	1.18 x 10^7	
N	3.64 x 10^6	0.013 3 (C, N, O)
O	2.21 x 10^7	
Ne	3.44 x 10^6	0.013 3
Na	6 x 10^4	
Mg	1.06 x 10^6	
Al	8.5 x 10^5	
Si	10^6(standard)	0.003 65 (Na–Ni)
S	5 x 10^5	
Ca	7.2 x 10^4	
Fe	8.3 x 10^5	
Ni	4.8 x 10^4	

2.3 Stellar Nucleosynthesis

According to the Big Bang Standard Model (BBSM), the only elements formed in large abundances in the Big Bang were hydrogen and helium, with a helium mass fraction of 0.235. Traces of ^{3}He (1.2 10^{-5}) and D (5-8 10^{-5}) were also present. Once a star is born, stellar nucleosynthsis takes place in its interior, transforming hydrogen and helium into heavier elements. In the same process, D is destroyed and converted into ^{3}He. Stellar nucleosynthesis enables the formation of heavy elements up to iron. Heavier elements are not synthetized in stars, but are produced in explosive burning processes like supernovae. Stellar nucleosynthesis is responsible for the cosmic abundances of the elements, as we observe them today in the Universe (Table 2).

2.4 The Interstellar Medium

The physical properties of cold molecular clouds (density, temperature) have been studied in detail from radio techniques, in particular H and CO mapping. The development of millimetre and submillimetre heterodyne spectroscopy has led to the discovery of a large number of interstellar molecules. About 80 different species have been found, some molecules having as many as 15 atoms. In addition, complex carbon molecules (possibly PAHs, i.e. Polycyclic Aromatic Hydrocarbons), bearing several tens of atoms, have been also detected in the ISM through their infrared signature. The solid phase is also present, in the form of small carbon aggregates and silicate grains, covered with organic refractory mantles or icy mantles. These results have demonstrated that a complex carbon chemistry does take place everywhere in the Universe, and all these elements, in gaseous and solid forms, are present in the interstellar cloud from which new stars are born (Table 3).

3 Early Theories of Solar-System Formation

The search for understanding how the solar system formed started with the Copernican model. In spite of the pioneering efforts of Aristarchos of Samos (280 BC), the heliocentric system was not recognized until the XVth century. Following the earlier work of Nicholas of Cusa (1401-1464), Nicholas Copernicus (1473-1543) set the basis of the heliocentric system in his famous book, "De revolutionibus orbium celestium, L VI", published in 1543. This new view of the world, rejected for a long time by the clerical authorities, was later supported by famous astronomers like Galileo Galilei (1564-1642) and Johannes Kepler (1571-1630) who found the planetary laws describing motions around the Sun. Later, Isaac Newton, who discovered the law of universal gravitation, provided the theoretical basis of the Copernican model.

Table 3. Interstellar molecules

Simple Hydrides, Sulfides, Halides, and Related Molecules				
H_2	CO	NH_2	CS	N_2Cl
HCl	SiO	SiH_4	SiS	AlCl
H_2O	SO_2	CC	H_2	KCl
	OCS	CH_4	PN	AlF
Nitriles, Acetylene Derivatives, and Related Molecules				
HCN	$HC{\equiv}C\text{-}CN$	$H_3C\text{-}C{\equiv}C\text{-}CN$	$H_3C\text{-}CH_2\text{-}CN$	$H_2C{\equiv}CH_2$
H_3CCN	$H(C{\equiv}C)_2\text{-}CN$	$H_3C\text{-}C{\equiv}CH$	$H_2C{=}CH\text{-}CN$	$HC{\equiv}CH$
CCCO	$H(C{\equiv}C)_3\text{-}CN$	$H_3C\text{-}(C{\equiv}C)_2\text{-}H$	HNC	
CCCS	$H(C{\equiv}C)_4\text{-}CN$		HN=C=O	
$HC{\equiv}CCHO$	$H(C{\equiv}C)_5\text{-}CN$		HN=C=S	
H_3CNC				
Aldehydes, Alcohols, Ethers, Ketones, Amides, and Related Molecules				
$H_2C{=}O$	H_3COH	HO-CH=O	H_2CNH	
$H_2C{=}S$	$H_3C\text{-}CH_2\text{-}OH$	$H_3C\text{-}O\text{-}CH{=}O$	H_3CNH_2	
$H_3C\text{-}CH{=}O$	H_3CSH	$H_3C\text{-}O\text{-}CH_3$	H_2NCN	
$NH_2\text{-}CH{=}O$	$(CH_3)_2CO$?	$H_2C{=}C{=}O$		
Cyclic Molecules				
C_3H_2	SiC_2	C_3H		
Ions				
CH^+	HCO^+	$HCNH^+$		
HN_2^+	$HOCO^+$	SO^+		
Radicals				
OH	C_3H	CN	HCO	C_2S
CH	C_4H	C_3N	NO	NS
C_2H	C_5H	H_2CCN	SO	
	C_6H			

3.1 Turbulence Model

René Descartes (1596-1650) was the first scientist and philosopher to address the question of solar system formation. In his "Theory of vortices" (1644), assuming that space is filled with ether, he described the nebular disk as being composed of eddies of a whole range of sizes. The model was only qualitative, and failed to explain the origin of the ecliptic plane. The concepts of friction and turbulence, however, were reconsidered in the XXth century by several authors, including von Weizsäker, Kuiper and Whipple.

3.2 The Nebular Theory

The first scientists two introduce the concept, now widely accepted, of the primordial nebula, were Emmanuel Kant (1724-1804) and Pierre-Simon Laplace (1749-1827). In 1755, Kant suggested for the first time that the proto-Sun was formed due to gravitational concentration of a fraction of dense material. He also suggested that the circular and coplanar planetary orbits were the result of their formation within a flattened disk in rotation around the Sun. In 1796, Laplace introduced the idea of a common origin for the proto-Sun and the proto-planetary disk, both resulting from the contraction of the primordial nebula.

The nebular theory had the main advantage of accounting for most of the orbital properties of the planets. However, a major question remained unsolved at that time: 60% of the angular momentum of the solar system is contained in Jupiter; only 25% lies in the Sun, which still contains 99% of the total mass. One would expect the Sun to rotate much faster. How was the angular momentum transferred to large heliocentric distances? In 1924, von Weizsäcker suggested that the angular momentum transfer might be due to turbulent viscosity. Cameron, in 1960, proposed thermal convection as a mechanism to support this viscosity. In the 1960s, Hoyle and Schatzman suggested that the Sun's rotation could be slowed down by the effect of its magnetic field and by transport of matter through the solar wind.

Two classes of nebular models were developed. In the massive nebula model (Cameron), a viscous disk of about 1 solar mass is formed, a large fraction of which (85%) is swept away by the solar wind in about 10^5 years, and planets formed from the gaseous nebula by gravitational instabilities. In the low-mass nebula model (Safronov, Hayashi, Elmegreen...), the mass of the disk is only about 10^{-2} solar mass. Condensation takes place during the cooling of the disk, leading to the formation of planetesimals which accrete to form small bodies and planets. The low-mass nebula model is now accepted as the baseline model for the formation model of the solar system.

3.3 The Tidal Theories

The first tidal theory was proposed in 1745 by Buffon (1707-1788) who suggested that a collision with a comet could be responsible for extracting from the Sun the matter which formed the planetary system. At that time the true nature of comets, as well as their sizes and masses, were unknown. Later Jeans, in 1917, developed this theory using this time another star instead of the comet. Later other scientists (Lyttletown, Alfven, Schmidt) proposed the capture of interstellar matter by the Sun.

Tidal theories were ruled out when new measurements from meteorites allowed the dating of solar system objects. Using isotopic ratios of radiogenic elements (in particular ^{40}K and ^{87}Rb which desintegrate respectively in ^{40}Ar and ^{87}Sr with time constants of 1.7 10^{10} y and 7 10^{10} y), the age of the solar system was found to be 4.55 (±/–0.10) 10^9 years, implying that the Sun and the solar system did form at the same time. Another important measurement was the first detection of a deuterated molecule (CH_3D) in Jupiter in 1972, by Beer. The inferred D/H value was about 2 10^{-5}, i.e. significantly more than in the Sun: as mentioned above, deuterium has been very rapidly destroyed in the Sun, as it is in any star by nucleosynthesis. This result demonstrated that the planets are not made of solar material.

4 A Plausible Model of Solar System Formation

Using our knowledge about the age of the solar system (similar to that of the Sun) and on the basis of many observations of young stars, it is possible to build a plausible model of the different stages of solar system evolution. The model starts with the collapse of a rotating cloud with a mass of about 1 solar mass, leading to the formation of a protoplanetary disk of about 0.01 to 0.05 solar masses.

4.1 The Condensation Sequence

Matter contained in the original collapsing cloud must have been in the form of gas and dust. During the formation of the protoplanetary disk, grains evaporated in the vicinity of the Sun. A condensation sequence occurred during the cooling of the nebula, and as a function of the heliocentric distance, the temperature decreasing from about 2000 K near the Sun down to about 50 K at large heliocentric distances. The sequence led to the condensation of Al, Ti, Ca, Mg, Si, Fe, Na, S... An important observational fact is that the abundances measured in primitive meteorites are in excellent agreement with the abundances found in phases stable at a temperature of 1400 K.

During the cooling of the protoplanetary disk, new solid particles were formed by condensation. They were mixed with the gas and, due to turbulence in the nebula, they probably grew by grain-molecule collisions and sticking of the molecule to the grain, possibly forming fluffy aggregates. Because of their higher density with respect to the gas, these growing particles must have settled toward the ecliptic plane. The effect of inelastic collisions in the equatorial plane was then likely to decrease the thickness of the disk and to accelerate the accretion process. Planetesimals were thus formed in the plane of the disk and their orbits were mostly circular (Fig. 1).

4.2 Terrestrial Planets and Giant Planets

The accretion theory provides a natural explanation for the basic difference between terrestrial planets and giant planets. At small heliocentric distance, only metals and silicates are in solid form. Since heavy elements are less abundant than lighter ones, the matter available for accretion is less abundant

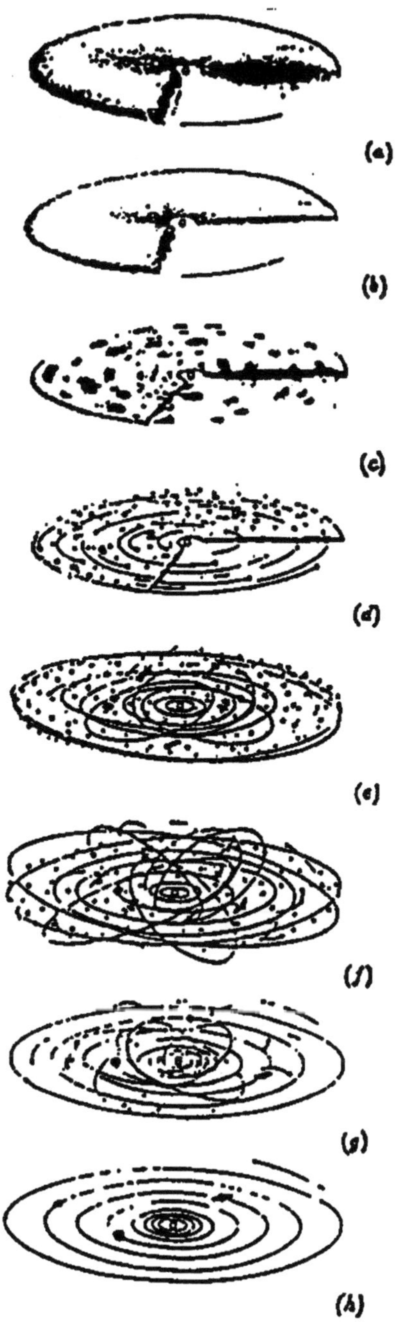

Fig. 1. Process of planetary formation. (a) and (b): formation of a disk; (c) formation of dust condensations; (d) and (e) formation of planetesimals; (f) appearance of embryos at separate orbits; (g) final state of planetary accumulation; (h) the contemporary solar system (from T.V. Ruzmaikina, in "Encyclopedia of Planetary Sciences", J.S. Shirley and R.W. Fairbridge, edts., Chapman and Hall, 1997

than at larger distances from the Sun. As a consequence, relatively small and dense bodies are formed. In contrast, at large heliocentric distances where the temperature is lower, most of the elements including C, N and O are in solid form (H_2O, CO_2, CH_4, NH_3 ices). As a result, large cores can be formed. Theoretical models predict that once a core reaches a mass of about 15 terrestrial masses, its gravity field is sufficient for the surrounding nebula (mostly composed of hydrogen and helium) to collapse. This explains why the giant planets, far from the Sun, have a large mass and a low density (Fig. 2).

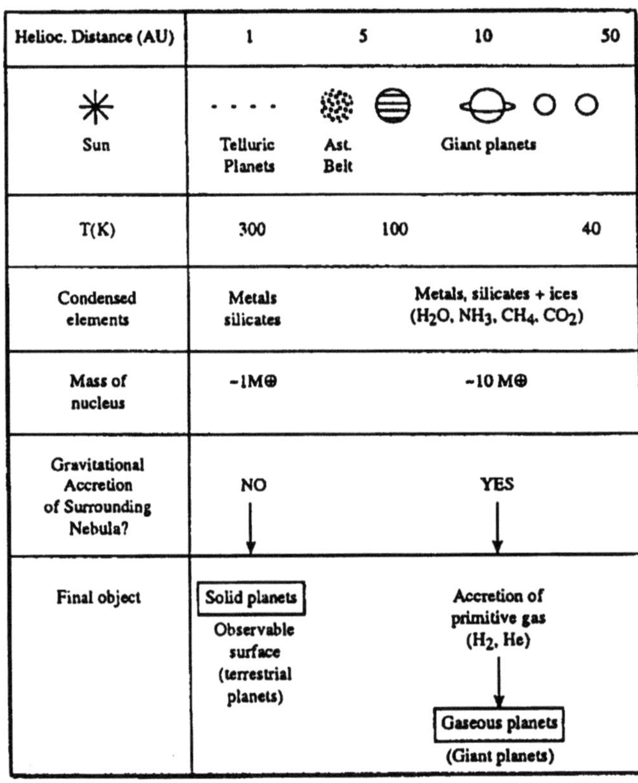

Fig. 2. The condensation sequence

4.3 Accretion Time Scales

Estimates of accretion times are provided by numerical simulations. Terrestrial planets could have formed in 10^7–10^8 years. Their accretion would have been in two steps: first, a fast runaway growth of embryos up to 3 10^4 years, followed by a slower accretion up to 10^8 years. In the case of the giant planets, Jupiter could have formed in 10^6 years, Saturn in about 2 10^6 years, Uranus in 10^7 years and Neptune in about 2 10^7 years. In the meantime, if we assume a phase of high activity (T-Tauri phase), for the early Sun most of the gas of the primordial nebula is expected to have been swept away after about 10^6 years. In such a scenario, Jupiter (and, to a lesser extent, Saturn) would have formed before the dissipation of the nebular gas, but not Uranus and Neptune. This might explain why Uranus and Neptune have a much lower fraction of gas as compared to their initial core (Table 4). This formation scenario thus accounts for the subdivision of the giant planets into two categories, the gaseous giants (Jupiter and Saturn) the initial cores of which are less than 10% of their total masses, and the icy giants (Uranus and Neptune) the initial cores of which are more than half of their total masses.

4.4 A Test for the Nucleation Model: C/H and D/H in the Giant Planets

In the massive nebula model developed, in particular, by Cameron, giant planets formed by gravitational instabilities within the primordial nebula. In contrast, in the low-mass nebula model, as described above, planets are formed from the accretion of planetesimals. A major diagnostic in favor of the nucleation model has been provided by the measurement of abundance ratios in the giant planets, in particular C/H and D/H.

If giant planets were formed from a gravitational instability, their interiors should be homogeneous and the abundance ratios in their atmospheres should reflect the values of the primordial nebula. C/H and D/H, in particular, should be the same for all four planets and equal to their values in the primorial nebula. The situation, however, is different in the nucleation model. For all giant planets, the initial core is about 15 terestrial masses, but the mass of gas is, relatively to the core, much higher for Jupiter than for Uranus and Neptune. Because the initial core is mostly made of CH_4 and other ices, the C/H ratio is much higher in the core than in the nebular gas. The same situation occurs for D/H, because D/H is significantly enriched in ices, due to lowtemperature ion-molecule reactions. As a result, in the nucleation model, both C/H and D/H are expected to increase from Jupiter to Neptune, while in the homogeneous model, they should be constant for the four planets.

C/H has been measured in the four giant planets from visible and near-infrared ground-based spectroscopy, and from Voyager IR spectroscopy. In the case of Jupiter, an accurate determination has been recently given by the Galileo probe. The C/H enrichment factor with respect to the solar value is 3 for Jupiter, 6 for Saturn, about 30 for Uranus and possibly 60 for Neptune. D/H has been measured in Jupiter by the Galileo probe, and in the four giant planets by the ISO satellite. The D/H ratio is about 2 10^{-5} in Jupiter and Saturn (in agreement with the primordial value) and about 6 10^{-5} in Uranus and Neptune. These results clearly demonstrate the validity of the nucleation model, which is now generally accepted.

4.5 Isotopic Anomalies in Meteorites: Evidence for Inhomogeneities in the Primordial Nebula

As mentioned above, the laboratory analysis of meteorites and lunar samples has provided added significantly to our knowledge of the age of the solar system. Another very important result was the evidence for some heterogeneity in the primordial nebula, i.e. the presence of "presolar" grains, with an isotopic composition different from that of the rest of the nebula, which were not vaporized at the time of the gravitational collapse.

This discovery was made from the measurement of oxygen isotopic anomalies in some meteorites. In the case of terrestrial samples, all abundance measurements of ^{16}O, ^{17}O and ^{18}O can be explained with the assumption of constant initial $^{17}O/^{16}O$ and $^{18}O/^{16}O$ ratios , equal to the solar value. However, in the case of the Allende meteorite, the parent of which is believed to be a primitive asteroid, the measured abundances of ^{16}O, ^{17}O and ^{18}O could only be explained by the presence of grains containing only ^{16}O. Such grains could not originate from the primordial nebula.

Another major isotopic anomaly was measured in the case of ^{26}Mg. Some samples contain an overabundance of ^{26}Mg as compared to the other isotopes of magnesium, proportional to their aluminium abundance. The ^{26}Mg excess is likely to come from the radioactive decay of ^{26}Al. However, the decay period for ^{26}Al disintegration is extremely short (700000 years). This implies that the time separating the formation of ^{26}Al, by nucleosynthesis in a supernova, from the condensation of the meteoritic grain, must have been of that order. As a consequence, the formation of the solar system may have occurred shortly after the explosion of a nearby supernova (and may in fact have been triggered by this explosion).

Following the discovery of the ^{26}Al anomaly, other isotopic anomalies have been found for other elements, in many different meteorites. In particular, an excess of ^{22}Ne was found, probably originating from the disintegration of ^{22}Na. This radiogenic element, which is formed in novae explosions, has a time decay of only 2.6 years.

All these results tend to indicate that "presolar grains", some of which had different nucleosynthesis origins, were present in a primordial nebula which was not fully vaporized in the collapse phase, or which was complemented by interstellar material of different nucleosynthetic origin during the accretion phase.

4.6 The Role of Collisions

All models of solar system formation recognize the major role played by collisions at all stages of the solar system evolution. Collisions between grains and molecules were first responsible for the growth of planetesimals. Later, at the end of the T-Tauri phase, where most of the gas and small particles left in the primordial nebula were dissipated, the newly-formed planets and small bodies were bombarded by particles of all sizes, as demonstrated by the numerous crater impacts still present on the surfaces of bare objects (Moon, Mercury, etc.).

Collisions between large bodies may also have been important in the evolution of some objects. The Earth-Moon system is believed to have originated from a collision between the Earth and a Mars-size planet. Fragments of the Earth's mantle would have been dissipated in the Earth's orbit, together with fragments of the collider. All debris would have later accreted again in orbit to form the present Moon.

Collisions with large objects are also a possible explanation for the retrograde rotation of Venus and the tilted rotation axis of Uranus. Another possible explanation, however, involves the chaotic behaviour of the planets' rotation. In any case, the event responsible for the present orientation of Uranus' rotation axis must have occurred early in its history, before the formation of its planetary system, since the satellite system of Uranus is, as for the other giant planets, in the planet's equatorial plane.

We still see the effects of collisions in a more recent past. Collisions with large meteorites are possibly responsible for major changes in the history of the past terrestrial climate. Finally, even more recently, in July 1994, we observed the collision of a comet with Jupiter. The probability of such an event is probably one every few hundred years.

Table 4. Mass fractions in the initial cores of the giant planets

Planet	Total Mass (terrestrial mass)	Fraction of core in mass(*)
Jupiter	318	0.03 - 0.05
Saturn	95	0.10 - 0.16
Uranus	15	0.66 - 1.0
Neptune	17	0.59 - 0.88

(*) assuming $M_{core} = 10 - 15$ terrestrial masses

5 Planetary Atmospheres

5.1 Primary and Secondary Atmospheres

It has been seen that the giant planets accreted their atmospheres from the collapse of the surrounding nebula, mostly composed of hydrogen and helium. Such an atmosphere is called "primary" as it basically reflects the primordial

Table 5. Gravitational escape from planets and satellites

Element	V_e(km/s)	T_{ex}*	m_c
Mercury	4.2	600	21
Venus	10.3	~3000	17
Earth	11.2	1500	7.4
Mars	5.0	~500?	12
Jupiter	61	~1000?	0.11
Saturn	37	~1000?	0.3
Uranus	22	~500?	0.6
Neptune	25	500?	0.5
Moon	2.4	300	32
Io	2.3	~800	93
Ganymede	2.8	110	9
Callisto	2.3	110	13
Titan	2.8	190	20

* is the present exospheric temperature.

solar composition. We have noted, however, that there is a significant difference between the gaseous giants, Jupiter and Saturn, whose initial cores were only a small fraction of their total masses, and the icy giants, Uranus and Neptune, whose initial cores were more than half of their total masses (Table 4).

In the case of the terrestrial planets, the gravity field generated by their cores was never sufficient to accrete the surrounding nebula. Their atmospheres, called "secondary", have two possible origins: outgassing from the interior (through volcanism) and meteoritic impacts.

5.2 Stability of a Molecule in a Planetary Atmosphere

Under which conditions can a molecule escape a planetary atmosphere? This is possible if its thermal velocity V_t is greater than the escape velocity V_e of the planet. The escape velocity is defined by the following equation:

$$\mathrm{mV_e^2/2 = mMG/R} \tag{1}$$

where m is the mass of the molecule, M the mass of the planet, R its radius, G the gravitational constant.

The thermal velocity of the molecule is defined by

$$\mathrm{mV_T^2/2 = 3kT/2} \tag{2}$$

where k is the Boltzmann constant and T is the temperature.

Spitzer (1952) has calculated that the condition of stability of a molecule over the lifetime of the solar system is

$$\mathrm{V_T < 0.2V_e} \tag{3}$$

As a result, a molecule is stable if its mass is greater than the critical mass m_c such that

$$m_c = 3kT/0.04V_e^2 \quad (4)$$

As a consequence, a molecule will escape more easily if m is small, M is low and T is high. Note that T is a parameter which depends upon the altitude in the planetary atmosphere, and which may have changed over the lifetime of the solar system. For estimating m_c, exospheric temperatures should be used; they may be significantly higher than surface temperatures.

Table 5 gives estimates of m_c for planets and major satellites. It can be seen that, in the case of the giant planets, m_c is lower than 1. This means that even hydrogen cannot escape from the gravity fields of the giant planets over the solar system lifetime. In the case of terrestrial planets, all light elements have escaped, especially hydrogen and helium.

5.3 Chemical Composition of Planetary Atmospheres

We have seen that the atmospheres of the terrestrial planets are dominated by N_2, CO_2 and oxygen compounds, while the ones of the giant planets are dominated by hydrogen and helium, with minor constituents in a reduced form. This basic difference can be understood in the light of their formation scenario.

In which form were carbon and nitrogen in the primordial nebula? According to models of thermodynamical equilibrium, the relative abundances of CH_4, CO, N_2 and NH_3 are defined by the following chemical reactions:

$$\begin{aligned} CH_4 + H_2O &\rightarrow CO + 3H_2 \\ 2NH_3 &\rightarrow N_2 + 3H_2 \end{aligned} \quad (5)$$

which evolve toward (CH_4, NH_3) at low temperature and toward (CO, N_2) at high temperature. This is consistent with the chemical composition of the giant planets where CH_4 and NH_3 are present. In contrast, CO and N_2 preferentially formed in the vicinity of the terrestrial planets, while H_2 escaped from their atmospheres.

The cooling process of the primordial nebula may also have had a dynamical effect on the composition of planetary and satellite atmospheres. Indeed, if the temperature of the nebula dropped rapidly as compared to the chemical reaction times, CO and N_2 may have been quenched in the outer solar system and may still be present at large heliocentric distances, except in the vicinity of the giant planets where the pressure was high enough for thermochemical equilibrium to take place. In this scenario, the composition of Saturn's satellite Titan, dominated by N_2 and CH_4, could be explained by the presence of CH_4 and NH_3 in the Saturnian subnebula, where NH_3 would have been subsequently converted into N_2 by photolysis.

References

1. S.K. Atreya: *Atmospheres and ionospheres of the outer planets and their satellites*, (Springer- Verlag 1986)
2. S.K. Atreya, et al., edts.: *Origin and evolution of planetary and satellite atmospheres*, (University of Arizona Press 1989)
3. J.T. Bergstrahl, et al., edts.: *Uranus*, (University of Arizona Press 1989)
4. P. Cassen, D.S. Woolum: *The origin of the solar system*, in "Encyclopedia of the Solar System", P. R. Weissman et al., edts., (Academic Press 1999)
5. D.P. Cruikshank, et al., edts.: *Neptune and Triton*, (University of Arizona Press 1996)
6. Th. Encrenaz, et al.: *The solar system*, (Springer- Verlag 1995)
7. Th. Encrenaz: *Gaseous envelopes of solar-system bodies: remote sensing spectroscopy in the infrared and millimeter range*, in "Infrared Astronomy from Space", F. Melchiorri and P. Encrenaz, edts., Scuola Superiore G. Reiss Romoli, 1992, pp. 1-90
8. T. Gehrels, M.S. Matthews, edts.: *Saturn*, (University of Arizona Press 1984)
9. J.S. Lewis: *Physics and chemistry of the solar system*, (Academic Press 1995)
10. D. Morrisson, T. Owen: *The planetary system*, (Addison-Wesley 1996)
11. T.V. Ruzmaikina: in "Encyclopedia of Planetary Sciences", J. S. Shirley and R. W. Fairbridge, edts., (Chapman and Hall 1997)
12. S.R. Taylor: *Solar system evolution*, (Cambridge 1992)

Dynamics of the Solar System

Carl D. Murray

Astronomy Unit, Queen Mary, University of London,
Mile End Road, London E1 4NS, UK

Abstract. The basic properties of the two- and three-body problems are reviewed together with applications to observed phenomena in the solar system. A study of the properties of the planetary disturbing function is presented showing how the dynamics of resonances can be understood using a pendulum approximation. Newtonian gravitation also permits chaotic motion. The properties of chaotic systems are examined together with applications to the long-term dynamical evolution of objects in the solar system. Planetary ring systems are used to illustrate particular aspects of the nature of resonant phenomena in the solar system.

1 Introduction

The planets, satellites, asteroids, comets and meteoroids that go to make up the solar system all respond to each other's gravitational attraction resulting in an incredibly intricate interacting system. At the heart of such a newtonian system is an inverse square law of force with its many ramifications. Many of these can be classified and quantified by an examination of a series expansion of the perturbing potential experienced by one body due to another as both orbit a central mass. This is the basis for the study of resonant phenomena in the solar system. Curiously, although Newton could be regarded as the champion of a deterministic universe, his vision of a clockwork solar system seems paradoxically at odds with the current realisation that chaos has played an important role in determining the dynamical structure of our solar system.

We start in Section 2 with an examination of the basic properties of the two-body problem whereby the gravitational interaction of two masses results (in the case of the majority of solar system objects) in elliptical motion. We then progress in Section 3 to study the restricted three-body problem and show how it can be applied to improve our understanding of the dynamics of various asteroids and satellites. Any study of resonant and long-term dynamics requires an understanding of the disturbing function and this is explored in Section 4 leading on to a study of resonance in Section 5. The role of chaos is examined in Section 6 while Section 7 deals with the application of many of these techniques to understanding some of the phenomena seen in planetary ring systems. This entire article is based on the more detailed discussion of solar system dynamics given by Murray & Dermott [1].

2 The Two-Body Problem

By observing the motion of the planets, Johannes Kepler had deduced his three laws of planetary motion:

- The planets move in elliptical paths with the Sun at one focus of the ellipse
- The line from the Sun to a planet sweeps out equal areas in equal times
- The square of the orbital period of a planet is proportional to the cube of its orbital distance

Kepler's laws were empirically derived at the end of the sixteenth century and although he had devised a magnetic vortex theory to explain his observations, it was Isaac Newton who was the first to show that an inverse square law of force gave rise to the elliptical motion. In this section we show how Kepler's laws can be derived from Newton's law of gravitation and investigate the basic properties of elliptical orbits.

2.1 Equations of Motion

Consider two masses, m_1 and m_2 with position vectors $\boldsymbol{r}_1$ and $\boldsymbol{r}_2$ referred to some fixed origin O. Let $\boldsymbol{r} = \boldsymbol{r}_2 - \boldsymbol{r}_1$ denote the relative position of m_2 with respect to m_1. Given that each mass is subjected to an inverse square gravitational force due to the other, it can easily be shown that $\boldsymbol{r}$ satisfies the vector differential equation

$$\ddot{\boldsymbol{r}} + \mu \frac{\boldsymbol{r}}{r^3} = 0 \tag{1}$$

where $\mu = \mathcal{G}(m_1 + m_2)$ and $\mathcal{G}$ is the universal gravitational constant.

Taking the vector product of $\boldsymbol{r}$ with Eq. (1) we have $\boldsymbol{r} \times \ddot{\boldsymbol{r}} = 0$ which can be integrated directly to give

$$\boldsymbol{r} \times \dot{\boldsymbol{r}} = \boldsymbol{h} \tag{2}$$

where $\boldsymbol{h}$ is a constant vector which has to be perpendicular to both $\boldsymbol{r}$ and $\dot{\boldsymbol{r}}$. Therefore the motion of m_2 about m_1 is confined to a plane perpendicular to the direction defined by $\boldsymbol{h}$. Another implication is that $\boldsymbol{r}$ and $\dot{\boldsymbol{r}}$ must also lie in the same plane (see Fig. 1). Equation (1) is called the *angular momentum integral.*

Because $\boldsymbol{r}$ and $\dot{\boldsymbol{r}}$ have to lie the same plane (the *orbit plane*) we now consider the motion in that plane. We make use of a polar coordinate system (r, θ) with origin at the mass m_1 and a reference direction corresponding to $\theta = 0$. Let $\hat{\boldsymbol{r}}$ and $\hat{\boldsymbol{\theta}}$ denote unit vectors along and perpendicular to the

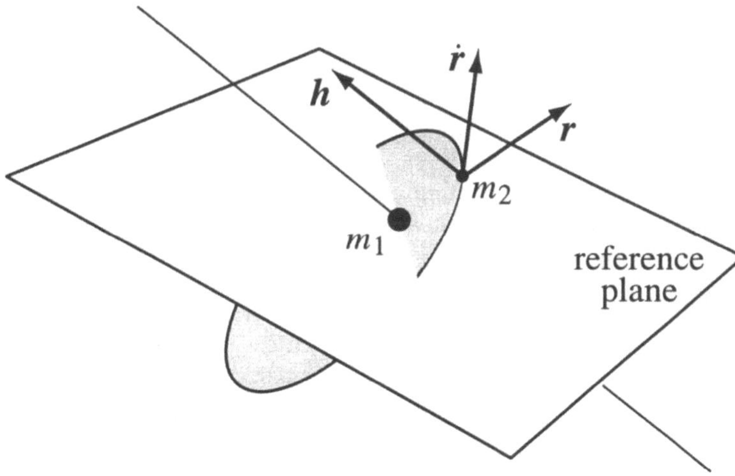

Fig. 1. The relationship between the position vector $\boldsymbol{r}$, the velocity vector $\dot{\boldsymbol{r}}$ and the angular momentum vector $\boldsymbol{h}$.

radius vector respectively. Using standard results the position, velocity and acceleration vectors can be written in polar coordinates as

$$\boldsymbol{r} = r\hat{\boldsymbol{r}} \tag{3}$$

$$\dot{\boldsymbol{r}} = \dot{r}\hat{\boldsymbol{r}} + r\dot{\theta}\hat{\boldsymbol{\theta}} \tag{4}$$

$$\ddot{\boldsymbol{r}} = (\ddot{r} - r\dot{\theta}^2)\hat{\boldsymbol{r}} + \left[\frac{1}{r}\frac{\mathrm{d}}{\mathrm{d}t}\left(r^2\dot{\theta}\right)\right]\hat{\boldsymbol{\theta}}\,. \tag{5}$$

Substituting the expression for $\dot{\boldsymbol{r}}$ given in Eq. (3) back into Eq. (2) gives $\boldsymbol{h} = r^2\dot{\theta}\hat{\boldsymbol{z}}$, where $\hat{\boldsymbol{z}}$ is a unit vector perpendicular to the plane of the orbit forming the customary right-handed triad with $\hat{\boldsymbol{r}}$ and $\hat{\boldsymbol{\theta}}$. Therefore the magnitude of the angular momentum vector is given by

$$h = r^2\dot{\theta}\,. \tag{6}$$

Now consider the motion of m_2 in a time interval δt. Let its polar coordinates at time t be $(r,\ \theta)$. At time $t+\delta t$ its polar coordinates are $(r+\delta r,\ \theta+\delta\theta)$. Therefore the area swept out by the radius vector in the time interval δt is

$$\delta A \approx \frac{1}{2}r\,(r+\delta r)\sin\delta\theta \approx \frac{1}{2}r^2\delta\theta \tag{7}$$

where we have neglected second- and higher-order terms in the small quantities. Dividing each side by δt and taking the limit as $\delta t \to 0$ gives

$$\dot{A} = \frac{1}{2}r^2\dot{\theta} = \frac{1}{2}h\,. \tag{8}$$

Because h is a constant this implies that equal areas are swept out in equal times. Therefore Eq. (8) is the mathematical form of Kepler's second law of planetary motion.

2.2 Orbital Position and Velocity

Substituting the expression for $\ddot{\mathbf{r}}$ from Eq. (5) into Eq. (1) and comparing the $\hat{\mathbf{r}}$ components gives

$$\ddot{r} - r\dot{\theta}^2 = -\frac{\mu}{r^2}. \tag{9}$$

We need to solve this equation and find r as a function of θ in order to determine the path of m_2 with respect to m_1. The standard way of doing this is to make the substitution $u = 1/r$ and eliminate the time by making use of Eq. (6). By differentiating r with respect to time, we obtain

$$\dot{r} = -\frac{1}{u^2}\frac{\mathrm{d}u}{\mathrm{d}\theta}\dot{\theta} = -h\frac{\mathrm{d}u}{\mathrm{d}\theta} \tag{10}$$

$$\ddot{r} = -h\frac{\mathrm{d}^2u}{\mathrm{d}\theta^2}\dot{\theta} = -h^2u^2\frac{\mathrm{d}^2u}{\mathrm{d}\theta^2} \tag{11}$$

and hence Eq. (9) can be written

$$\frac{\mathrm{d}^2u}{\mathrm{d}\theta^2} + u = \frac{\mu}{h^2}. \tag{12}$$

This is a second-order, linear differential equation with a general solution

$$u = \frac{\mu}{h^2}\left[1 + e\cos(\theta - \varpi)\right] \tag{13}$$

where e (an amplitude) and ϖ (a phase) are two constants of integration. Substituting back gives

$$r = \frac{p}{1 + e\cos(\theta - \varpi)}. \tag{14}$$

This is the general equation of a conic in polar coordinates where e is the *eccentricity* and p is the *semilatus rectum* given by

$$p = h^2/\mu. \tag{15}$$

Although four possible conics are possible (circle, ellipse, parabola and hyperbola) here we are only concerned with the properties of the ellipse and the special case of a circle. For an ellipse $0 < e < 1$ and $p = a(1 - e^2)$ where a is a constant referred to as the *semi-major axis* of the ellipse. For a circle $e = 0$.

The positions and velocities of the planets imply that the path of any planet about the Sun is elliptical and closed in inertial space (see Fig. 2). Therefore we have shown that Kepler's first law of planetary motion is a natural consequence of the inverse square law of force. Note that the mass m_1 lies at one focus of the ellipse while the other focus is empty.

Most permanent members of the solar system have $e \ll 1$. Planetary exceptions are Pluto ($e = 0.25$) and Mercury ($e = 0.21$). Nereid, a moon of

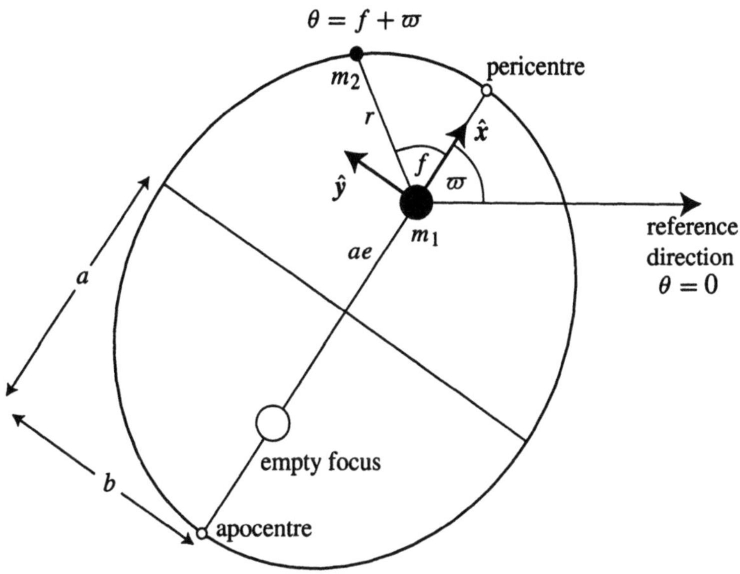

Fig. 2. The geometry of the ellipse of semi-major axis a, semi-minor axis b, eccentricity e and longitude of pericentre ϖ.

Neptune, has the largest eccentricity ($e = 0.75$) of any known natural satellite. We are concentrating on elliptical motion and in this case the quantities a and e are related by

$$b^2 = a^2(1 - e^2) \tag{16}$$

where b is the *semi-minor axis* of the ellipse (see Fig. 2). For completeness, we can now write

$$r = \frac{a(1 - e^2)}{1 + e\cos(\theta - \varpi)} . \tag{17}$$

The angle θ is called the *true longitude*. A simple inspection of Eq. (17) shows that the minimum and maximum values of the orbital radius occur at $r_p = a(1-e)$ and $r_a = a(1+e)$ with $\theta = \varpi$ and $\theta = \varpi + \pi$ respectively. These points are called the *pericentre* (or *periapse*) and the *apocentre* (or *apoapse*) respectively. The distance of either focus from the centre of the ellipse is ae.

The angle ϖ is called the *longitude of pericentre*. Although this is a constant for the two-body problem, it can vary with time when there are additional perturbations (e.g. motion around an oblate planet, the perturbing effect of another planet, etc). It is usually more convenient to refer the angular coordinate to the pericentre rather than the arbitrary reference line. This leads to the introduction of the *true anomaly*, $f = \theta - \varpi$ (see Fig. 2). The equation of the ellipse can now be written as

$$r = \frac{a(1 - e^2)}{1 + e\cos f} . \tag{18}$$

The area swept out by a radius vector in the course of one orbital period T is simply the area $A = \pi ab$ enclosed by the ellipse. But from Eq. (8) this must equal $hT/2$ and hence

$$T^2 = \frac{4\pi^2}{\mu}a^3 . \tag{19}$$

This is the mathematical formulation of Kepler's third law of planetary motion. Note that the orbital period is a function of μ and a only.

If any solar system object (e.g. a planet or asteroid) has a satellite, then observations of its distance and period of the satellite can be used with Kepler's third law to derive an estimate the mass of the object. Let m_{c}, m and m' denote the mass of the Sun, the object and the object's satellite respectively with similar definitions for the semi-major axes and orbital periods. We can use Eq. (19) to write

$$\frac{m+m'}{m_{\mathrm{c}}+m} \approx \frac{m}{m_{\mathrm{c}}} = \left(\frac{a'}{a}\right)^3 \left(\frac{T}{T'}\right)^2 \tag{20}$$

where we have made the reasonable assumptions that $m' \ll m$ and $m \ll m_{\mathrm{c}}$. Therefore, if a, a', T and T' are known from observations, then m/m_{c} can be found. Therefore the mass of the object can be estimated from the orbital properties of its satellite.

Figure 3 shows an image of the asteroid (243) Ida and its moon Dactyl taken by the *Galileo* spacecraft on its way to Jupiter. Ida has dimensions of $56 \times 24 \times 21$ km while Dactyl is about 1.4 km across. Observations of Dactyl's motion and estimates of Ida's shape from the *Galileo* images have resulted in an estimated density of 2.6 ± 0.5 g cm^{-3} for Ida [2].

Because the angle θ covers 2π radians in one orbital period we can define the *mean motion*, n as

$$n = \frac{2\pi}{T} . \tag{21}$$

Note that although n is a constant, this same is not true of the angular velocity $\dot{\theta} = \dot{f}$.

We can derive another constant of two-body motion by taking the scalar product of $\dot{\boldsymbol{r}}$ with Eq. (1) and using the expressions for $\boldsymbol{r}$ and $\dot{\boldsymbol{r}}$ given in Eqs. (3) and (4). This produces a scalar equation which can be integrated to give

$$\frac{1}{2}v^2 - \frac{\mu}{r} = C \tag{22}$$

where $v^2 = \dot{\boldsymbol{r}} \cdot \dot{\boldsymbol{r}}$ is the square of the velocity and C is a constant of the motion. Thus the two-body problem has four constants of the motion: the energy integral, C, and the three components of the angular momentum integral, $\boldsymbol{h}$.

From the definition of $\dot{\boldsymbol{r}}$ given in Eq. (4) we can write

$$v^2 = \dot{\boldsymbol{r}} \cdot \dot{\boldsymbol{r}} = \dot{r}^2 + r^2\dot{f}^2 . \tag{23}$$

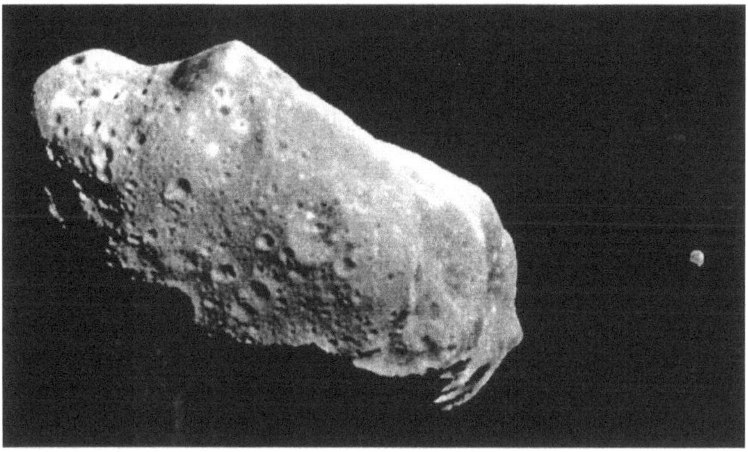

Fig. 3. An image of the asteroid (243) Ida and its moon Dactyl taken by the *Galileo* spacecraft on 28th August 1993. *(Image courtesy of NASA/JPL.)*

By finding expressions for $\dot{r}$ and $r\dot{f}$ it can be shown that this can be written as

$$v^2 = \mu \left(\frac{2}{r} - \frac{1}{a} \right) . \tag{24}$$

Hence the energy constant, C, can be written as

$$C = -\frac{\mu}{2a} . \tag{25}$$

Therefore the energy of an elliptical orbit is a function of its semi-major axis alone and is independent of the eccentricity.

2.3 Kepler's Equation

In finding the relationship between r and θ in the previous section we managed to eliminate the time t from the problem. Therefore, we now know that the path is elliptical but we still need to establish the whereabouts of the orbiting object at any given time. What we really want is an angle that is a linear function of time so that we can calculate time averages of various quantities. Such an angle is the *mean amonaly*, M, defined by

$$M = n(t - \tau) \tag{26}$$

where the constant τ is the time of pericentre passage. The mean anomaly has no simple geometrical interpretation but it is related to another angle, the eccentric anomaly E, which does.

Consider a circumscribed circle, of radius a, concentric to a keplerian ellipse of semi-major axis a and eccentricity e (see Fig. 4). A line perpendicular to the major axis of the ellipse is extended through the point on the orbit and intersects the circle. The eccentric anomaly, E, is defined to be the angle between the major axis of the ellipse and the radius from the centre to the intersection point on the circumscribed circle (see Fig. 4).

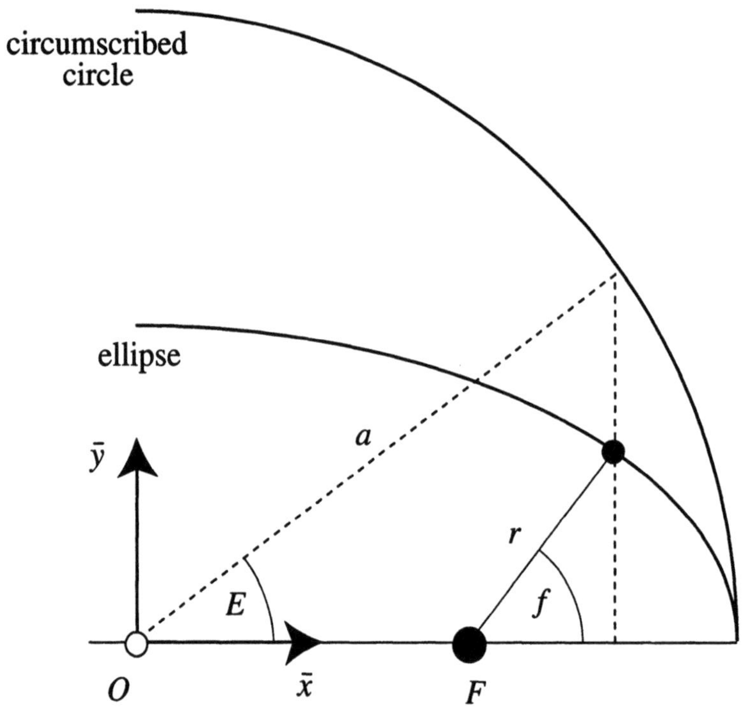

Fig. 4. The relationship between the true anomaly f and the eccentric anomaly E.

Using the equation of a centred ellipse in rectangular coordinates,

$$(\bar{x}/a)^2 + (\bar{y}/b)^2 = 1 \tag{27}$$

and the geometry shown in Fig. 4, we can write $\bar{x} = a \cos E$ and hence $\bar{y}^2 = b^2 \sin^2 E$. From the definition of b in Eq. (16) we have $\bar{y} = a\sqrt{1-e^2} \sin E$. Thus the projections of r in the horizontal and vertical directions are

$$x = a(\cos E - e)\,, \tag{28}$$

$$y = a\sqrt{1-e^2} \sin E \tag{29}$$

from which we obtain

$$r = a(1 - e \cos E)\,. \tag{30}$$

By deriving expressions for $r\dot{f}$ and using Eqs. (23) and (24) we can write

$$\dot{r}^2 = n^2a^3\left(\frac{2}{r} - \frac{1}{a}\right) - \frac{n^2a^4(1-e^2)}{r^2}\,. \tag{31}$$

Hence

$$\dot{r} = \frac{na}{r}\sqrt{a^2e^2 - (r-a)^2}\,. \tag{32}$$

By making the substitution $r - a = -ae\cos E$ we can rewrite this as

$$\dot{E} = \frac{n}{1 - e\cos E}\,. \tag{33}$$

This can be easily integrated to give

$$n(t-\tau) = E - e\sin E \tag{34}$$

where we have taken τ to be the constant of integration and used the appropriate boundary condition $E = 0$ when $t = \tau$ (see Fig. 4). Hence, from our definition of M in Eq. (26) we have

$$M = E - e\sin E\,. \tag{35}$$

This is *Kepler's equation* and its solution is the key to finding the orbital position at any given time.

At this stage it is useful to define another angle, the *mean longitude*, λ by

$$\lambda = M + \varpi\,. \tag{36}$$

Therefore λ is also a linear function of time and, as with M, it has no geometrical interpretation, except in the special case of a circular orbit. Note that all longitude angles (i.e. θ, ϖ and λ) are defined with respect to a common, arbitrary reference direction (see Fig. 2).

As pointed out by Colwell [3], papers about the solution of Kepler's equation have been published in almost every decade since 1650. Because it is transcendental in M, Kepler's equation cannot be solved directly apart from trivial cases. It is possible to derive series solutions in terms of powers of e but these are not usually convergent or at best slowly convergent for large values of e. Danby [4] gives a variety of numerical methods for solving Kepler's equation. The most common and easiest to implement is the standard Newton-Raphson method. By defining the function $f(E)$ to be

$$f(E) = E - e\sin E - M \tag{37}$$

we can solve Kepler's equation by finding the root of the non-linear equation, $f(E) = 0$. The relevant implementation of the Newton-Raphson iteration scheme is

$$E_{i+1} = E_i - \frac{f(E_i)}{f'(E_i)}, \qquad i = 0, 1, 2, \ldots \tag{38}$$

where $f'(E_i) = \mathrm{d}f(E_i)/\mathrm{d}E_i = 1 - e\cos E_i$. Because $E = M$ when $e = 0$, it is customary to take $E_0 = M$ as an initial estimate provided e is small.

2.4 The Orbit in Space

The plane of the orbit of one mass about another is the plane perpendicular to the angular momentum vector (see Fig. 1). We have shown how the values of $\boldsymbol{r} = (x, y)$ and $\dot{\boldsymbol{r}} = (\dot{x}, \dot{y})$ define a unique orbit and a location on that orbit by means of the three constants a, e, ϖ and the variable f. Because motion in the solar system is not confined to a single plane, we now consider the motion in three dimensions (see Fig. 5).

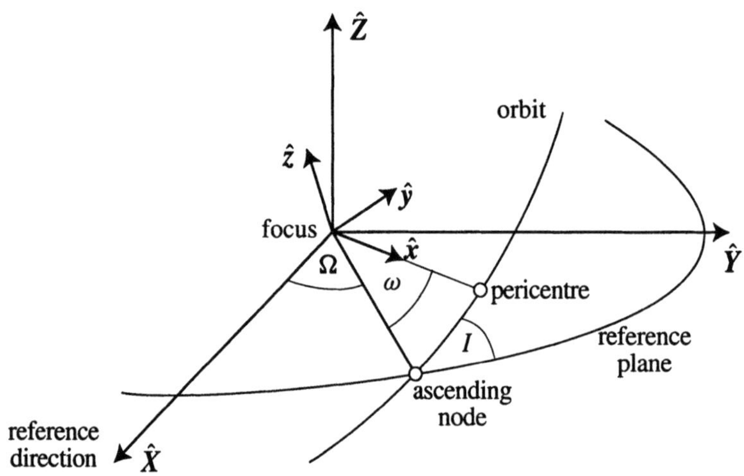

Fig. 5. Three dimensional orbital motion showing the orientation angles I (inclination), ω (argument of pericentre) and Ω (longitude of ascending node).

Let an arbitrary point in space have position vector, $\boldsymbol{r} = (x, y, z) = x\hat{\boldsymbol{x}} + y\hat{\boldsymbol{y}} + z\hat{\boldsymbol{z}}$ with respect to a coordinate system referred to an orbit plane. Here the x-axis lies along the major axis of the ellipse in the direction of pericentre, the y-axis is perpendicular to the x-axis and lies in the orbital plane, while the z-axis is mutually perpendicular to the other two forming a right-handed triad (see Fig. 5).

In order to refer the coordinates in the orbit reference system to some standard frame we have to introduce three additional angles. These are the inclination, I, the argument of pericentre, ω, and the longitude of ascending node, Ω as shown in Fig. 5. The standard reference plane has its X-axis in a particular direction with its Y-axis defining the plane and its Z-axis perpendicular to both of them. For the motion of objects around the Sun it is customary to use a Sun-centred, or *heliocentric* coordinate system where the reference plane is the plane of the Earth's orbit (the *ecliptic*) and the reference line (the X-axis) is in the direction of the *vernal equinox*, along the line of intersection of the plane of the Earth's equator and the ecliptic. The angle the orbit plane makes with the reference plane is I. The point in both planes where the orbit crosses the reference plane moving from below to above the plane is called the *ascending node* while the angle between the reference line and the radius vector to the ascending node is Ω. The angle between this same radius vector and the pericentre of the orbit is ω.

Note that I has to be in the range $0 \leq I \leq 180°$. If $I < 90°$ the motion is *prograde* while if $I \geq 90°$ it is *retrograde*. We can make use of the previously defined longitude of pericentre, ϖ, to write

$$\varpi = \Omega + \omega \tag{39}$$

even though we are now dealing with the inclined case and the angles Ω and ω lie in different planes.

In order to transform from the (x, y, z), orbital plane system to the general (X, Y, Z) reference system we have to carry out three rotations:

- a rotation about the z-axis through an angle ω so that the x-axis coincides with the line of nodes,
- a rotation about the x-axis through an angle I so that the two planes are coincident and
- a rotation about the z-axis through an angle Ω.

These transformations can be represented by three, 3×3 rotation matrices, denoted by $\boldsymbol{P}_1$, $\boldsymbol{P}_2$ and $\boldsymbol{P}_3$ respectively with elements

$$\boldsymbol{P}_1 = \begin{pmatrix} \cos\omega & -\sin\omega & 0 \\ \sin\omega & \cos\omega & 0 \\ 0 & 0 & 1 \end{pmatrix}, \tag{40}$$

$$\boldsymbol{P}_2 = \begin{pmatrix} 1 & 0 & 0 \\ 0 & \cos I & -\sin I \\ 0 & \sin I & \cos I \end{pmatrix}, \tag{41}$$

$$\boldsymbol{P}_3 = \begin{pmatrix} \cos\Omega & -\sin\Omega & 0 \\ \sin\Omega & \cos\Omega & 0 \\ 0 & 0 & 1 \end{pmatrix}. \tag{42}$$

Consequently, if we restrict ourselves to objects in the orbit plane (i.e. $z = 0$) we can write

$$\begin{pmatrix} X \\ Y \\ Z \end{pmatrix} = \boldsymbol{P}_3\boldsymbol{P}_2\boldsymbol{P}_1 \begin{pmatrix} r\cos f \\ r\sin f \\ 0 \end{pmatrix} \tag{43}$$

$$= r \begin{pmatrix} \cos\Omega\cos(\omega + f) - \sin\Omega\sin(\omega + f)\cos I \\ \sin\Omega\cos(\omega + f) + \cos\Omega\sin(\omega + f)\cos I \\ \sin(\omega + f)\sin I \end{pmatrix}. \tag{44}$$

Note that because rotations always preserve lengths, the values of a, e and all distances are unchanged by the transformations.

3 The Three-Body Problem

In the previous section we saw how the motion of objects in the solar system can be described, at any instant, by means of their position on an elliptical orbit. However, although the two-body problem provides a good approximation to the motion of such objects, it is important to realise that everything perturbs everything else in the solar system. Consequently we have to consider the effects of at least one additional body. We are fortunate in that the observed orbital parameters (e.g. small eccentricities and inclinations) and hierarchical structure (e.g. Sun-planet-moon) that exist in the solar system allow us to make a number of useful approximations. In this section we study the properties of the three-body problem with particular emphasis on the approximations whereby (i) two of the masses move in circular orbits about their common centre of mass and (ii) the third mass is too small to affect the motion of the other two. This is the circular, restricted three-body problem.

3.1 Equations of Motion

Consider axes ξ and η in the inertial frame referred to the centre of mass, O, of the system (see Fig. 6). Let the ξ axis lie along the line from m_1 to m_2 at time $t = 0$ with the η axis perpendicular to it and in the orbital plane of the two masses. Let the coordinates of the two masses in this reference frame be (ξ_1, η_1) and (ξ_2, η_2). Because of their assumed circular motion the two masses have a constant separation and the same angular velocity about each other and their common centre of mass. Let the unit of mass be chosen such that

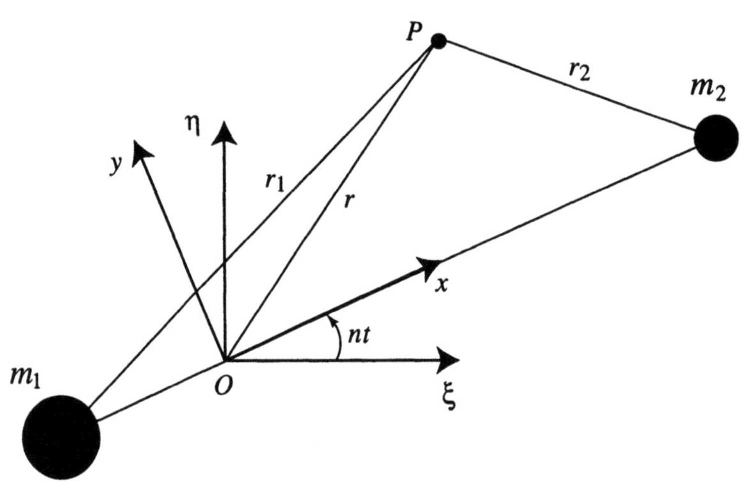

Fig. 6. The relationship between the (ξ, η) and (x, y) coordinate systems for the particle at point P. The origin, O, of both systems is located at the centre of mass.

$\mu = \mathcal{G}(m_1 + m_2) = 1$. If we assume that $m_1 > m_2$ and define

$$\bar{\mu} = \frac{m_2}{m_1 + m_2} \tag{45}$$

then in this system of units the two masses are $\mu_1 = \mathcal{G}m_1 = 1 - \bar{\mu}$ and $\mu_2 = \mathcal{G}m_2 = \bar{\mu}$ where $\bar{\mu} < \frac{1}{2}$. The unit of length is chosen such that the constant separation of the two masses is unity. It then follows that the common mean motion, n, of the two masses is also unity.

In our system the equations of motion of the particle are

$$\ddot{\xi} = \mu_1 \frac{\xi_1 - \xi}{r_1^3} + \mu_2 \frac{\xi_2 - \xi}{r_2^3}, \tag{46}$$

$$\ddot{\eta} = \mu_1 \frac{\eta_1 - \eta}{r_1^3} + \mu_2 \frac{\eta_2 - \eta}{r_2^3}, \tag{47}$$

$$\tag{48}$$

where, from Fig. 6,

$$r_1^2 = (\xi_1 - \xi)^2 + (\eta_1 - \eta)^2, \tag{49}$$

$$r_2^2 = (\xi_2 - \xi)^2 + (\eta_2 - \eta)^2. \tag{50}$$

Now consider a coordinate system rotating at a uniform rate n in the positive direction (see Fig. 6). The direction of the x-axis is chosen such that the two masses always lie along it with coordinates $(x_1, y_1) = (-\mu_2, 0)$ and $(x_2, y_2) = (\mu_1, 0)$. Hence

$$r_1^2 = (x + \mu_2)^2 + y^2, \tag{51}$$

$$r_2^2 = (x - \mu_1)^2 + y^2, \tag{52}$$

where (x, y) are the coordinates of the particle with respect to the rotating system. These coordinates are related to the coordinates in the inertial system by a simple rotation through an angle nt. It can easily be shown that the equations of motion in the rotating system can be written as

$$\ddot{x} - 2n\dot{y} = \frac{\partial U}{\partial x}, \tag{53}$$

$$\ddot{y} + 2n\dot{x} = \frac{\partial U}{\partial y}, \tag{54}$$

where $U = U(x, y)$ is a scalar function of position given by

$$U = \frac{n^2}{2}(x^2 + y^2) + \frac{\mu_1}{r_1} + \frac{\mu_2}{r_2}. \tag{55}$$

3.2 The Jacobi Integral

Multiplying Eq. (53) by $\dot{x}$, Eq. (54) by $\dot{y}$ and adding gives

$$\dot{x}\ddot{x} + \dot{y}\ddot{y} = \frac{\partial U}{\partial x}\dot{x} + \frac{\partial U}{\partial y}\dot{y} = \frac{\mathrm{d}U}{\mathrm{d}t} \tag{56}$$

which can be integrated to give

$$v^2 = 2U - C_\mathrm{J} \tag{57}$$

where C_J is a constant of integration and $v^2 = \dot{x}^2 + \dot{y}^2$. This constant, purely a function of the position and velocity of the particle, can be written as

$$C_\mathrm{J} = n^2(x^2 + y^2) + 2\left(\frac{\mu_1}{r_1} + \frac{\mu_2}{r_2}\right) - \dot{x}^2 - \dot{y}^2. \tag{58}$$

Therefore $2U - v^2 = C_\mathrm{J}$ is a constant of the motion. This is the *Jacobi integral* or *Jacobi constant* and it is the only integral of the circular, restricted three-body problem. Although it cannot be used to provide an exact solution for the orbital motion, it can be used to determine regions from which the particle is excluded. Consider the locations where the velocity of the particle is zero. In this case we have

$$n^2(x^2 + y^2) + 2\left(\frac{\mu_1}{r_1} + \frac{\mu_2}{r_2}\right) = C_\mathrm{J}. \tag{59}$$

This defines a set of curves for particular values of C_J. These curves, known as *zero velocity curves* are bounds on the motion of the particle. Figure 7 shows examples of these curves for a mass $\mu_2 = 0.2$ where we have taken the mean motion, n to be unity. From Eq. (57) it is clear that we must always have $2U \geq C_\mathrm{J}$ for real values of v. Thus Eq. (59) defines the boundary curves of regions where particle motion is not possible, i.e. excluded regions. Hence, although we cannot solve the restricted three-body problem to find the motion of the particle for arbitrary starting conditions, the existence of the Jacobi integral does allow us, in certain circumstances, to find regions of space where the particle cannot be. The result can easily be extended to three dimensions.

Consider the implications of the shaded areas shown in Fig. 7. If a particle with $C_\mathrm{J} = 3.9$ is in orbit in the unshaded region around the mass m_1 in Fig. 7a, then it can never orbit the mass m_2 or escape from the system since it would have to cross the excluded region to do so. Similarly in Fig. 7b for $C_\mathrm{J} = 3.7$, if the particle is orbiting the mass m_1 then it is possible that it could eventually orbit the mass m_2, but it could never escape from the system. This is the concept of *Hill's stability*.

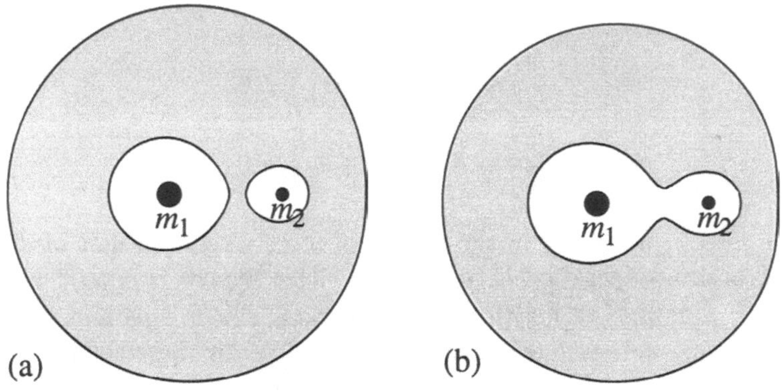

Fig. 7. Two sets of zero velocity curves for the case when $\mu_2 = 0.2$. The values are (a) $C_J = 3.9$ and (b) $C_J = 3.7$. The shaded areas denote regions from which the particle is excluded.

3.3 Lagrangian Equilibrium Points

It is possible to find a number of special solutions to the circular restricted three-body problem by searching for points where the particle has zero velocity and zero acceleration in the rotating frame. These are equilibrium points of the system. We can write

$$\mu_1 r_1^2 + \mu_2 r_2^2 = x^2 + y^2 + \mu_1\mu_2 \tag{60}$$

and hence

$$U = \mu_1\left(\frac{1}{r_1} + \frac{r_1^2}{2}\right) + \mu_2\left(\frac{1}{r_2} + \frac{r_2^2}{2}\right) - \frac{1}{2}\mu_1\mu_2 \,. \tag{61}$$

Note that r_1 and r_2, unlike x and y, are always positive quantities.

Now consider the equations of motion, Eqs. (53) and (54), with $\ddot{x} = \ddot{y} = \dot{x} = \dot{y} = 0$ (i.e. zero velocity and zero acceleration). To find the locations of the equilibrium points we must solve the simultaneous non-linear equations,

$$\frac{\partial U}{\partial x} = \frac{\partial U}{\partial r_1}\frac{\partial r_1}{\partial x} + \frac{\partial U}{\partial r_2}\frac{\partial r_2}{\partial x} = 0\,, \tag{62}$$

$$\frac{\partial U}{\partial y} = \frac{\partial U}{\partial r_1}\frac{\partial r_1}{\partial y} + \frac{\partial U}{\partial r_2}\frac{\partial r_2}{\partial y} = 0\,. \tag{63}$$

These equations reduce to

$$\mu_1\left(-\frac{1}{r_1^2} + r_1\right)\frac{x + \mu_2}{r_1} + \mu_2\left(-\frac{1}{r_2^2} + r_2\right)\frac{x - \mu_1}{r_2} = 0\,, \tag{64}$$

$$\mu_1\left(-\frac{1}{r_1^2} + r_1\right)\frac{y}{r_1} + \mu_2\left(-\frac{1}{r_2^2} + r_2\right)\frac{y}{r_2} = 0\,. \tag{65}$$

These have the trivial solutions

$$\frac{\partial U}{\partial r_1} = \mu_1 \left(-\frac{1}{r_1^2} + r_1 \right) = 0, \tag{66}$$

$$\frac{\partial U}{\partial r_2} = \mu_2 \left(-\frac{1}{r_2^2} + r_2 \right) = 0, \tag{67}$$

which gives $r_1 = r_2 = 1$ in the system of units where the unit of distance is the constant separation of m_1 and m_2. This implies $(x + \mu_2)^2 + y^2 = 1$ and $(x - \mu_1)^2 + y^2 = 1$ with the two solutions $x = \frac{1}{2} - \mu_2$ and $y = \pm\frac{\sqrt{3}}{2}$. Since $r_1 = r_2 = 1$, each of the two points defined by these equations forms an equilateral triangle with the masses m_1 and m_2. These are the so-called *triangular Lagrangian equilibrium points* usually denoted L_4 $(y > 0)$ and L_5 $(y < 0)$.

It is clear from Eq. (65) that $y = 0$ is a simple solution of Eq. (63), implying that there are additional equilibrium points along the x-axis satisfy Eq. (62). These are the *colinear Lagrangian equilibrium points* denoted by L_1, L_2 and L_3. The L_1 point lies between the masses m_1 and m_2, the L_2 point lies outside the mass m_2 and the L_3 point lies on the negative x-axis close to the unit radius. If we define

$$\alpha = \left(\frac{\mu_2}{3\mu_1} \right)^{1/3} \tag{68}$$

then it can be shown (see, for example, [1]) that L_1 is located at a distance

$$r_2 = \alpha - \frac{1}{3}\alpha^2 - \frac{1}{9}\alpha^3 - \frac{23}{81}\alpha^4 + \mathcal{O}(\alpha^5) \tag{69}$$

from the mass m_2. Similarly L_2 is located at a distance

$$r_2 = \alpha + \frac{1}{3}\alpha^2 - \frac{1}{9}\alpha^3 - \frac{31}{81}\alpha^4 + \mathcal{O}(\alpha^5) \tag{70}$$

from the mass m_2. Therefore, to $\mathcal{O}(\alpha)$, the L_1 and L_2 points are equidistant from m_2 on either side of it. The distance of L_3 from m_1 can be written as (see [1])

$$r_1 = 1 - \frac{7}{12}\left(\frac{\mu_2}{\mu_1}\right) + \frac{7}{12}\left(\frac{\mu_2}{\mu_1}\right)^2 - \frac{13223}{20736}\left(\frac{\mu_2}{\mu_1}\right)^3 + \mathcal{O}\left(\frac{\mu_2}{\mu_1}\right)^4 . \tag{71}$$

Note that as $\mu_2 \to 0$, L_1 and L_2 become more equidistant from the mass m_2 (at separations of α) while L_3 moves towards a unit distance from m_1. Figure 8 shows a sample of zero velocity curves for the case $\mu_2 = 0.01$ together with the location of the five Lagrangian equilibrium points.

It is not enough to know the locations of the five equilibrium points — we also need to know their stability properties. These can be investigated

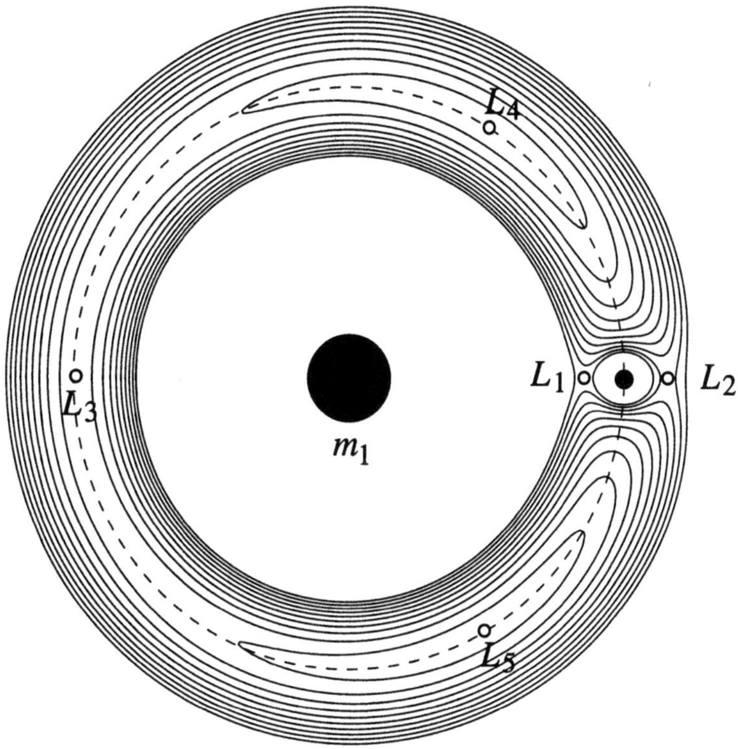

Fig. 8. A selection of zero velocity curves for the case $\mu_2 = 0.01$ showing the corresponding locations of the Lagrangian equilibrium points (open circles).

by using the standard technique of considering a small displacement from an equilibrium position and solving the linearised equations of motion that describe the resulting path taken by the particle with respect to the equilibrium point. Such an anlaysis (see, for example, [1]) shows that L_4 and L_5 are linearly stable provided

$$\mu_2 \leq \frac{27 - \sqrt{621}}{54} \approx 0.0385 . \tag{72}$$

This condition is satisfied by all Sun-planet and planet-satellite pairs in the solar system with the single exception of the Pluto-Charon system ($\mu_2 \sim 0.1$). The type of paths (in the rotating frame) of particles started close to the equilibrium points show evidence of two frequencies. The first is almost the same as the keplerian frequency of m_2 about m_1. The second is a smaller frequency associated with librational motion around the equilibrium point. However, a similar analysis for the colinear equilibrium points, L_1, L_2 and L_3 shows that they are all linearly unstable.

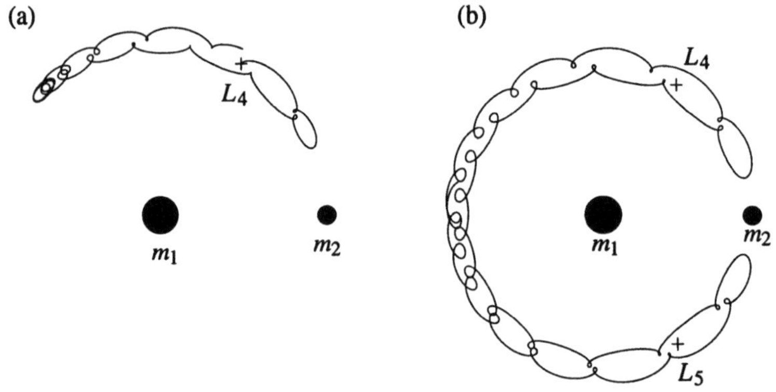

Fig. 9. (a) A tadpole orbit librating about the L_4 point in a system with $\mu_2 = 0.001$. (b) A horseshoe orbit encompassing the L_3, L_4 and L_5 points in a system with $\mu_2 \approx 0.001$. The figures are derived from [1].

3.4 Tadpoles and Horseshoes

Particles started in low-eccentricity, low-inclination orbits close to the semi-major axis of m_2 can exhibit what are called *tadpole* and *horseshoe* orbits. The names are derived from the shape of their associated zero velocity curves. Figure 9 shows an example of each type of orbit; each was integrated numerically using a starting condition close to the L_4 point of the system. The tadpole orbit shown in Fig. 9a encompasses the L_4 point. For a larger initial displacement a horseshoe orbit is possible as shown in Fig. 9b. Note the 'loops' in each path. These are nothing more than the effect of a keplerian eccentricity when viewed in the rotating frame. If the initial displacement results in an initial semi-major axis that is outside the region enclosed between the L_1 and L_2 points (i.e. more than a distance $\sim \alpha$ from the unit semi-major axis) then the orbit will not turn but rather circulate past the mass m_2.

It is important to remember that the orbital elements that were constant in the two-body problem are subject to change due to the perturbations in the three-body problem. This is clearly seen in Fig. 10 for three separate trajectories, each of which was numerically integrated for 100 orbital periods. The trajectories represent typical examples of a small libration amplitude tadpole, a large amplitude tadpole and a horseshoe orbit. Note that the change in semi-major axis is significant in each case.

3.5 Trojan Asteroids and Satellites

The behaviour of objects moving around the stable equilibrium points was known to Lagrange in the eighteenth century. However, actual examples of such objects were not found until the twentieth century. In 1906 the first example was the asteroid (588) Achilles, discovered to be librating around

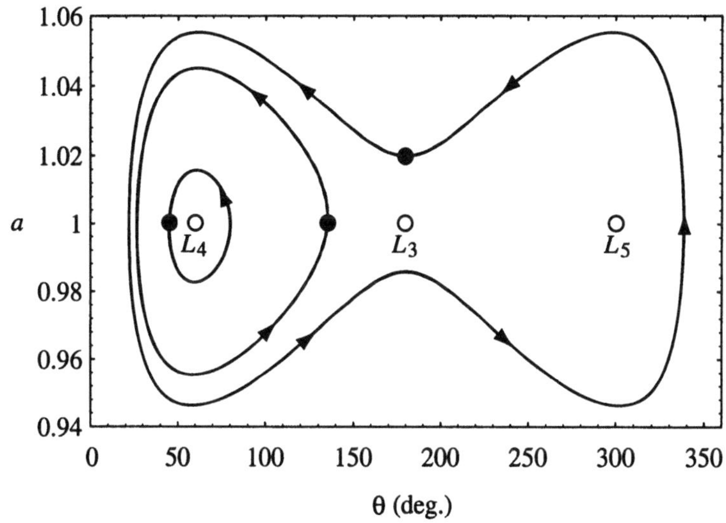

Fig. 10. The variation of semi-major axis a with the angle θ around the orbit for three trajectories with $\mu_2 = 10^{-3}$. The starting positions are indicated by filled circles.

the L_4 point in the Sun–Jupiter system. By the end of 1997 a total of 405 Trojan asteroids had been discovered librating about Jupiter's L_4 and L_5 points. The amplitudes of libration can exceed 30° but the mean value of the amplitude is 14°.

Figure 11 shows the distribution of the Sun–Jupiter Trojan asteroids in December 1997. The projected positions onto the plane of the ecliptic are shown in Fig. 11a, together with the orbit and position of Jupiter with respect to the Sun. Although there are two distinct clusterings about the triangular points, large libration amplitudes are evident. Figure 11b shows the vertical extent of the Trojan groups.

Jupiter does not have a monopoly on Trojan asteroids. The first Sun–Mars Trojan, (5261) Eureka, librating about the L_5 point, was discovered in 1990 [5]. Asteroid (3753) Cruithne, originally designated 1986TO, is involved in an unusual horseshoe libration in the Sun–Earth system [6].

These are all examples of librational motion about the L_4 and L_5 points of Sun–planet systems. However, we can also consider motion in the vicinity of the triangular points of a planet–satellite system. These are usually referred to as *co-orbital satellites*, or *Trojan satellites*. The first co-orbital satellites were discovered in 1980 using ground-based CCD observations of the Saturn system. The three objects lie in the orbits of Tethys and Dione and all were directly imaged by the Voyager spacecraft during their Saturn flybys in 1980 and 1981. Telesto and Calypso lie close to (within $\sim 2°$ and $\sim 4°$ respectively) the L_4 and L_5 points of Tethys while Helene lies at most $\sim 17°$ from Dione's L_4 point.

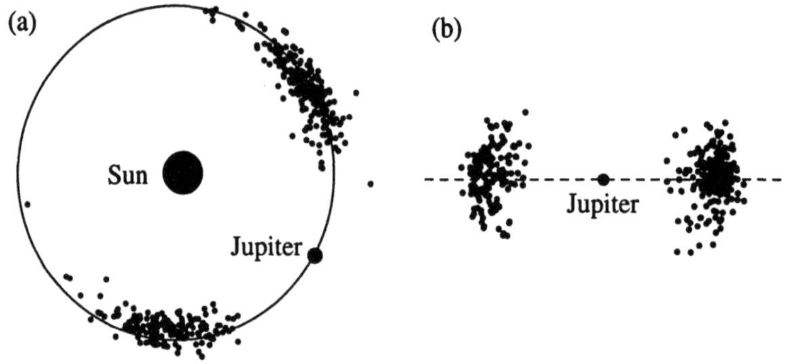

Fig. 11. The distribution of asteroids in the vicinity of the orbit of Jupiter on Julian Date 2450800.5 (18 December 1997). (a) The positions of the asteroids projected on to the plane of the ecliptic. (b) The vertical distribution of the same asteroids viewed along the Jupiter–Sun line. The dashed line denotes the plane of Jupiter's orbit.

3.6 Janus and Epimetheus

As well as the tadpole type orbits being followed by Telesto, Calypso and Helene in the saturnian system, there is at least one example of a horseshoe configuration, albeit slightly different from the classical model previously discussed. This is the co-orbital pair Janus and Epimetheus. In 1980, the year of their discovery, Janus was observed to have a semi-major axis $a_{\rm J} = 151472$ km while Epimetheus, the smaller satellite, had $a_{\rm E} = 151422$ km, i.e. an orbital separation of only 50 km. They have approximate mean diameters of 175 km and 105 km respectively and were $\sim 180°$ apart in February 1980. A simple analysis would suggest a collision in 1982. However, it was quickly realised that the orbits are performing a variation on the horseshoe configuration of the circular restricted problem.

In a reference frame centred on Saturn and rotating with the average mean motion of either satellite, Janus and Epimetheus each librates on its own horseshoe path about longitudes 180° apart. If $W_{\rm J}$ and $W_{\rm E}$ are the average widths of the librational arcs of Janus and Epimetheus respectively, then by assuming circular orbits for each satellite and conservation of the total orbital angular momentum it is easy to show that

$$m_{\rm J} W_{\rm J} = m_{\rm E} W_{\rm E}. \tag{73}$$

Figure 12 shows the paths of the satellites in the frame rotating with the average mean motion of either satellite. The actual half-widths of the Janus and Epimetheus arcs are 10 km and 40 km respectively and each orbit has a mean semi-major axis of 150,432 km. The satellites never pass one another because every 4 years, as they approach one another, their mutual gravitational perturbations cause an exchange of angular momentum, leading to the satellite on the outer path moving to an inner one and vice versa.

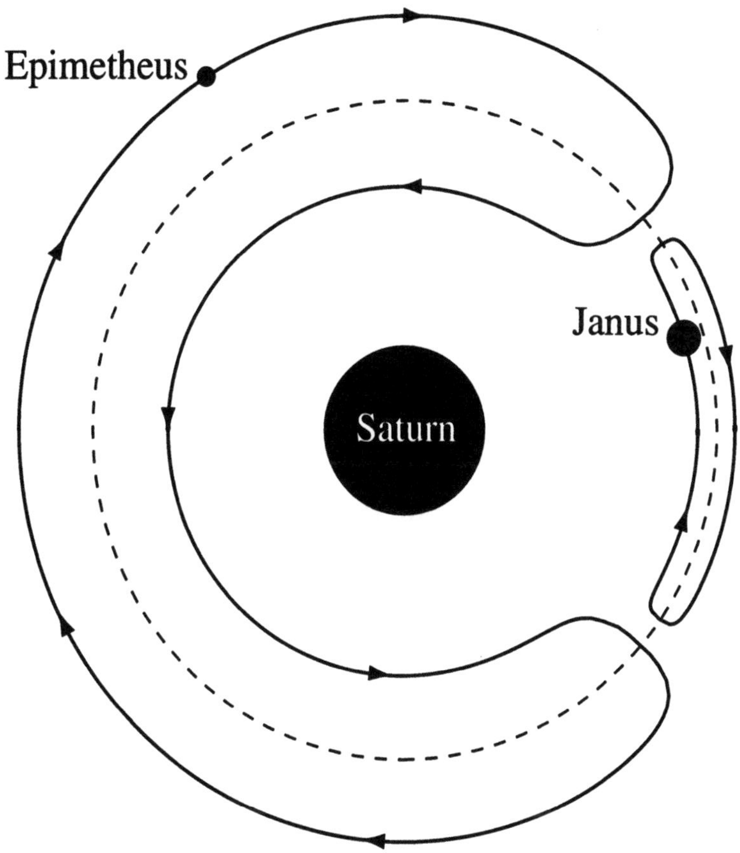

Fig. 12. A schematic diagram showing the paths of Janus and Epimetheus in a frame rotating with the average mean motion of either satellite. The relative radial extent of the librational arcs are exaggerated by a factor $\sim 10^3$; the ratio of the radial widths of the arcs is equal to the Janus–Epimetheus mass ratio (~ 0.25). (Taken from [1].)

Dermott & Murray [7] showed that observations of the motions of Janus and Epimetheus could be used to determine the sum of the masses as well as their ratio. From new observational data Nicholson *et al.* [8] showed that Janus and Epimetheus can approach one another to within 5.64° and that the resulting masses yield densities of 0.65 ± 0.08 g cm^{-3} and 0.63 ± 0.11 g cm^{-3} respectively. Given that these are probably icy bodies, such low values point towards the possibility that these and other satellites of Saturn may be made of a form of porous ice. Therefore knowledge of the dynamics of satellite orbits can lead directly to constraints on their internal properties.

4 The Disturbing Function

In our investigation of the three-body problem in Section 3 we only considered orbits that lay close (in terms of semi-major axis) to that of the secondary mass. In this section we discuss the more general problem of analysing the effect of a perturbing secondary mass on the keplerian motion of an object in orbit around a primary mass. In order to achieve this we have to consider the additional gravitational potential over and above the central mass potential produced by a perturbing body. This is commonly called the disturbing function. As we shall see, an understanding of the disturbing function and its properties is the key to understanding a variety of phenomena in the solar system.

4.1 Perturbing Potential

Consider a central mass m_c and two orbiting masses, m and m', with position vectors $\boldsymbol{r}$ and $\boldsymbol{r}'$ with respect to m_c (see Fig. 13). By formulating the equations of motion in an inertial frame it can easily be shown that the motion of m and m' can be described by the solution to the equations

$$\ddot{\boldsymbol{r}} + \mathcal{G}(m_\mathrm{c}+m)\frac{\boldsymbol{r}}{r^3} = \mathcal{G}m'\left(\frac{\boldsymbol{r}'-\boldsymbol{r}}{|\boldsymbol{r}'-\boldsymbol{r}|^3} - \frac{\boldsymbol{r}'}{r'^3}\right), \tag{74}$$

$$\ddot{\boldsymbol{r}}' + \mathcal{G}(m_\mathrm{c}+m')\frac{\boldsymbol{r}'}{r'^3} = \mathcal{G}m\left(\frac{\boldsymbol{r}-\boldsymbol{r}'}{|\boldsymbol{r}-\boldsymbol{r}'|^3} - \frac{\boldsymbol{r}}{r^3}\right). \tag{75}$$

These in turn can be written in terms of the gradient of potential functions U, U', $\mathcal{R}$ and $\mathcal{R}'$ as

$$\ddot{\boldsymbol{r}} = \nabla(U+\mathcal{R}), \tag{76}$$

$$\ddot{\boldsymbol{r}}' = \nabla'(U'+\mathcal{R}'), \tag{77}$$

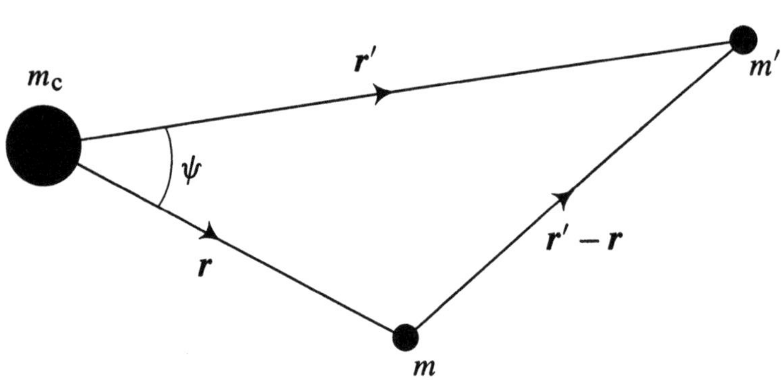

Fig. 13. The position vectors, $\boldsymbol{r}$ and $\boldsymbol{r}'$ of two masses, m and m', with respect to a central mass, m_c.

where

$$U = \mathcal{G}\frac{(m_c + m)}{r}, \quad (78)$$

$$U' = \mathcal{G}\frac{(m_c + m')}{r'}, \quad (79)$$

$$\mathcal{R} = \frac{\mathcal{G}m'}{|\boldsymbol{r}' - \boldsymbol{r}|} - \mathcal{G}m'\frac{\boldsymbol{r}\cdot\boldsymbol{r}'}{r'^3}, \quad (80)$$

$$\mathcal{R}' = \frac{\mathcal{G}m}{|\boldsymbol{r} - \boldsymbol{r}'|} - \mathcal{G}m\frac{\boldsymbol{r}\cdot\boldsymbol{r}'}{r^3}. \quad (81)$$

Here $\mathcal{R}$ is the disturbing function experienced by the mass m due to m' and $\mathcal{R}'$ is the disturbing function experienced by the mass m' due to m. The first term on the right-hand side of Eqs. (80) and (81) is called the *direct term* and the second term is called the *indirect term*.

While it is relatively trivial to express $\mathcal{R}$ or $\mathcal{R}'$ in terms of cartesian coordinates, the real task is to derive expressions for the disturbing function in terms of the orbital elements of both bodies. This is dealt with in some detail by Murray & Dermott [1] who provide an expansion complete to fourth-order in the individual orbital elements.

By writing

$$\mathcal{R} = \frac{\mu'}{a'}\mathcal{R}_\mathrm{D} + \frac{\mu'}{a'}\alpha\mathcal{R}_\mathrm{E}, \quad (82)$$

$$\mathcal{R}' = \frac{\mu}{a'}\mathcal{R}_\mathrm{D} + \frac{\mu}{a'}\frac{1}{\alpha^2}\mathcal{R}_\mathrm{I}, \quad (83)$$

where now $\mu = \mathcal{G}m$, $\mu' = \mathcal{G}m'$ and

$$\mathcal{R}_\mathrm{D} = \frac{a'}{|\boldsymbol{r}' - \boldsymbol{r}|}, \quad (84)$$

$$\mathcal{R}_\mathrm{E} = -\left(\frac{r}{a}\right)\left(\frac{a'}{r'}\right)^2 \cos\psi, \quad (85)$$

$$\mathcal{R}_\mathrm{I} = -\left(\frac{r'}{a'}\right)\left(\frac{a}{r}\right)^2 \cos\psi, \quad (86)$$

where now $\alpha = a/a' < 1$ is the ratio of the semi-major axes and ψ is the angle between the position vectors (see Fig. 13). In these expressions $\mathcal{R}_\mathrm{D}$ is derived from the direct part of the disturbing function, $\mathcal{R}_\mathrm{E}$ comes from the indirect part due to an external perturber while $\mathcal{R}_\mathrm{I}$ comes from the indirect part for an internal perturber. Therefore we can use an expansion of $\mathcal{R}_\mathrm{D}$ to obtain the direct part of either $\mathcal{R}$ or $\mathcal{R}'$.

4.2 Literal Expansion to Second Order

Murray & Dermott [1] give an explicit second-order expansion of $\mathcal{R}_\mathrm{D}$, $\mathcal{R}_\mathrm{E}$ and $\mathcal{R}_\mathrm{I}$ in terms of the orbital elements. The expansion of $\mathcal{R}_\mathrm{D}$ makes use of

Laplace coefficients $b_s^{(j)}(\alpha)$ which are defined as

$$\frac{1}{2}b_s^{(j)}(\alpha) = \frac{1}{2\pi}\int_0^{2\pi} \frac{\cos j\psi \,\mathrm{d}\psi}{(1 - 2\alpha\cos\psi + \alpha^2)^s} \tag{87}$$

where s is a half-integer (i.e. $s = \frac{1}{2}, \frac{3}{2}, \frac{5}{2}, \ldots$). This can also be written as a hypergeometric series in α of the form

$$\frac{1}{2}b_s^{(j)}(\alpha) = \frac{s(s+1)\ldots(s+j-1)}{1\cdot 2\cdot 3\ldots j}\alpha^j \times \left[1 + \frac{s(s+j)}{1(j+1)}\alpha^2 + \frac{s(s+1)(s+j)(s+j+1)}{1\cdot 2(j+1)(j+2)}\alpha^4 + \ldots\right]. \tag{88}$$

The expansion of $\mathcal{R}_\mathrm{D}$ complete to second-order in the eccentricities and inclinations is

$$\begin{aligned}\mathcal{R}_\mathrm{D} = {} & (f_0 + f_1(e^2 + e'^2) + f_2(s^2 + s'^2))\cos[j\lambda' - j\lambda] \\ & + f_3 ee'\cos[j\lambda' - j\lambda + \varpi' - \varpi] + f_4 ss'\cos[j\lambda' - j\lambda + \Omega' - \Omega] \\ & + f_5 e\cos[j\lambda' + (1-j)\lambda - \varpi] + f_6 e'\cos[j\lambda' + (1-j)\lambda - \varpi'] \\ & + f_7 e^2\cos[j\lambda' + (2-j)\lambda - 2\varpi] \\ & + f_8 ee'\cos[j\lambda' + (2-j)\lambda - \varpi' - \varpi] \\ & + f_9 e'^2\cos[j\lambda' + (2-j)\lambda - 2\varpi'] \\ & + f_{10} s^2\cos[j\lambda' + (2-j)\lambda - 2\Omega] \\ & + f_{11} ss'\cos[j\lambda' + (2-j)\lambda - \Omega' - \Omega] \\ & + f_{12} s'^2\cos[j\lambda' + (2-j)\lambda - 2\Omega']\end{aligned} \tag{89}$$

where $s = \sin\frac{1}{2}I$, $s' = \sin\frac{1}{2}I'$ and

$$f_0 = \frac{1}{2}b_{1/2}^{(j)} \tag{90}$$

$$f_1 = \frac{1}{8}\left[-4j^2 + 2\alpha D + \alpha^2 D^2\right] b_{1/2}^{(j)} \tag{91}$$

$$f_2 = -\frac{1}{4}\left(\alpha b_{3/2}^{(j-1)} + \alpha b_{3/2}^{(j+1)}\right) \tag{92}$$

$$f_3 = \frac{1}{4}\left[2 + 6j + 4j^2 - 2\alpha D - \alpha^2 D^2\right] b_{1/2}^{(j+1)} \tag{93}$$

$$f_4 = \alpha b_{3/2}^{(j+1)} \tag{94}$$

$$f_5 = \frac{1}{2}\left[-2j - \alpha D\right] b_{1/2}^{(j)} \tag{95}$$

$$f_6 = \frac{1}{2}\left[-1 + 2j + \alpha D\right] b_{1/2}^{(j-1)} \tag{96}$$

$$f_7 = \frac{1}{8}\left[-5j + 4j^2 - 2\alpha D + 4j\alpha D + \alpha^2 D^2\right] b_{1/2}^{(j)} \tag{97}$$

$$f_8 = \frac{1}{4}\left[-2+6j-4j^2+2\alpha D-4j\alpha D-\alpha^2D^2\right] b_{1/2}^{(j-1)} \tag{98}$$

$$f_9 = \frac{1}{8}\left[2-7j+4j^2-2\alpha D+4j\alpha D+\alpha^2D^2\right] b_{1/2}^{(j-2)} \tag{99}$$

$$f_{10} = \frac{1}{2}\alpha b_{3/2}^{(j-1)} \tag{100}$$

$$f_{11} = -\alpha b_{3/2}^{(j-1)} \tag{101}$$

$$f_{12} = \frac{1}{2}\alpha b_{3/2}^{(j-1)} . \tag{102}$$

In these expressions D denotes the differential operator $\mathrm{d}/\mathrm{d}\alpha$.
The expressions for $\mathcal{R}_\mathrm{E}$ and $\mathcal{R}_I$ are

$$\begin{aligned}
\mathcal{R}_\mathrm{E} = & \left(-1+\frac{1}{2}e^2+\frac{1}{2}e'^2+s^2+s'^2\right)\cos[\lambda'-\lambda] \\
& - ee'\cos[2\lambda'-2\lambda-\varpi'+\varpi] - 2ss'\cos[\lambda'-\lambda-\Omega'+\Omega] \\
& - \frac{1}{2}e\cos[\lambda'-2\lambda+\varpi] + \frac{3}{2}e\cos[\lambda'-\varpi] - 2e'\cos[2\lambda'-\lambda-\varpi'] \\
& - \frac{3}{8}e^2\cos[\lambda'-3\lambda+2\varpi] - \frac{1}{8}e^2\cos[\lambda'+\lambda-2\varpi] \\
& + 3ee'\cos[2\lambda-\varpi'-\varpi] - \frac{1}{8}e'^2\cos[\lambda'+\lambda-2\varpi'] \\
& - \frac{27}{8}e'^2\cos[3\lambda'-\lambda-2\varpi'] - s^2\cos[\lambda'+\lambda-2\Omega] \\
& + 2ss'\cos[\lambda'+\lambda-\Omega'-\Omega] - s'^2\cos[\lambda'+\lambda-2\Omega']
\end{aligned} \tag{103}$$

and

$$\begin{aligned}
\mathcal{R}_\mathrm{I} = & \left(-1+\frac{1}{2}e^2+\frac{1}{2}e'^2+s^2+s'^2\right)\cos[\lambda'-\lambda] \\
& - ee'\cos[2\lambda'-2\lambda-\varpi'+\varpi] - 2ss'\cos[\lambda'-\lambda-\Omega'+\Omega] \\
& - 2e\cos[\lambda'-2\lambda+\varpi] + \frac{3}{2}e'\cos[\lambda-\varpi'] - \frac{1}{2}e'\cos[2\lambda'-\lambda-\varpi'] \\
& - \frac{27}{8}e^2\cos[\lambda'-3\lambda+2\varpi] - \frac{1}{8}e^2\cos[\lambda'+\lambda-2\varpi] \\
& + 3ee'\cos[2\lambda-\varpi'-\varpi] - \frac{1}{8}e'^2\cos[\lambda'+\lambda-2\varpi'] \\
& - \frac{3}{8}e'^2\cos[3\lambda'-\lambda-2\varpi'] - s^2\cos[\lambda'+\lambda-2\Omega] \\
& + 2ss'\cos[\lambda'+\lambda-\Omega'-\Omega] - s'^2\cos[\lambda'+\lambda-2\Omega'] .
\end{aligned} \tag{104}$$

It is worthwhile examining some of the properties of the disturbing function expansion because of its importance in later parts of this article. The expansion of $\mathcal{R}$ (or $\mathcal{R}'$) has the general form

$$\mathcal{R} = \frac{\mu'}{a'}\left(\mathcal{R}_\mathrm{D}+\alpha\mathcal{R}_\mathrm{E}\right) = \frac{\mu'}{a'}\sum_j S(\alpha,e,e',I,I')\cos\varphi \tag{105}$$

where j is an integer and φ is a combination of longitudes that can be written as

$$\varphi = j_1\lambda' + j_2\lambda + j_3\varpi' + j_4\varpi + j_5\Omega' + j_6\Omega \tag{106}$$

where the j_i are all integers. Each cosine argument, φ, satisifes the *d'Alembert relation* whereby the sum of the coefficients of the various longitudes in the argument is zero; i.e. $\sum_{i=1}^{6} j_i = 0$. In fact, this allows us to identify a valid argument in the series expansion to any order. The second point to note is that the absolute value of the coefficient of ϖ', ϖ, Ω' and Ω (i.e. $|j_3|$, $|j_4|$, $|j_5|$ and $|j_6|$) is equal to the *lowest* power of e', e, s' and s, respectively, that occurs in the accompanying term. All the arguments can be classified according to the absolute value of the sum of the coefficients of the mean longitudes, λ' and λ in the cosine argument. For example, the expansion of $\mathcal{R}_{\mathrm{D}}$ given in Eq. (89) has three zeroth-order arguments, two first-order arguments and six second-order arguments. Therefore an expansion that includes Nth-order arguments will involves Nth-order powers of the eccentricities and inclinations. Note too that there are no Laplace coefficients in the expansions for the indirect terms $\mathcal{R}_{\mathrm{E}}$ and $\mathcal{R}_{\mathrm{I}}$. Furthermore, in these two expansions each argument is explicit and does not involve the integer j.

4.3 Lagrange's Planetary Equations

In order to make use of the disturbing function we need to be able to calculate the changes in orbital elements to which they give rise. To do this we make use of *Lagrange's planetary equations*. These express the rates of change of a, e, I, ϖ, Ω and the mean longitude at epoch, ε, and are given by

$$\frac{\mathrm{d}a}{\mathrm{d}t} = \frac{2}{na}\frac{\partial\mathcal{R}}{\partial\varepsilon} \tag{107}$$

$$\frac{\mathrm{d}e}{\mathrm{d}t} = -\frac{\sqrt{1-e^2}}{na^2e}\left(1-\sqrt{1-e^2}\right)\frac{\partial\mathcal{R}}{\partial\varepsilon} - \frac{\sqrt{1-e^2}}{na^2e}\frac{\partial\mathcal{R}}{\partial\varpi} \tag{108}$$

$$\frac{\mathrm{d}\varepsilon}{\mathrm{d}t} = -\frac{2}{na}\frac{\partial\mathcal{R}}{\partial a} + \frac{\sqrt{1-e^2}\left(1-\sqrt{1-e^2}\right)}{na^2e}\frac{\partial\mathcal{R}}{\partial e} + \frac{\tan\frac{1}{2}I}{na^2\sqrt{1-e^2}}\frac{\partial\mathcal{R}}{\partial I} \tag{109}$$

$$\frac{\mathrm{d}\Omega}{\mathrm{d}t} = \frac{1}{na^2\sqrt{1-e^2}\sin I}\frac{\partial\mathcal{R}}{\partial I} \tag{110}$$

$$\frac{\mathrm{d}\varpi}{\mathrm{d}t} = \frac{\sqrt{1-e^2}}{na^2e}\frac{\partial\mathcal{R}}{\partial e} + \frac{\tan\frac{1}{2}I}{na^2\sqrt{1-e^2}}\frac{\partial\mathcal{R}}{\partial I} \tag{111}$$

$$\frac{\mathrm{d}I}{\mathrm{d}t} = \frac{-\tan\frac{1}{2}I}{na^2\sqrt{1-e^2}}\left(\frac{\partial\mathcal{R}}{\partial\varepsilon} + \frac{\partial\mathcal{R}}{\partial\varpi}\right) - \frac{1}{na^2\sqrt{1-e^2}\sin I}\frac{\partial\mathcal{R}}{\partial\Omega}\,. \tag{112}$$

Consider the expression for $\dot{\varepsilon}$ given in Eq. (109). The first term on the right-hand side contains a factor $\partial\mathcal{R}/\partial a$. Therefore, because a occurs both explicitly (in the Laplace coefficients) and implicitly (in cosine arguments because $\lambda = nt + \varepsilon$) in the expansion, this gives rise to the time, t, occurring as a factor when the partial derivative is taken. This can be overcome by defining a new mean longitude at epoch, ε^* by

$$\frac{d\varepsilon^*}{dt} = \frac{d\varepsilon}{dt} + t\frac{dn}{dt}. \tag{113}$$

Hence,

$$\frac{d\lambda}{dt} = n + \frac{d\varepsilon^*}{dt} \tag{114}$$

and

$$\lambda = \int n\, dt + \varepsilon^* . \tag{115}$$

This can also be written as

$$\lambda = \rho + \varepsilon^* \tag{116}$$

where

$$\frac{d\rho}{dt} = n \tag{117}$$

$$\frac{d^2\rho}{dt^2} = \frac{dn}{dt} = -\frac{3}{2}\frac{n}{a}\frac{da}{dt} \tag{118}$$

or

$$\frac{d^2\rho}{dt^2} = -\frac{3}{a^2}\frac{\partial \mathcal{R}}{\partial \varepsilon}. \tag{119}$$

The end result is that we should consider any derivatives $\partial/\partial\varepsilon$, such as those that occur in the expressions for $\dot{a}$, $\dot{e}$ and $\dot{I}$, to mean $\partial/\partial\lambda$. In practice the variation of ε can usually be neglected because it is a small effect.

4.4 Secular, Resonant and Short-Period Terms

The key point about the use of Lagrange's equations is that we do not normally make use of all the terms in $\mathcal{R}$. Rather, we identify those terms in the expansion of $\mathcal{R}$ that are likely to be important for the particular problem of interest. In that sense we invoke the *averaging principle* whereby we assume that all the non-important, short-period terms will have zero time-averaged effect. The key orbital elements in this process are the semi-major axes, a and a'.

We have shown that each cosine argument, φ, in $\mathcal{R}$ is written as a linear combination of the angles λ', λ, ϖ', ϖ, Ω' and Ω. In the unperturbed problem the mean longitudes, λ' and λ, increase linearly at rates n' and n respectively. On the other hand, all the other angles are constant in the unperturbed problem. Therefore, when considering the perturbed system λ' and λ are rapidly varying quantities, while all the other angles are slowly varying. Therefore, any arguments which do not involve mean longitudes are slowly varying. These give rise to long-period or *secular* terms. This does not imply that all other arguments are of short period.

Consider a general argument, φ, of the form given in Eq. (106). We can write $j_1\lambda' + j_2\lambda \approx (j_1 n' + j_2 n)t +$ constant and hence, if a and a' are such that

$$j_1 n' + j_2 n \approx 0 \tag{120}$$

then this argument also has a period longer than either orbital period. Equation (120) is satisfied when there is a commensurability (i.e. a simple numerical relationship) between the two mean motions or orbital periods. These arguments give rise to *resonant* terms in the expansion. If we consider the semi-major axes the equivalent condition is

$$a \approx \left(|j_1|/|j_2|\right)^{\frac{2}{3}} a' . \tag{121}$$

Therefore resonant terms are localised in semi-major axis. While a particular combination of angles may be slowly varying at one semi-major axis of the perturbed body, the same combination could be varying rapidly at another. In contrast the secular terms can be considered as global.

Any argument which is neither secular nor resonant is considered to give rise to a *short-period* term. The application of the averaging principle allows us to ignore the infinite number of short-period terms in the expansion and accept that the dynamics is dominated by the appropriate secular and resonant terms.

At this point it is worth formalising a procedure for determining the appropriate averaged term, $\langle\mathcal{R}\rangle$ or $\langle\mathcal{R}'\rangle$ in the disturbing function.

- Decide which combination of angles, φ, is applicable to the problem at hand. This requires knowledge of the physical problem.
- Determine the 'order', $N = |j_1 + j_2|$, of the argument (see Eq. (106)).
- By looking at the appropriate order terms in the expansion of $\mathcal{R}_{\mathrm{D}}$, determine the value of the integer j which gives agreement with the desired argument, φ.
- Calculate the combination of Laplace coefficients for that value of j to give the explicit form of the term of interest, $\langle\mathcal{R}_{\mathrm{D}}\rangle$ say.
- Decide whether an external or an internal perturbation is being considered. This is determined by the nature of the problem.
- If the perturbation is external, then look at the appropriate order terms in the expansion of the indirect part, $\mathcal{R}_{\mathrm{E}}$ and isolate a matching argument, if it exists, and read off the corresponding indirect term $\langle\mathcal{R}_{\mathrm{E}}\rangle$.
- If the perturbation is internal, then look at the appropriate order terms in the expansion of the indirect part, $\mathcal{R}_{\mathrm{I}}$ and isolate a matching argument, if it exists, and read off the corresponding indirect term $\langle\mathcal{R}_{\mathrm{I}}\rangle$.
- If the perturbation is external then

$$\langle\mathcal{R}\rangle = \frac{\mu'}{a'}\left(\langle\mathcal{R}_{\mathrm{D}}\rangle + \alpha\langle\mathcal{R}_{\mathrm{E}}\rangle\right). \tag{122}$$

- If the perturbation is internal then

$$\langle \mathcal{R}' \rangle = \frac{\mu}{a} \left(\alpha \langle \mathcal{R}_{\mathrm{D}} \rangle + \frac{1}{\alpha} \langle \mathcal{R}_{\mathrm{I}} \rangle \right) . \tag{123}$$

Before proceeding it is worthwhile stating the lowest-order form of Lagrange's equations. Considering only the variation in a, e, ϖ and Ω, it is easy to show from Eqs. (107)–(112) that

$$\frac{\mathrm{d}a}{\mathrm{d}t} = \frac{2}{na} \frac{\partial \langle \mathcal{R} \rangle}{\partial \lambda} \tag{124}$$

$$\frac{\mathrm{d}e}{\mathrm{d}t} = -\frac{1}{na^2 e} \frac{\partial \langle \mathcal{R} \rangle}{\partial \varpi} \tag{125}$$

$$\frac{\mathrm{d}\varpi}{\mathrm{d}t} = +\frac{1}{na^2 e} \frac{\partial \langle \mathcal{R} \rangle}{\partial e} \tag{126}$$

$$\frac{\mathrm{d}\Omega}{\mathrm{d}t} = +\frac{1}{na^2 \sin I} \frac{\partial \langle \mathcal{R} \rangle}{\partial I} \tag{127}$$

where $\langle \mathcal{R} \rangle$ is the averaged part of the disturbing function for an external perturber and we have implemented the use of λ instead of ε as discussed above.

As an example consider an asteroid's motion at 3.27 AU under the perturbing effect of Jupiter on a fixed orbit; we will take Jupiter's inclination to be zero. Because Jupiter's semi-major axis is 5.20 AU we have, $(3.27/5.20)^{3/2} \approx 0.499$. Hence, $2n' \approx n$ and we would expect resonant terms to be important. Therefore, as well as the secular terms, we also need to consider those terms in the expansion of the disturbing function which contain $2\lambda' - \lambda$; i.e. the resonant terms for this location.

Inspection of Eq. (89) shows that in a second-order expansion there are two terms in $\langle \mathcal{R}_{\mathrm{D}} \rangle$ that have a cosine argument containing $2\lambda' - \lambda$ for specific values of j. The relevant direct part of the averaged disturbing function is

$$\begin{aligned} \langle \mathcal{R}_{\mathrm{D}} \rangle = C_0 + C_1 (e^2 + e'^2) + C_2 s^2 + C_3 e e' \cos(\varpi - \varpi') \\ + C_4 e \cos(2\lambda' - \lambda - \varpi) + C_5 e' \cos(2\lambda' - \lambda - \varpi') \end{aligned} \tag{128}$$

where the constants are given by

$$C_0 = \frac{1}{2} b_{1/2}^{(0)}(\alpha) \tag{129}$$

$$C_1 = \frac{1}{8} \left[2\alpha D + \alpha^2 D^2 \right] b_{1/2}^{(0)}(\alpha) \tag{130}$$

$$C_2 = -\frac{1}{2} \alpha b_{3/2}^{(1)}(\alpha) \tag{131}$$

$$C_3 = \frac{1}{4} \left[2 - 2\alpha D - \alpha^2 D^2 \right] b_{1/2}^{(1)}(\alpha) \tag{132}$$

$$C_4 = \frac{1}{2} \left[-4 - \alpha D \right] b_{1/2}^{(2)}(\alpha) \tag{133}$$

$$C_5 = \frac{1}{2} \left[3 + \alpha D \right] b_{1/2}^{(1)}(\alpha) . \tag{134}$$

Note that we only require partial derivatives of $\langle\mathcal{R}\rangle$ with respect to λ, e, ϖ and I and so the terms in e'^2 and s'^2 as well as the first term in Eq. (129) are effectively additive constants and can be ignored. The second of the two resonant arguments makes no contribution to $\dot{e}$, $\dot{\varpi}$ and $\dot{\Omega}$, but does contribute a term to $\dot{a}$. Inspection of Eq. (103) shows that there is also a $-2\alpha e'$ contribution to the same argument from the indirect part.

Application of the lowest order form of Lagrange's equations given in Eqs. (124)–(127) gives

$$\begin{aligned}\frac{\mathrm{d}a}{\mathrm{d}t} &= 2n\alpha a(m'/m_c)C_4 e\sin(2\lambda'-\lambda-\varpi)\\ &\quad+2n\alpha a(m'/m_c)\left(C_5-2\alpha\right)e'\sin(2\lambda'-\lambda-\varpi') \qquad (135)\\ \frac{\mathrm{d}e}{\mathrm{d}t} &= n\alpha(m'/m_c)C_3 e'\sin(\varpi-\varpi')\\ &\quad+n\alpha(m'/m_c)C_4\sin(2\lambda'-\lambda-\varpi) \qquad (136)\\ \frac{\mathrm{d}\varpi}{\mathrm{d}t} &= n\alpha(m'/m_c)\left[2C_1+C_3(e'/e)\cos(\varpi-\varpi')\right]\\ &\quad+n\alpha(m'/m_c)(C_4/e)\cos(2\lambda'-\lambda-\varpi) \qquad (137)\\ \frac{\mathrm{d}\Omega}{\mathrm{d}t} &= n\alpha(m'/m_c)(C_2/2) \qquad (138)\end{aligned}$$

for the variations in a, e, ϖ and Ω due to the secular terms (those involving C_1, C_2 and C_3) and the 2:1 resonance terms (those involving C_4 and C_5).

If we consider approximate solutions for these equations we obtain

$$\begin{aligned}a &= a_0-\frac{2n\alpha a(m'/m_c)C_4 e}{2n'-n-\dot{\varpi}}\left[\cos(2\lambda'-\lambda-\varpi)-\cos(\lambda_0+\omega_0)\right]\\ &\quad-\frac{2n\alpha a(m'/m_c)(C_5-2\alpha)e'}{2n'-n}\left[\cos(2\lambda'-\lambda-\varpi')-\cos\lambda_0\right] \qquad (139)\\ e &= e_0-\frac{n\alpha}{\dot{\varpi}}(m'/m_c)C_3 e'\left[\cos\varpi_0-\cos\varpi\right]\\ &\quad+\frac{n\alpha(m'/m_c)C_4}{2n'-n-\dot{\varpi}}\left[\cos(2\lambda'-\lambda-\varpi)-\cos(\lambda_0+\omega_0)\right] \qquad (140)\\ \varpi &= \varpi_0+n\alpha(m'/m_c)2C_1 t\\ &\quad+\frac{n\alpha(m'/m_c)(C_4/e)}{2n'-n-\dot{\varpi}}\left[\sin(2\lambda'-\lambda-\varpi)+\sin(\lambda_0+\omega_0)\right] \qquad (141)\\ \Omega &= \Omega_0+n\alpha(m'/m_c)(C_2/2)t\,. \qquad (142)\end{aligned}$$

In order to derive these solutions we have assumed that the only time varying quantities on the right-hand side of the equations for $\dot{a}$, $\dot{e}$ and $\dot{\varpi}$ are in the cosine arguments, and that ϖ increases linearly with time at a constant rate $\dot{\varpi}$ determined by secular theory. However, they predict that there is no secular change in a while e undergoes a combined secular and resonant change. Both ϖ and Ω should vary linearly with time with ϖ subjected to an additional sinusoidal variation due to the resonance.

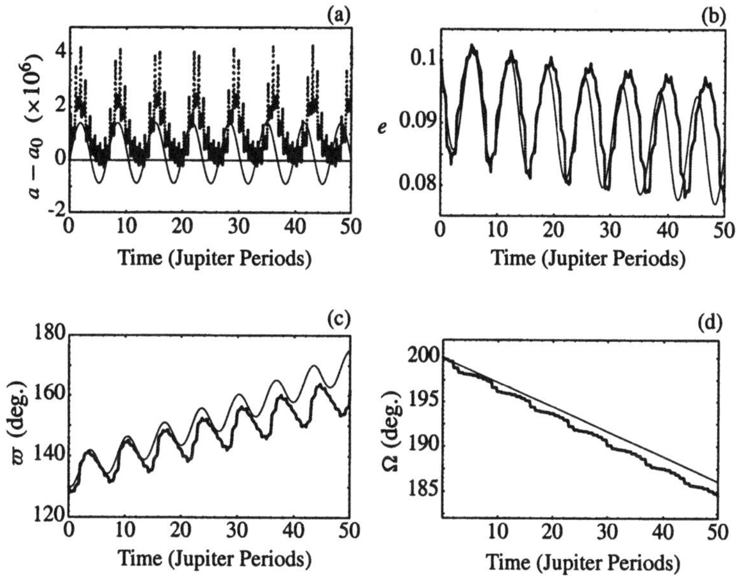

Fig. 14. A comparison of the results of a full numerical integration (points or thick line) with predictions from analytical theory (thin line) for the variation of (a) semi-major axis, (b) eccentricity, (c) longitude of perihelion, and (d) longitude of ascending node for a test particle near the 2:1 resonance undergoing resonant and secular perturbations from Jupiter. Taken from [1].

Figure 14 shows the result of a full integration of the equations of motion and a comparison with the predicted variations from the combined secular and resonant theory outlined above. The calculations were done with fixed elements $a' = 1$, $e' = 0.048$, $\varpi' = 0$, $I' = 0$, $m'/m_c = 1/1047.355$ for the perturber (Jupiter) and starting values $a_0 = 0.6$, $e_0 = 0.1$, $\varpi_0 = 130°$, $\Omega_0 = 200°$, $\lambda_0 = 300°$ and $\lambda' = 0°$. The relevant constants are $C_1 = 0.314001$, $C_2 = -1.25600$, $C_3 = -0.447005$, $C_4 = -1.04332$ and $C_5 = 1.55230$. Examination of Fig. 14 shows that there is good agreement between the predictions and the numerical results, with the amplitudes and frequencies of the variations in a, e and ϖ being close to their predicted values. We would expect there to be some differences, partly due to our approximate form of Lagrange's equations and partly due to the fact that in order to integrate the differential equations we took the quantities a and e on the right-hand side of Eqs. (135)–(138) as well as in Eqs. (129)–(134) to be constant, whereas it is clear that they vary.

Note from Eqs. (139)–(142) that all the amplitudes due to the resonant terms contain a divisor of the form $2n' - n - \dot{\varpi}$, i.e. the time derivative of the resonant argument $2\lambda' - \lambda - \varpi$. This implies that we should expect large changes in the elements as the exact resonance (the location in semi-major axis where $2n' - n - \dot{\varpi} = 0$) is approached. However, in such circumstances the assumptions of our simple analytical model break down and a more careful approach is required.

4.5 The Effect of Planetary Oblateness

Lagrange's equations are equally valid in cases where the disturbing function arises from a non-spherical or *oblate* central object. Here we summarise how such additional terms lead to changes in the keplerian ellipse of the two-body orbit. Although there is no need to include non-spherical terms when considering the motion of planets around the Sun, their inclusion is essential when studying the motion of planetary satellites and ring particles.

It can be shown from potential theory that the gravitational potential experienced by a satellite orbiting a non-spherical planet of mass m_c and radius R_p can be written as

$$V = -\frac{\mathcal{G}m_c}{r}\left[1 - \sum_{i=2}^{\infty} J_i (R_p/r)^i P_i(\sin\alpha)\right] \tag{143}$$

where in this case α is the latitude of the satellite, r is its orbital radius, $P_i(\sin\alpha)$ is the Legendre polynomial of degree i in $\sin\alpha$ and the J_i are dimensionless coefficients which characterise the size of the non-spherical components of the potential. If i is even then the J_i are called the zonal harmonic coefficients.

Murray & Dermott [1] show how motion under this potential gives rise to three frequencies n (the mean motion), κ (the radial frequency) and ν (the vertical frequency) given by

$$n^2 = \frac{\mathcal{G}m_p}{a^3}\left[1 + \frac{3}{2}J_2\left(\frac{R_p}{a}\right)^2 - \frac{15}{8}J_4\left(\frac{R_p}{a}\right)^4\right] \tag{144}$$

$$\kappa^2 = \frac{\mathcal{G}m_p}{a^3}\left[1 - \frac{3}{2}J_2\left(\frac{R_p}{a}\right)^2 + \frac{45}{8}J_4\left(\frac{R_p}{a}\right)^4\right] \tag{145}$$

$$\nu^2 = \frac{\mathcal{G}m_p}{a^3}\left[1 + \frac{9}{2}J_2\left(\frac{R_p}{a}\right)^2 - \frac{75}{8}J_4\left(\frac{R_p}{a}\right)^4\right] \tag{146}$$

where terms up to and including J_4 have been included. Note that if $J_2 = J_4 = 0$ then $n^2 = \kappa^2 = \nu^2 = n_0^2$ where $n_0 = \sqrt{\mathcal{G}m_p/a^3}$ is the keplerian mean motion of the satellite around a point mass planet. If we consider the case of the mean motion, n, it is clear that the inclusion of the extra terms means that for a given semi-major axis the satellite moves faster than the rate expected at that location if the motion was purely keplerian. Therefore, since the observable quantity for a satellite is usually n, the semi-major axis is not that determined from Kepler's third law. Instead it is necessary to solve Eq. (144), a non-linear equation in a.

The inclusion of the J_{2i} terms and the resulting small differences in the three frequencies is that the orbit is no longer closed and the pericentre and node are no longer fixed. In fact

$$\dot{\varpi} = n - \kappa \tag{147}$$

$$\dot{\Omega} = n - \nu \tag{148}$$

and, to $\mathcal{O}(R_p/a)^4$ we have

$$\dot{\varpi} = +n_0 \left[\frac{3}{2} J_2 \left(\frac{R_p}{a} \right)^2 - \frac{15}{4} J_4 \left(\frac{R_p}{a} \right)^4 \right] \tag{149}$$

$$\dot{\Omega} = -n_0 \left[\frac{3}{2} J_2 \left(\frac{R_p}{a} \right)^2 - \frac{9}{4} J_2^2 \left(\frac{R_p}{a} \right)^4 - \frac{15}{4} J_4 \left(\frac{R_p}{a} \right)^4 \right] . \tag{150}$$

We can also use Lagrange's equations to derive expressions for $\dot{\varpi}$ and $\dot{\Omega}$. To second order in e and I this gives

$$\dot{\varpi} = +n \left[\frac{3}{2} J_2 \left(\frac{R_p}{a} \right)^2 - \frac{9}{8} J_2^2 \left(\frac{R_p}{a} \right)^4 - \frac{15}{4} J_4 \left(\frac{R_p}{a} \right)^4 \right] \tag{151}$$

$$\dot{\Omega} = -n \left[\frac{3}{2} J_2 \left(\frac{R_p}{a} \right)^2 - \frac{27}{8} J_2^2 \left(\frac{R_p}{a} \right)^4 - \frac{15}{4} J_4 \left(\frac{R_p}{a} \right)^4 \right] . \tag{152}$$

Note the differences with Eqs. (150) and (151) because of our use of n rather than n_0 outside the brackets.

5 The Dynamics of Resonance

A resonance can occur when the ratio of two periods or frequencies in a system is close to a rational number. In the solar system the frequencies could be two mean motions (in the case of an orbit-orbit resonance) or a spin frequency and a mean motion (in the case of a spin-orbit resonance). However, as we shall see, the proximity to a rational is a necessary but not sufficient condition for resonance to occur. In the context of orbit-orbit resonance a more precise definition is that an exact resonance exists when the time derivative of a particular cosine argument in the expansion of the disturbing function is zero.

5.1 Resonance in the Solar System

The work of Roy & Ovenden [9], Goldreich [10] and others has shown that there are more simple ratios of pairs of mean motions in the solar system than one would expect by chance. The reason for this preference for commensurability in satellite systems is probably tidal evolution (see Sect. 5.6)

Table 1. Planetary and satellite resonances in the solar system

System	Resonant argument	Amplitude	Period (y)
Planets			
Neptune–Pluto	$3\lambda' - 2\lambda - \varpi'$	86°	19,857
Jupiter			
Io–Europa	$2\lambda' - \lambda - \varpi$	1°	
Io–Europa	$2\lambda' - \lambda - \varpi'$	3°	
Europa–Ganymede	$2\lambda' - \lambda - \varpi$	3°	
Saturn			
Mimas–Tethys	$4\lambda' - 2\lambda - \Omega' - \Omega$	43.6°	71.8
Enceladus–Dione	$2\lambda' - \lambda - \varpi$	0.297°	11.1
Titan–Hyperion	$4\lambda' - 3\lambda - \varpi'$	36.0°	1.75

whereas with planetary orbits it should be realised that orbital migration is possible in the early history of the solar system (see, for example, [11]).

The actual resonances known to exist between planets or between satellites are listed in Table 1. There are a number of points to note: The only planetary resonance is the 3:2 resonance between Neptune and Pluto. While it has been known for more than 200 years that Jupiter and Saturn are close to a 5:2 resonance, the time derivative of the relevant resonant angle does not change sign and hence circulates rather than librates. The jovian system has resonances involving three out of the four Galilean satellites. The Io–Europa 2:1 resonance is especially significant because it leads directly to the tidal heating of Io resulting in its spectacular active volcanism. In the saturnian system (see colour plate XXX) there are three pairs of satellite in resonance. One resonant argument, that of the Mimas–Tethys pair, involves the ascending nodes of both satellites and hence their inclinations are contained in the associated term. Curiously there are no known resonances between satellites in the uranian system, even though it has at least 18 known moons. This may be explained by the fact that chaos plays a slightly different role in this system.

Another important point to note about all the resonances listed in Table 1 is that the integers involved in the resonant argument are all small and, with one exception, the order of the resonant argument is always one. Therefore first-order resonances predominate in the solar system. This can be understood by recalling that the terms associated with an Nth-order argument will be of Nth order in the eccentricities or inclinations. Therefore in the planar case first-order arguments are associated with strong, first-order terms in e or e'.

5.2 The Geometry of Resonance

In order to understand the possible significance of a simple numerical relationship between two orbital periods, consider the two examples illustrated in Fig. 15. Let an asteroid in an elliptical orbit have an orbital period exactly one half that of Jupiter (assumed to be moving in a circular co-planar orbit). Note that at this stage we are not considering any perturbations by Jupiter on the asteroid; we are merely interested to se the dynamical consequences of their relative geometries.

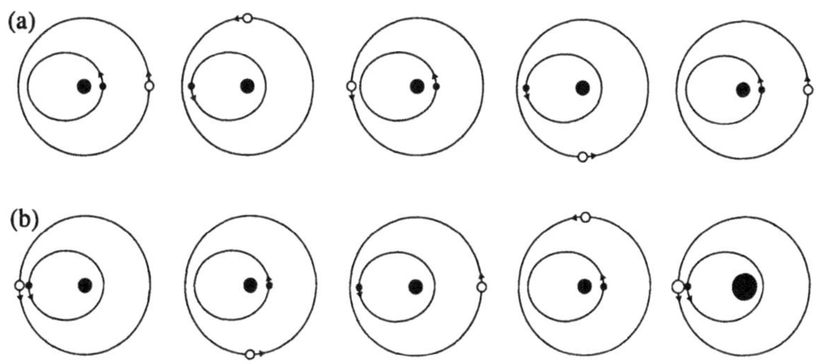

Fig. 15. The relative positions of Jupiter (white circle) in a circular orbit and an asteroid (small filled circle) in an elliptical orbit for the (a) stable and (b) unstable configurations of a 2:1 resonance.

We consider two possible starting configurations. In Fig. 15a the asteroid and Jupiter are such that at time $t = 0$, Jupiter and the asteroid are at conjunction and the asteroid is at the perihelion of its orbit. Because the objects are in a 2:1 resonance, the asteroid will complete two periods for every one period of Jupiter. Each frame in Fig. 15a advances the time by one quarter of a Jupiter period. In the second frame the asteroid is now at the aphelion of its orbit, and Jupiter has completed $\frac{1}{4}$ of an orbit. Note that Jupiter is not nearby when the asteroid is at this dangerous position. Similarly, when Jupiter reaches this position one quarter of a Jupiter period later, the asteroid is back at its perihelion. One step later the asteroid returns to the danger point but Jupiter is not nearby; the original configuration is repeated one frame later. Therefore large perturbations from Jupiter at the asteroid's aphelion are avoided by the resonance mechanism. This is an example of a stable equilibrium configuration between Jupiter and the asteroid. Conversely, starting Jupiter and the asteroid at conjunction at the asteroid's aphelion (Fig. 15b), would lead to an unstable equilibrium configuration, where damaging close approaches would be repeated every Jupiter period.

Note that the asteroid in the 2:1 resonance goes through perihelion and aphelion twice for every Jupiter period. Therefore, in a frame rotating at Jupiter's constant angular velocity (i.e. its mean motion because we have assumed that the orbit is circular) the path has to have two minima and two maxima separations from the Sun. Figure 16 shows the paths in the rotating frame for the 2:1, 3:2, 4:3, 5:4 and 6:5 resonances for eccentricities in the range $0.1 \leq e \leq 0.4$.

It is clear from Fig. 16 that for these first-order interior resonances the path in the rotating frame develops loops as e increases. For the $p+1 : p$

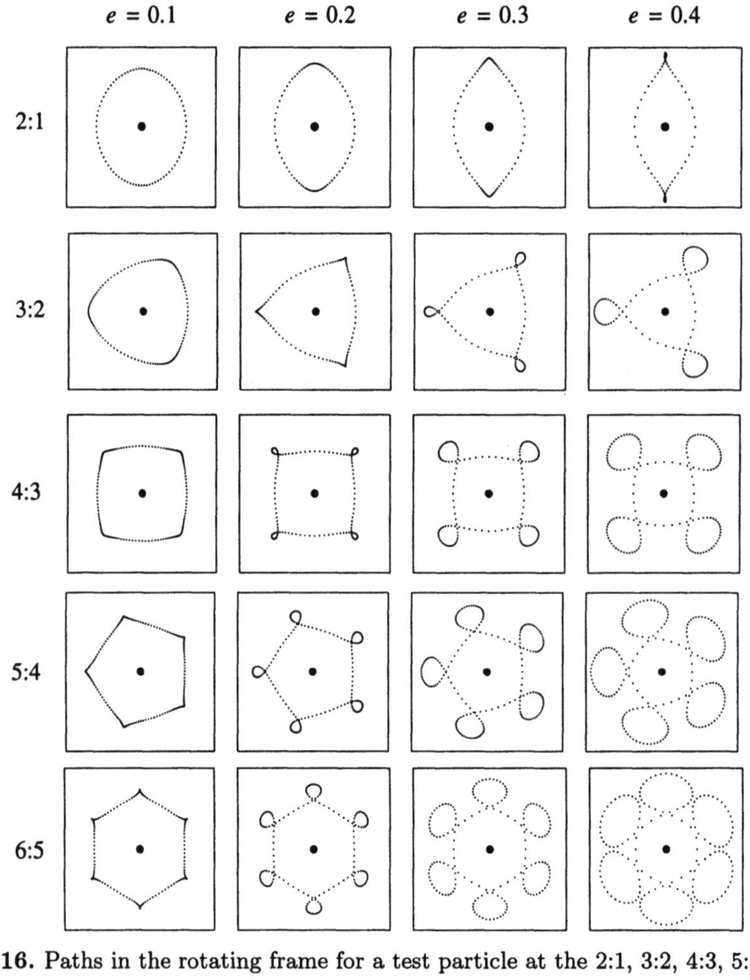

Fig. 16. Paths in the rotating frame for a test particle at the 2:1, 3:2, 4:3, 5:4 and 6:5 interior resonances for values of the eccentricity, $e = 0.1$, 0.2, 0.3 and 0.4. The positions of the particle along each path are drawn at equal time intervals.

resonance there are always $p+1$ loops. These arise because of the use of a rotating reference frame. As the eccentricity of the test particle increases its angular velocity at aphelion gets smaller. Beyond some critical eccentricity the particle's angular velocity will be less than that of Jupiter and so in the rotating frame it will appear to be going backwards, the size of the loop increasing as e increases. Murray & Dermott [1] give examples of such behaviour for external as well as second-order resonances.

5.3 The Pendulum Model

Consider the case of the planar, circular, restricted three-body problem where a body on an external orbit perturbs an inner body of negligible mass, both objects moving in the reference plane. We have already seen that the general term of the averaged expansion reduces to

$$\langle \mathcal{R} \rangle = \frac{\mathcal{G}m'}{a'} \left[f_{\rm sec}(\alpha) e^2 + f_{\rm res}(\alpha) e^{|j_4|} \cos\varphi \right] \tag{153}$$

where

$$\varphi = j_1\lambda' + j_2\lambda + j_4\varpi\,. \tag{154}$$

Note that because $j_1 + j_2 + j_4 = 0$ to satisfy the d'Alembert relation, we must have $|j_4| = 1$ for a first-order resonance ($|j_1 + j_2| = 1$) and $|j_4| = 2$ for a second-order resonance ($|j_1 + j_2| = 2$). The corresponding equations of motion derived from Lagrange's equations are

$$\dot{n} = 3 j_2 \mathcal{C}_{\rm res} n e^{|j_4|} \sin\varphi \tag{155}$$

$$\dot{e} = j_4 \mathcal{C}_{\rm res} e^{|j_4|-1} \sin\varphi \tag{156}$$

$$\dot{\varpi} = 2\mathcal{C}_{\rm sec} + |j_4| \mathcal{C}_{\rm res} e^{|j_4|-2} \cos\varphi \tag{157}$$

where we have neglected the variation of the mean longitude at epoch. The constants arising from the resonant and secular parts of the disturbing function are given by

$$\mathcal{C}_{\rm res} = \frac{\mathcal{G}m'}{na^2a'} f_{\rm res}(\alpha) = \left(\frac{m'}{m_{\rm c}}\right) n\alpha f_{\rm res}(\alpha) \tag{158}$$

$$\mathcal{C}_{\rm sec} = \frac{\mathcal{G}m'}{na^2a'} f_{\rm sec}(\alpha) = \left(\frac{m'}{m_{\rm c}}\right) n\alpha f_{\rm sec}(\alpha) \tag{159}$$

respectively, where $m_{\rm c}$ is the mass of the central body and we have made use of Kepler's third law to write $\mathcal{G} = n^2a^3/m_{\rm c}$. Note with reference to the functions f_i defined in Eqs. (91)–(102) that $f_{\rm sec} = f_1$ and $f_{\rm res} = f_5$ (for a first-order resonance) or $f_{\rm res} = f_7$ (for a second-order resonance). Here it is assumed that we are incorporating any indirect terms into $f_{\rm res}$.

Given that the orbit of the external body is fixed we can write

$$\dot{\varphi} = j_1 n' + j_2(n + \dot{\epsilon}) + j_4\dot{\varpi}\,. \tag{160}$$

Neglecting the $\dot{\varepsilon}$ term and differentiating again with respecting to the time gives

$$\ddot{\varphi} = j_2\dot{n} + j_4\ddot{\varpi} \tag{161}$$

where we have used the fact that $\dot{n}' = 0$. If we write

$$\dot{\varpi} = 2\mathcal{C}_{\rm sec} + |j_4|\mathcal{C}_{\rm res}G(e)\cos\varphi \tag{162}$$

where $G(e) = e^{|j_4|-2}$ then

$$\ddot{\varpi} = |j_4|\mathcal{C}_{\rm res}\left(\frac{\mathrm{d}G(e)}{\mathrm{d}e}\dot{e}\cos\varphi - G(e)\dot{\varphi}\sin\varphi\right). \tag{163}$$

From Eqs. (159) and (160) we see that $\mathcal{C}_{\rm res}$ and $\mathcal{C}_{\rm sec}$ contain a factor of $m'/m_{\rm c}$ which is usually a small quantity. Therefore the $\dot{e}$ and $\dot{\varphi}$ terms in Eq. (164) introduce another factor of $m'/m_{\rm c}$ and hence the contribution of $\ddot{\varpi}$ to $\ddot{\varphi}$ can be neglected in most circumstances. However, in the case of first-order resonances (i.e. those for which $|j_4| = 1$) there can be a significant contribution to the precession of ϖ because $G(e) = 1/e$ and e is a small quantity. In these circumstances the $\ddot{\varpi}$ term can contribute to $\ddot{\varphi}$. Note that in either case $\ddot{\varphi}$ no longer contains a contribution from the secular part of the disturbing function.

If we neglect the $\ddot{\varphi}$ contribution the variation of the resonant variable is described by the equation

$$\ddot{\varphi} = -\omega_0^2\sin\varphi \tag{164}$$

where we are taking ω_0 to be a constant given by

$$\omega_0^2 = -3j_2^2\mathcal{C}_{\rm res}ne^{|j_4|}\,. \tag{165}$$

Note that we are assuming that n and e are approximately constant for the purposes of calculating ω_0.

Equation (165) is a pendulum equation with the centre of libration depending on the sign of $\mathcal{C}_{\rm res}$. Because f_5 is always negative for an odd-order resonance the centre of stable libration is $\varphi = 0$. Conversely, because f_7 is always positive for an even-order resonance, the centre of stable libration is $\varphi = \pi$. The general solution can be described either as a circulation or a libration of φ, with the type of motion depending on the total energy per unit mass, E, of the system. This is given by

$$E = \frac{1}{2}\dot{\varphi}^2 + 2\omega_0^2\sin^2\frac{1}{2}\varphi\,. \tag{166}$$

The various types of motion are classified by the value of E. This is illustrated in Fig. 17 where we plot the potential energy and three possible values of the total energy.

- If $E = E_1 > E_3$ the motion of φ is *unbounded* corresponding to *circulation* of the angle φ.

- If $E = E_2 < E_3$ the motion of φ is *bounded* corresponding to *oscillation* or *libration* of the angle φ.
- If $E = E_3$ motion occurs on the *separatrix* which divides the circulation regime from the libration regime.

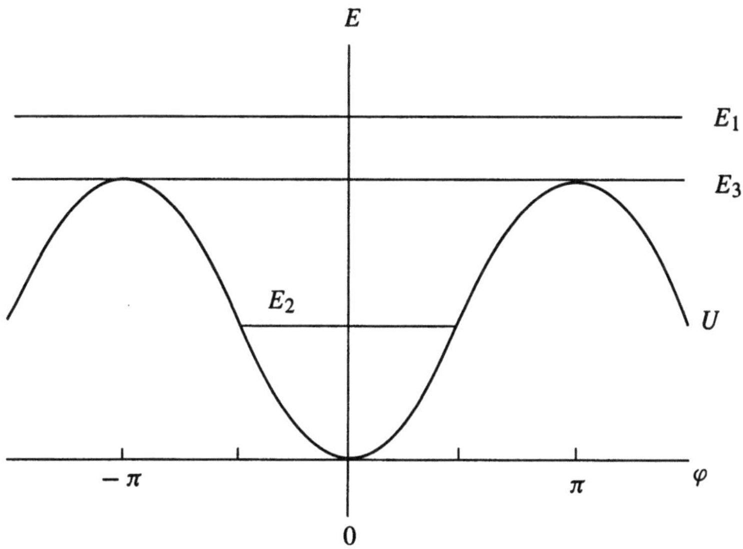

Fig. 17. The potential energy, U, as a function of φ, compared with three possible values of the total energy, E.

When φ is librating the period of libration is given by (see [1])

$$T_{\mathrm{lib}} = \frac{1}{\omega_0} \int_0^{2\pi} \frac{\mathrm{d}\theta}{\left(1 - (E/2\omega_0^2)\sin^2\theta\right)^{1/2}} = \frac{4}{\omega_0} K\left(\frac{E}{2\omega_0^2}\right) . \tag{167}$$

Here $K(x)$ is the complete elliptical integral of the first kind. Note that $K(0) = \pi/2$ and hence, when E is small the corresponding small amplitude librations about the equilibrium point will have a period

$$\lim_{\varphi_0 \to 0} T_{\mathrm{lib}} = \frac{2\pi}{\omega_0} . \tag{168}$$

This is equivalent to using the approximation $\sin\varphi \approx \varphi$ that leads to simple harmonic motion about the quilibrium point. Note also that $T_{\mathrm{lib}} \to \infty$ as the separatrix is approached.

5.4 Libration Width

By deriving an analytical model of resonance we can estimate the variation in orbital parameters caused by individual resonances without resorting to numerical integration. In many applications it is necessary to calculate the extent of the libration in semi-major axis (or mean motion) for an object in resonance. This could be used, for example, to relate the known libration width with an observed phenomenon such as a ring feature, or a gap in the asteroid distribution.

Consider Eq. (167) using our simple pendulum model. It is clear from Fig. 17 that the energy associated with maximum libration occurs when $\dot{\varphi} = 0$ at $\varphi = \pm\pi$. This implies

$$E_{\max} = -6 j_2^2 \mathcal{C}_{\text{res}} n e^{|j_4|}. \tag{169}$$

Now let $E = E_{\max}$ and consider the variation of φ given by

$$\dot{\varphi} = \pm j_2 \left(12 |\mathcal{C}_{\text{res}}| n e^{|j_4|}\right)^{1/2} \cos \frac{1}{2}\varphi . \tag{170}$$

We can relate the variation of φ to the variation in n by means of Eqs. (156) and (171) giving

$$\mathrm{d}n = 3 j_2 \mathcal{C}_{\text{res}} n e^{|j_4|} \frac{\sin \varphi}{\dot{\varphi}} \mathrm{d}\varphi = \pm \left(3 |\mathcal{C}_{\text{res}}| n e^{|j_4|}\right)^{1/2} \sin \frac{1}{2}\varphi \, \mathrm{d}\varphi . \tag{171}$$

Integration gives

$$n = n_0 \pm \left(12 |\mathcal{C}_{\text{res}}| n e^{|j_4|}\right)^{1/2} \cos \frac{1}{2}\varphi \tag{172}$$

and therefore the maximum change in the mean motion is

$$\delta n_{\max} = \pm \left(12 |\mathcal{C}_{\text{res}}| n e^{|j_4|}\right)^{1/2} \tag{173}$$

which occurs when $\varphi = 0$. We can use Kepler's third law to calculate the equivalent maximum change in semi-major axis. This gives

$$\delta a_{\max} = \pm \left(\frac{16}{3} \frac{|\mathcal{C}_{\text{res}}|}{n} e^{|j_4|}\right)^{1/2} a . \tag{174}$$

This formula can be easily modified to include situations where the $\ddot{\varpi}$ term in the equation for $\ddot{\varphi}$ is non-negligible (see [1]).

5.5 Resonance Splitting

In our study of resonance using the pendulum model (Sect. 5.3 and 5.4) our starting point was the planar, circular restricted three-body problem. We also considered the lowest order terms in the expansion of the disturbing function. In those restrictions there was only one resonant argument, φ. However, if the perturbing object moves on an elliptical orbit in a different plane to the perturbed object, many additional resonant arguments are now possible, and the same is true if we include higher order expansions of the disturbing function. In each case the location of each exact resonance (i.e. the value of the semi-major axis for which $\dot{\varphi} = 0$) depends on the particular combination of angles. For example, if we are carrying out a second-order (in eccentricity and inclination) analysis of the 2:1 resonance we must consider the two first-order arguments

$$\varphi_1 = 2\lambda' - \lambda - \varpi \tag{175}$$

$$\varphi_2 = 2\lambda' - \lambda - \varpi' \tag{176}$$

as well as the six second-order arguments

$$\varphi_3 = 4\lambda' - 2\lambda - 2\varpi \tag{177}$$

$$\varphi_4 = 4\lambda' - 2\lambda - \varpi' - \varpi \tag{178}$$

$$\varphi_5 = 4\lambda' - 2\lambda - 2\varpi' \tag{179}$$

$$\varphi_6 = 4\lambda' - 2\lambda - 2\Omega \tag{180}$$

$$\varphi_7 = 4\lambda' - 2\lambda - \Omega' - \Omega \tag{181}$$

$$\varphi_8 = 4\lambda' - 2\lambda - 2\Omega' . \tag{182}$$

These are all at approximately the same semi-major axis determined by Kepler's third law and the fact that $n \approx 2n'$. However, their exact locations depend on the values of the quantities $\dot{\varpi}$, $\dot{\varpi}'$, $\dot{\Omega}$ and $\dot{\Omega}'$. Therefore it is clear that the resonances can be widely separated where the pericentre and node rates are large. This is the phenomenon of *resonance splitting* and it is particularly important in satellite systems where the planet's oblateness can dominate precession rates. If the resonances are sufficiently well separated then each can be treated individually, irrespective of the perturbing effects of the others.

Saturn's oblateness causes large rates of pericentre precession and nodal regression for objects orbiting close to the planet; the same is also true of Jupiter but for other planets the effect is less noticeable. Therefore the precession rate of the perturber and the perturbed objects (dominated by the planet's J_2) cause the numerous resonant arguments to become separated in semi-major axis. As an example Fig. 18 shows the location of the Mimas 6:4 and Tethys 3:1 resonances in the saturnian system, with e and I denoting the eccentricity and inclination of the object in resonance, with single and double primes denoting the equivalent values for Mimas and Tethys respectively. The resonant locations were calculated using values of $\dot{\varpi}$, $\dot{\Omega}$, $\dot{\varpi}'$ and $\dot{\Omega}'$ taken from Harper & Taylor [35]. The Mimas 3:2 first-order resonances involving e and e' are coincident with the Mimas 6:4 resonances involving e^2 and e'^2 because each of the two 6:4 arguments is simply twice the 3:2 arguments. Because Tethys is more distant than Mimas results in rates that are a factor five smaller ($\sim \pm 0.19°\mathrm{d}^{-1}$ versus $\sim \pm 1°\mathrm{d}^{-1}$). The reason for the gen-

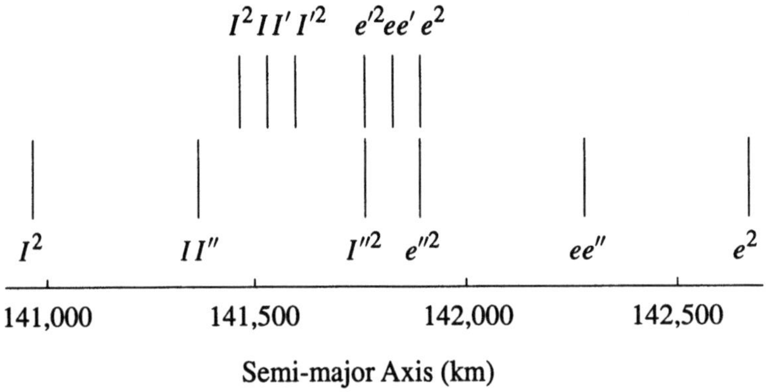

Fig. 18. The locations in semi-major axis of the Mimas 6:4 (upper lines) and Tethys 3:1 (lower lines) exact resonances. The appropriate terms in eccentricity and inclination associated with each argument are indicated.

eral proximity of the resonances of these satellites is the fact that Mimas and Tethys are involved in a 4:2 resonance with each other (see Table 1 above).

5.6 Resonant Encounters in Satellite Systems

It is likely that tidal evolution provides a mechanism for pairs of satellites to enter into a resonant configuration. It can be shown (see [1]) that the tides raised on a planet by a satellite of mass m and semi-major axis a moving in a prograde orbit cause the satellite's semi-major axis to change at a rate given by

$$\dot{a} = \frac{3k_2}{Q_\mathrm{p}}\left(\frac{\mathcal{G}}{m_\mathrm{p}}\right)^{1/2} R_\mathrm{p}^5 \frac{m}{a^{11/2}} \tag{183}$$

where m_p, R_p, Q_p and k_2 denote the mass, radius, tidal dissipation function and Love number of the planet, respectively. Therefore, provided the physical properties of the planet remain constant over time, $\dot{a}$ is purely a function of a and m. Now consider another satellite with semi-major axis a' and mass m' on an exterior orbit. The two satellites will be approaching one another if at a given time the ratio

$$\dot{a}/\dot{a}' = (m/m')(a'/a)^{11/2} \tag{184}$$

is greater than unity. If we let $\mathcal{N} = n'/n$ then the condition for capture can be written as $\dot{\mathcal{N}} = \mathrm{d}(n'/n)/\mathrm{d}t > 1$.

The equation for $\dot{a}$ can be solved to give

$$a(t) = \left[a_0^{13/2} - \frac{13}{2}Cm(t - t_0)\right]^{2/13} \tag{185}$$

where $a_0 = a(t_0)$ (the current value of a), and C (assumed to be constant) is a function of m_p, R_p, Q_p and k_2 (i.e. physical parameters of the planet). Although m_p and R_p are usually well known, Q_p and k_2 are more difficult to measure especially if one has to calculate their primordial values. Nevertheless, the fact that the semi-major axes of satellites vary with time means that resonances can be encountered.

All satellites in prograde orbits above the synchronous orbit will evolve outwards because of the tides they raise. It can be shown (see [1]) that in order for capture to take place the orbit of the inner satellite must be expanding faster than the orbit of the outer satellite and thus the condition is that the satellites are on converging orbits [13]. Once capture occurs into a stable orbit-orbit resonance, $\dot{\mathcal{N}} \approx 0$ and the orbits of both satellites expand at the same rate. Although angular momentum continues to be lost from the spin of the planet this is at the same rate as before the encounter, but now gravitational forces between the satellites due to the resonance act to transfer orbital angular momentum from the inner to the outer satellite allowing their orbits to expand together. Therefore the resonance is maintained and $\dot{\mathcal{N}} \approx 0$ is still satisfied. However, the forces also act to increase the eccentricities or inclinations of the satellites involved in the resonance at rates determined by the resonant argument and these increases can provide evidence for orbital evolution.

Figure 19 shows an example of possible changes in the semi-major axes of some of Saturn's satellites due to tidal evolution. In these plots time is to be thought of as an integral involving averaging of Q_p (see [13]). Therefore the zero point on the time axis should not be considered to be fixed.

We know that there are two resonances involving the satellites shown in Fig. 19: the Mimas–Tethys 4:2 and the Enceladus–Dione 2:1. The plots show that currently each of these pairs of satellites is evolving outwards at almost the same rate although they approached one another at a higher rate in the past. Therefore, the necessary condition for capture occurs in each case. However, if we take the example of Enceladus and Tethys we see that they have always evolved on diverging paths and could have passed through a number of first-order resonance without capture. Curiously Mimas and Enceladus are on clearly converging paths and yet have managed to escape resonant capture so far.

6 Chaos and Long-Term Evolution

The equations of motion of the two- and three-body problems have a common characteristic: They describe systems which are *deterministic* such that the current state of the system permits us to calculate its past and future state providing we know all the forces that are acting on it. Therefore, given the initial state of the system, we should be able to calculate its future state by obtaining solutions of the equations of motion. While this is true for the

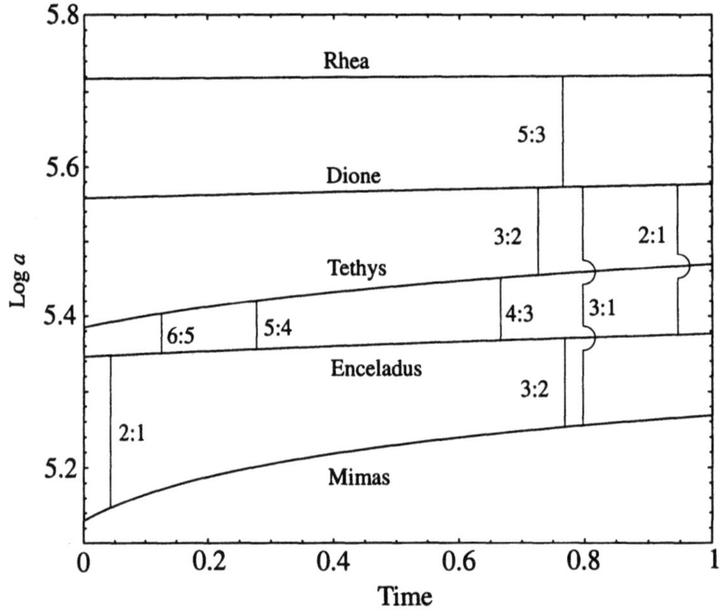

Fig. 19. Possible changes in the semi-major axes (measured in km) as a function of time for satellites in the inner part of the saturnian system. Some first- and second-order resonances between pairs of satellites are indicated.

two-body problem it is not true for the three-body problem because of the phenomenon called *chaos.*

In the late 19th century Henri Poincaré studied the mathematics of the three-body problem. His work hinted at the complicated nature of the motion that can arise for some starting conditions. The advent of fast digital computers combined with new observations and advances in theory means that we can now recognize that the phenomenon is widespread and that it has played an important role in determining the dynamical structure and evolution of the solar system.

For our purposes we can define chaos in the following way: An object in the solar system can be said to exhibit chaotic motion if its final dynamical state is sensitively dependent on its initial state. Because the measurement of any physical quantity has a built-in error, the lack of precision in starting conditions is transformed into an uncertainty in final conditions.

Figure 20 illustrates the result of integrating two nearby starting conditions for the orbit of a test particle perturbed by Jupiter in the planar, circular restricted problem. In each case the test particle passes close to Jupiter yet a difference of only 0.3° in initial longitude produces a dramatically different result. In this example a small change in starting conditions changes the geometry of the encounter and hence the size of the direct perturbation received from the planet.

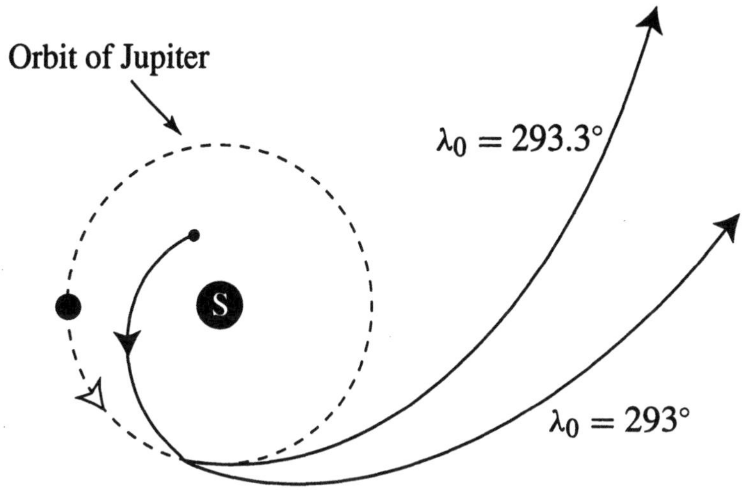

Fig. 20. The trajectories of two test particles started with the same a, e, ϖ but with different initial λ ($\lambda_0 = 293°$ and $\lambda_0 = 293.3°$).

6.1 Regular and Chaotic Orbits

We can demonstrate the differences between regular motion and chaotic motion by using numerical examples from the planar, circular restricted three-body problem to show their characteristics.

Figure 21 shows the evolution of e as a function of time for two different starting conditions but the same value of the Jacobi constant, $C_J = 3.07$, for the Sun-Jupiter mass ratio. In Fig. 21a we used $a_0 = 0.6944$ and $e_0 = 0.2065$. The plot shows a regular variation in the eccentricity in the range 0.206–0.248. At this location we would expect to see some effect of resonant perturbations because $(a/a_J)^{3/2} = 0.564 \approx 4/7$ and the orbit of the test particle is close to a 7:4 resonance with Jupiter. In Fig. 21b we used $a_0 = 0.6984$ and $e_0 = 0.1967$. These are only slightly different from the values used above and yet the nature of the variations in e are very different. Now the eccentricity undergoes irregular variations from 0.188 to 0.328; this is a chaotic trajectory with no obvious pattern to the variations in the orbital elements.

It is easy to visualise the values of x, y, $\dot{x}$ and $\dot{y}$ at any given time corresponding to a point in a four-dimensional space. However, because of the existence of the Jacobi constant in the circular restricted problem the trajectory of the particle in this space is confined to a surface. Therefore, for a given, fixed value of the Jacobi constant we only require three out of the four quantities in order to define the instantaneous orbit uniquely. This is because the other quantity can be determined, at least up to a sign change, by the equation defining the Jacobi constant (see Eq. (58)). For example, if x, y and $\dot{x}$ are our three quantities then $\dot{y}$ can be determined provided we know the value of C_J. If we now consider a plane, say $y = 0$, in the resulting

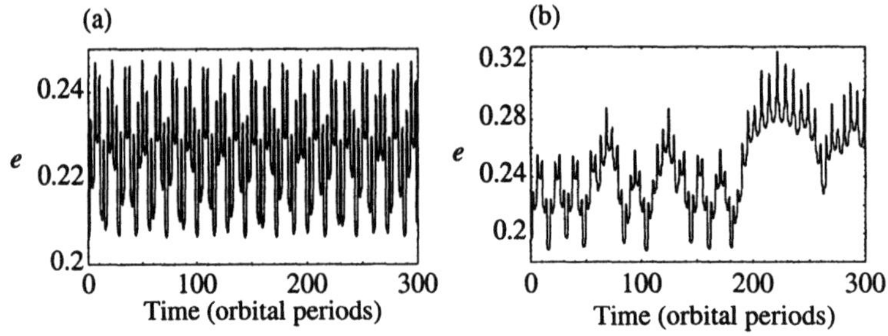

Fig. 21. The time evolution of the eccentricity for (a) a regular trajectory and (b) a chaotic trajectory in the circular restricted three-body problem.

three-dimensional space, the values of x and $\dot{x}$ can be plotted every time the particle has $y = 0$. The problem about the sign of $\dot{y}$ is removed by considering only those crossings with the same sign of $\dot{y}$. This is the method of the *Poincaré surface of section* and we can use it to illustrate the regular and chaotic regions in the circular restricted problem. The section is obtained by fixing a plane in the phase space and plotting the points when the trajectory intersects this plane in a particular direction. Note that as a result we do not plot the points at equal time intervals; we only plot a point when an intersection takes place.

Figure 22 shows the two surfaces of section corresponding to the two trajectories shown in Fig. 21. These were obtained by plotting x and $\dot{x}$ whenever $y = 0$ with $\dot{y} > 0$. In Fig. 22a there are three, distinct 'islands' the appearance of which is a characteristic of resonant motion. Note that rather than trace out one island at a time, successive points occur at each of the three island locations in turn, until they gradually appear to form three smooth curves. If x_0 was chosen to be at the centre of the island on the $\dot{x} = 0$ line, then the trajectory would appear as a succession of three points, one at the centre of each island in turn. The centre of each island corresponds to a starting condition that places the test particle at the middle of the resonance. These are said to be periodic points of the Poincaré map because the system returns to the same point every third time the trajectory crosses the plane. By moving the starting location further away from the centre the islands would get larger, corresponding to larger variations in e and a. Eventually some starting values would lead to trajectories not in resonant motion and these would no longer form distinct islands in the section plot.

The identification of the orbital evolution shown in Fig. 21b as chaotic becomes obvious from its Poincaré surface of section shown in Fig. 22b. Note that the orbit covers a larger region of phase space than in the regular example. Furthermore the points are beginning to fill an area of the phase space with a tendency for some points to 'stick' to the edge of the 7:4 and other resonances. This stickiness phenomenon is useful to define several empty regions, each of which can be associated with a resonance. The existence of

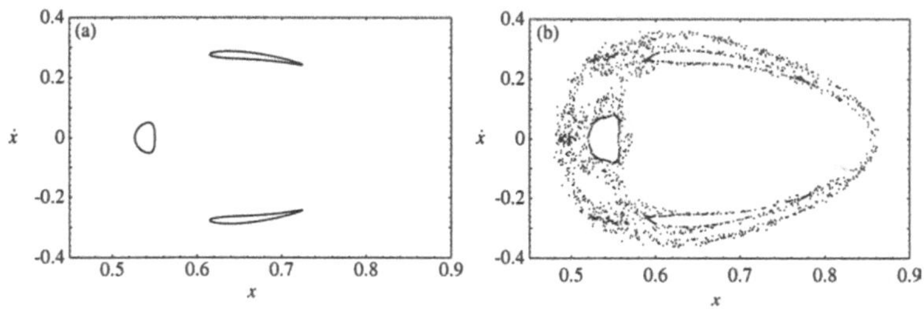

Fig. 22. Poincaré surfaces of section corresponding to the two orbits shown in Fig. 21.

such phenomena means that a chaotic trajectory may give the impression of being regular at particular times. In fact, chaotic behaviour can give the impression of regular motion for long periods of time making it difficult to detect in some circumstances.

We can make use of the divergence property of chaotic orbits to measure the *maximum Lyapounov characteristic exponent* (LCE) of a system. This gives a quantitative measure of the rate of divergence of nearby trajectories. It can be shown that a measurement of the local divergence of nearby trajectories leads to an estimate of the largest of a number of LCEs of the system.

Consider two orbits separated in phase space by a distance d_0 at time t_0 (see Fig. 23). Let d be the separation at time t. The orbit is chaotic if d is approximately related to d_0 by

$$d = d_0 \exp \gamma(t - t_0) \tag{186}$$

where γ is the maximum LCE. Note that we must have $\gamma > 0$ otherwise the trajectories would approach one another as t increased. We can estimate the value of γ from the results of a numerical integration by means of the relation

$$\gamma = \lim_{t \to \infty} \frac{\ln (d/d_0)}{t - t_0} . \tag{187}$$

The behaviour of γ as a function of time on a log–log scale usually reveals a striking difference between regular and chaotic trajectories. A regular orbit will have initial and final displacements close to one another ($d \approx d_0$) and hence a log–log plot would have a slope of -1. However, if the orbit is chaotic, then γ tends to a positive value. We will see an example of a plot of $\log \gamma$ as a function of $\log t$ in Sect. 6.4.

6.2 The Rotation of Hyperion

One of the first recognised examples of chaotic motion in the solar system concerns the rotational behaviour of the saturnian satellite Hyperion. Most

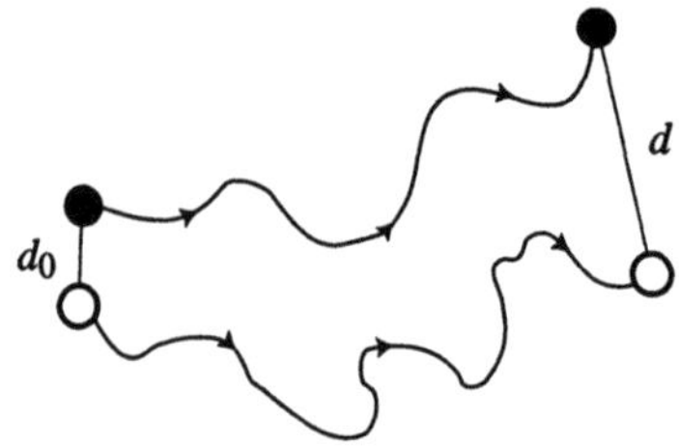

Fig. 23. Calculation of the maximum Lyapounov characteristic exponent by measuring the divergence of nearby trajectories.

natural satellites in the solar system have spin periods that are approximately equal to their orbital periods. This is known as synchronous rotation and such satellites always maintain the same face pointed towards the planet. This has not happened by chance– it is a consequence of tidal effects acting over the age of the solar system. Most satellites settle into this configuration on a timescale considerably less than the age of the solar system. However, small satellites orbiting far from a parent planet may not have had sufficient time to evolve into such a spin-orbit resonance. Hyperion has an unusual shape with approximate radial dimensions of 175 km × 120 km × 100 km and its orbital eccentricity is 0.1 with an orbital radius of 24.5 Saturn radii making it one of the most distant satellites of Saturn.

Initial observations of Hyperion's light curve suggested that it had a spin period of 13 d [14]. Because Hyperion's orbital period is 21.3 d, this implied a non-synchronous rotation. In order to derive the spin period a fixed rotation rate had been assumed throughout a 61 d interval during the Voyager 2 encounter. However, a paper by Wisdom, Peale & Mignard [15] suggested that Hyperion's rotation was chaotic and that a constant spin period could not be assumed.

The equation of motion that governs the rotation of a satellite is

$$\mathcal{C}\ddot{\theta} - \frac{3}{2}(\mathcal{B} - \mathcal{A})\frac{\mathcal{G}m_{\mathrm{p}}}{r^3}\sin 2\psi = 0 \qquad (188)$$

where r is the orbital radius, m_{p} is the mass of the planet, $\mathcal{A}$, $\mathcal{B}$ and $\mathcal{C}$ are the (assumed constant) principal moments of inertia of the satellite, θ is the angle the long axis of the satellite makes with the planet–pericentre line and $\psi = f - \theta$ where f is the true anomaly of the satellite. But r and ψ are both functions f which is a non-linear function of time. Consequently this equation is non-integrable. Wisdom, Peale and Mignard [15] carried out numerical and analytical studies of the solution to Eq. (189). Their results implied Hyperion's rotation was chaotic and underwent essentially random changes with time.

A surface of section using values appropriate for Hyperion, is shown in Fig. 24. This is the result of integrating a single trajectory for 20,000 orbital periods of Hyperion with the values of θ and $\dot{\theta}/n$ being plotted at each peri-

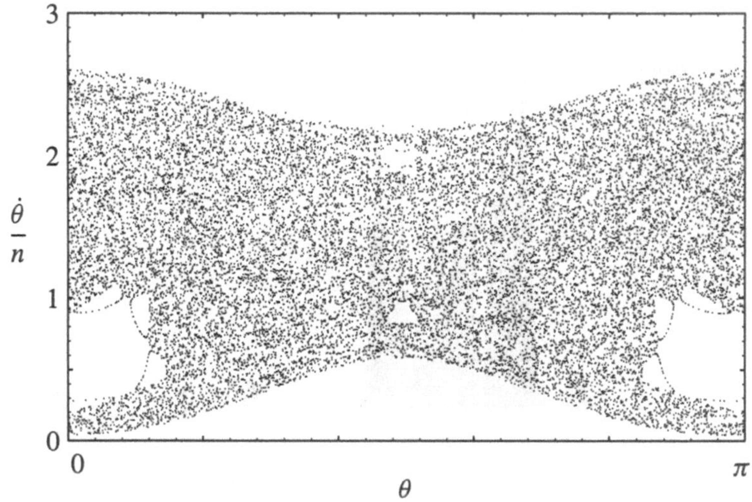

Fig. 24. Poincaré surface of section for the planar rotational behaviour of Hyperion based on a single starting condition. (Taken from [1].)

centre passage. If Hyperion happens to be trapped in one of the few islands corresponding to resonant motion, it could avoid the chaos but the evidence suggests that its rotation is chaotic. This seems to have been confirmed by ground-based observation [16]. However, the actual dynamics may be more complicated [17].

Hyperion's chaotic rotation is primarily a consequence of its unusual shape. The tides raised on Hyperion by Saturn should act to dampen the eccentricity of the Hyperion's orbit. However, Hyperion's eccentricity cannot damp because it is a forced eccentricity due to its 4:3 orbit-orbit resonance with the satellite Titan.

6.3 The Kirkwood Gaps

In 1867 Daniel Kirkwood [18] noticed that the distribution of asteroids was not random and that there was structure associated with jovian resonances. Nowadays using samples of several thousand asteroids it is easy to see this structure. Figure 25 shows a histogram of the distribution together with the marked locations of several strong jovian resonances. There are clear gaps at the 4:1, 3:1, 5:2 and 2:1 resonances but curiously there are also concentrations of asteroids at the 3:2 and 1:1 resonances. Objects at the 1:1 resonance are the Trojan asteroids discussed in Sect. 3.5 and their presence can be understood in terms of stable equilibrium points in the three-body problem. The gaps at other resonances, the concentration at the 3:2 resonance and the relatively abrupt cut-off beyond the 3:2 resonance pose more of a problem.

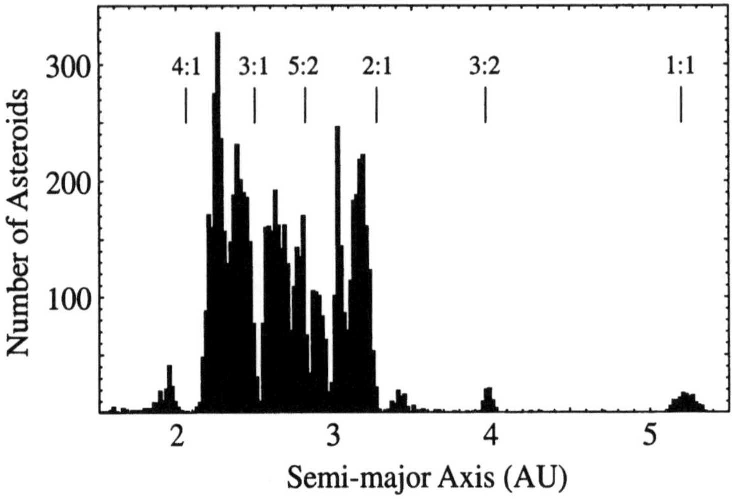

Fig. 25. Histogram of the distribution of numbered asteroids as a function of semi-major axis.)

If the Kirkwood gaps can be understood by studying the Sun-Jupiter-asteroid restricted three-body problem then a simple, long-term numerical integration of the equations of motion should be sufficient to discover any removal mechanism. However, our limited integrations at the 7:4 resonance showed that the central part of the resonance (i.e. each island in the surfaces of section) was regular in appearance. In fact, this is true for all resonances studied in the circular restricted three-body problem. Therefore the Kirkwood gaps cannot be explained by means of the planar, circular restricted problem alone.

Wisdom ([19], [20], [21]) showed that chaos had played an important role in the origin of the 3:1 Kirkwood gap. By deriving an algebraic mapping to speed up the numerical studies [19] he showed that for some starting conditions a test particle at the 3:1 resonance could achieve a large increase in its eccentricity. At its location ($a \approx 2.5$ AU) the asteroid would cross the orbit of Mars and eventually be removed by direct perturbations. Wisdom contended that although Jupiter causes the chaos, Mars actually removes the asteroid. He went on to show [20] that chaos occurred even in the planar case but that the crucial element was Jupiter's eccentricity. Murray & Fox [23] showed that the chaos was inherent in the averaged and full equations of motion. In further work Wisdom [21] provided a convincing analytical basis for the chaotic motion at the 3:1 resonance as well as showing [24] that chaos could also be involved in the delivery of meteorites to Earth.

The decline in asteroid numbers beyond the 3:2 resonance (at 3.97 AU) can also be explained by chaos. It is known that chaotic motion is associated with an overlap of adjacent resonances. Wisdom [24] derived a resonance overlap criterion and applied it to the asteroid belt showing that there should be a chaotic zone extending 0.9 AU inside Jupiter's orbit. This would result in a cleared zone in the asteroid belt beyond 4.3 AU in good agreement with observations (see Fig. 25).

Other jovian resonances have been investigated using a variety of numerical and analytical techniques. Murray [25] derived a map for first-order resonances and used it to investigate motion at the 2:1 and 3:2 resonances with Jupiter. However, he failed to demonstrate any fundamental difference between the two resonances which could account for a concentration of objects at the 3:2 and a lack of them at the 2:1. However, Wisdom [26] showed the presence of a large chaotic zone at the centre of the 2:1 resonance and the absence of such a zone at the 3:2 resonance. It remains to be seen whether or not chaos can explain all of the observed Kirkwood gaps but there are many indications that it has played a major role in this process.

6.4 The Stability of the Solar System

In the eighteenth century Pierre Simon de Laplace claimed to have shown that the solar system was stable based on his analysis of a secular perturbation theory. However, Laplace had to make a number of simplifying assumptions that are not strictly valid for the planets, and so his result cannot be considered as the final word on the subject.

With the availability of fast, cheap computers it is now possible to carry out numerical integrations of the full equations of motion of the planetary orbits for times approaching the age of the solar system. For example, a number of long-term integrations of the outer solar system have now been carried out ([27], [28], [29]). These clearly show that the orbit of Pluto, trapped in a 3:2 resonance with Neptune, is chaotic although there is no sign of it leaving the resonance. Figure 26 shows a plot of $\log \gamma$ as a function of $\log t$ for Pluto based on the results of a numerical investigation. Note that the slope is -1 for more that 10^7 years suggesting regular motion, even though it levels off after 10^9 years.

An alternative to full integration is to work with a system of averaged equations derived from the planetary disturbing function. This is always an approximation to the real system and can involve considerable algebraic calculations before any numerical integration can start. However, with typical step sizes of perhaps 500 y it is possible to integrate the solar system over several billion years. Laskar [30] used such a system of averaged equations to show that the inner planets are also chaotic with maximum LCE's of $\sim 10^{-6.7}$ y^{-1}.

None of the inetgrations show any sign of gross instabilities in the system over timescales comparable to the age of the solar system. This means that the planetary orbits are chaotic but that there is no sign of, for example, intersecting orbits. It means in practice that there are fundamental limits to our ability to predict the positions of the planets for long time intervals. This

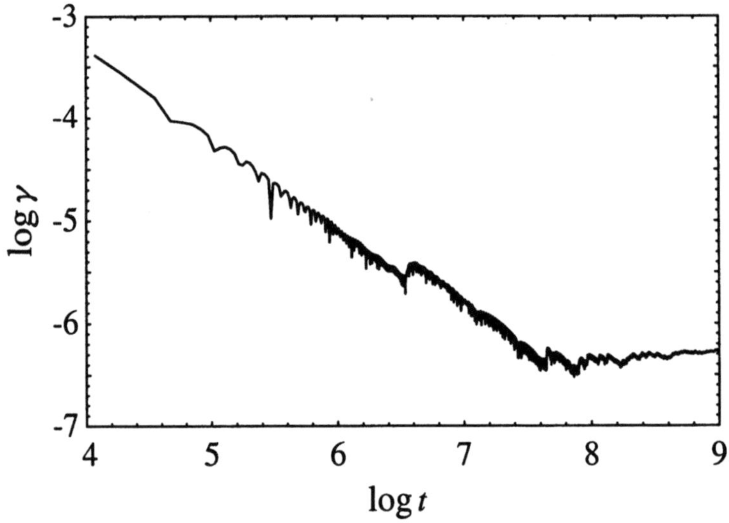

Fig. 26. Plot of $\log\gamma$ as a function of $\log t$ for Pluto based on the results of a numerical integration. The final value of γ gives a numerical estimate of Pluto's maximum Lyapounov exponent. (Taken from [1].)

is because any physical measurement has finite precision providing an inbuilt error in the calculations. This will propagate exponentially in a chaotic system. In the case of the Earth, for example, this means that an error of 1 cm in the position of the Earth today is sufficient to ensure that we cannot currently predict its location in 200 million years time.

In further work Laskar [31] used an averaging method to study the planetary orbits for 10 billion years into the past and 15 billion years into the future. The most dramatic result was the large chaotic variations in the orbit of Mercury ($0.1 < e < 0.5$, $8° < I < 21°$). Laskar shifted Earth's initial position by 150 m to find that Mercury's orbit could become almost hyperbolic 3.5 billion years in the future or 6.6 billion years in the past. However, over such long timescales it is important to question the validity of the physical model being used.

It now appears that the solar system is chaotic yet stable in the sense that the planets remain close to their current orbits for timescales approaching a billion years or more. So far an analytical proof of the solar system's stability or the origin of the chaos is as elusive as ever. Studies of the stability properties of other planetary systems have now begun [32].

7 Planetary Rings

The study of planetary ring systems provides one of the most interesting areas of research in solar system dynamics. All of the major planets possess ring systems and although each has its own peculiarities, some properties are common to all.

In this section we do attempt to provide a thorough study of the dynamics of planetary rings. A more complete analysis is given in Chapter 10 of [1] and the references therein while a summary appears in [33]. Here our interest is in applying the methods and techniques for studying resonance and chaos to try to understand some specific properties of ring systems.

7.1 Ring Systems

The chance discovery of the uranian ring system in 1977 showed that at least two of the outer planets possessed rings. In contrast to the bright, broad and bland rings of Saturn, the rings of Uranus were dark, narrow and sharp-edged. Two years later the *Voyager* spacecraft discovered the presence of a dusty ring of Jupiter. Meanwhile occultations of stars by Neptune were used to search for neptunian rings with only one in ten observations showing any detected features. The true nature of the neptunian ring system (optically thicker "arcs"of material orbiting within fainter rings) was only apparent after *Voyager 2* observations in 1989. Figure 27 shows a sample *Voyager* image of each ring system.

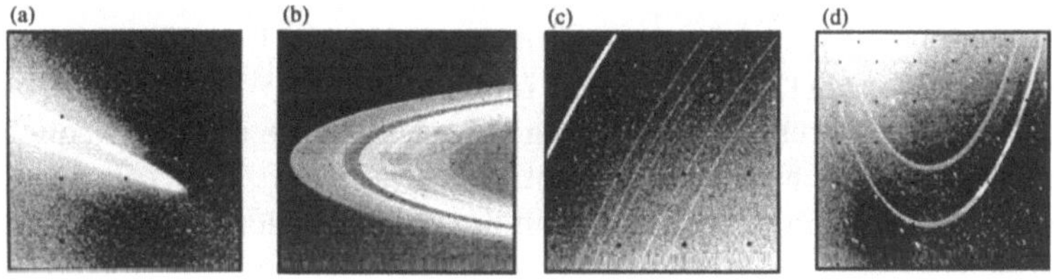

Fig. 27. Voyager images of the ring systems of (a) Jupiter, (b) Saturn, (c) Uranus and (d) Neptune. (Images courtesy of NASA/JPL.)

The relationship between the existence of rings and the presence of nearby small moons is evident in Fig. 28 where the rings and inner satellite systems of the major planets are drawn on a scale with the planetary radius the same for each planet.

It is clear from Fig. 28 that many small satellites lie close to the ring systems. There is good evidence that satellites such as Metis and Adrastea in the jovian system act as sources of ring material while in other cases, such as the saturnian pair Prometheus and Pandora on either side of the F ring, satellites influence nearby ring material. In another case the uranian satellites

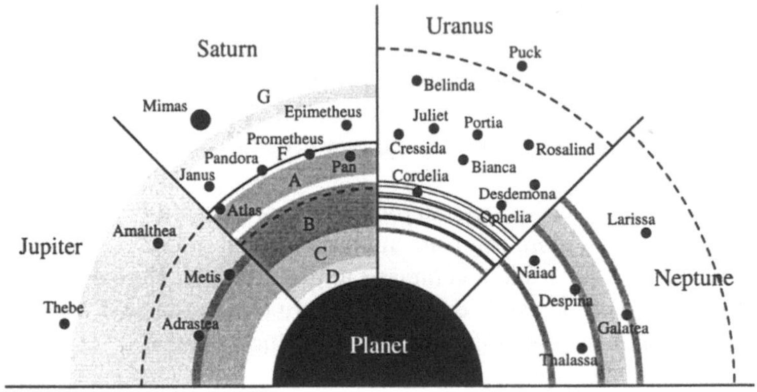

Fig. 28. The ring systems and associated small satellites of the outer planets shown in a scale of uniform planetary radius. The dashed curve denotes the location of the synchronous orbit for each planet. (Taken from [33]).

Cordelia and Ophelia on either side of the narrow ε ring help confine ring particles by means of resonances. The close connection between rings and small satellites suggests a common origin and evolution.

7.2 Types of Resonance

In Sect. 4.5 we noted that planetary oblateness leads to a radial frequency κ and a vertical frequency ν as well as a modification to the mean motion n in the motion of an orbiting object such as a ring particle. Now consider the effect of a perturbing satellite with its own set of frequencies n', κ' and ν' given by Eqs. (144)–(146) with a replaced by a'. We define the *pattern speed*, Ω_{ps}, of the satellite's perturbing potential as the angular frequency of a reference frame in which this potential is stationary. This will depend on the exact combination of frequencies under consideration; it may be written as

$$m\Omega_{\text{ps}} = mn' + k\kappa' + p\nu' = (m+k+p)n' - k\dot{\varpi}' - p\dot{\Omega}' \tag{189}$$

where m, k and p are integers and m is non-negative. A resonance will occur when an integer multiple of the difference between n and Ω_{ps} is equal to zero (for *corotation resonances*), or the natural frequency of the radial oscillations (for *eccentric* or *Lindblad resonances*) or vertical oscillations of the ring particle (for *vertical resonances*). We now consider each resonance type in turn. As we shall see below, these resonances can also be described using a disturbing function approach.

Corotation Resonances At a corotation resonance

$$m(n - \Omega_{\text{ps}}) = 0 \tag{190}$$

and the resonance condition becomes

$$(m+k+p)n' - mn - k\dot{\varpi}' - p\dot{\Omega}' = 0\,. \tag{191}$$

The resonant condition can also be written as $\dot{\varphi}_{\rm cr} = 0$ where $\varphi_{\rm cr}$ is the resonant angle given by

$$\varphi_{\rm cr} = j\lambda' + (k+p-j)\lambda - k\varpi' - p\Omega' \tag{192}$$

where $j = m+k+p$. This argument satisfies the d'Alembert relation and from the known properties of the disturbing function we know that $|p|$ must be even. Furthermore the associated resonance is one of order $|k+p|$. Therefore the 1:1 (or co-orbital) resonance, where $p = k = 0$, is a special case of a corotation resonance.

The maximum width in semi-major axis of a corotation resonance can be calculated using the pendulum model discussed in Sect. 5.4. The relevant part of the averaged disturbing function is

$$\mathcal{R} = \frac{\mathcal{G}m'}{a'} f_{\rm res}(\alpha) e'^{|k|} s'^{|p|} \cos\varphi_{\rm cr} \tag{193}$$

where $\alpha = a/a'$ and the exact form of $f_{\rm res}(\alpha)$ depends on the resonance in question. For example, for 3:2 corotation resonance $j = 3$ with $k = 1$ and $p = 0$ giving $f_{\rm res} = f_6$ (as given in Eq. (96)) with no indirect terms. Using the pendulum approach the maximum width, $W_{\rm cr}$, of a general corotation resonance can be written as a function of the magnitude of $\mathcal{R}$ as

$$W_{\rm cr} = 8\left(\frac{a\,|\mathcal{R}|}{3\mathcal{G}m_{\rm p}}\right)^{1/2} a\,. \tag{194}$$

Lindblad Resonances At a Lindblad resonance

$$m(n - \Omega_{\rm ps}) = \pm\kappa \tag{195}$$

where the upper and lower signs correspond to the inner (ILR) and outer (OLR) Lindblad resonance respectively. The choice of sign allows us to consider a ring particle that is orbiting inside or outside the orbit of the perturbing satellite. The resonance condition is

$$(m+k+p)n' - (m\mp 1)n - k\dot{\varpi}' \mp \dot{\varpi} - p\dot{\Omega}' = 0\,. \tag{196}$$

In terms of the resonant angle, $\varphi_{\rm lr}$, the resonance condition is $\dot{\varphi}_{\rm lr}$=0 where

$$\varphi_{\rm lr} = j\lambda' + (k+p\pm 1-j)\lambda - k\varpi' \mp \varpi - p\Omega' \tag{197}$$

and where, as before, $j = m+k+p$. This is a resonant argument of order $|k+p\perp 1|$.

When calculating the width of a Lindblad resonance the pendulum approach is not appropriate. We have to remember that we are dealing with an ensemble of ring particles responding to the perturbing potential. The Lindblad resonance induces a forced eccentricity on the ring particles such that, at a given semi-major axis, the particles move in *streamline* motion. The resulting pattern in the rotating frame gives the appearance of a wave on the ring. The magnitude of the forced eccentricity decreases as the distance from the exact resonance increases, with a phase change of 180° on either side of exact resonance. The width of the resonance is determined by the separation from the exact resonance such that the value of the forced eccentricity is just sufficient for the outer streamline to intersect the inner one (see the inner part of Fig. 29). The mechanism is discussed in Porco & Nicholson [34], Murray & Dermott [1] and Murray [33]. For the specific case of the $k = p = 0$ inner Lindblad resonance (ILR) it can be shown that the width of the resonance is given by

$$W_{\mathrm{lr},0} \approx 2.9(m'/m_{\mathrm{p}})^{1/2}a. \tag{198}$$

and this width is approximately the same for all first-order resonances.

Vertical Resonances At a vertical resonance

$$m(n - \Omega_{\mathrm{ps}}) = \pm\nu \tag{199}$$

where the upper and lower signs correspond to the inner (IVR) and outer (OVR) vertical resonances respectively. The resonance condition is

$$(m + k + p)n' - (m \mp 1)n - k\dot{\varpi}' \mp \dot{\Omega} - p\dot{\Omega}' = 0\,. \tag{200}$$

In terms of the resonant angle, φ_{vr}, the resonance condition is $\dot{\varphi}_{\mathrm{vr}}=0$ where

$$\varphi_{\mathrm{vr}} = j\lambda' + (k + p \pm 1 - j)\lambda - k\varpi' - p\Omega' \mp \Omega \tag{201}$$

where, as before, $j = m+k+p$. This is a resonant argument of order $|k+p\pm1|$.

In the case of vertical resonances there is no mechanism analagous to that of the Lindblad resonance in a coplanar ring system. There is a forced inclination but in the real situation this does not rise to infinity as exact resonance is approached. A more careful analysis shows that a pendulum-like equation of motion is required to understand the resonance mechanism.

7.3 Location of Resonances

In our discussion of resonance splitting in Sect. 5.5 it was clear that more than one exact resonance can exist close to the nominal location of a resonance. Thus, while the approximate semi-major axis of an internal $p + q : p$ resonance is given by $a = [p/(p + q)]^{2/3}a'$, where a' is the semi-major axis

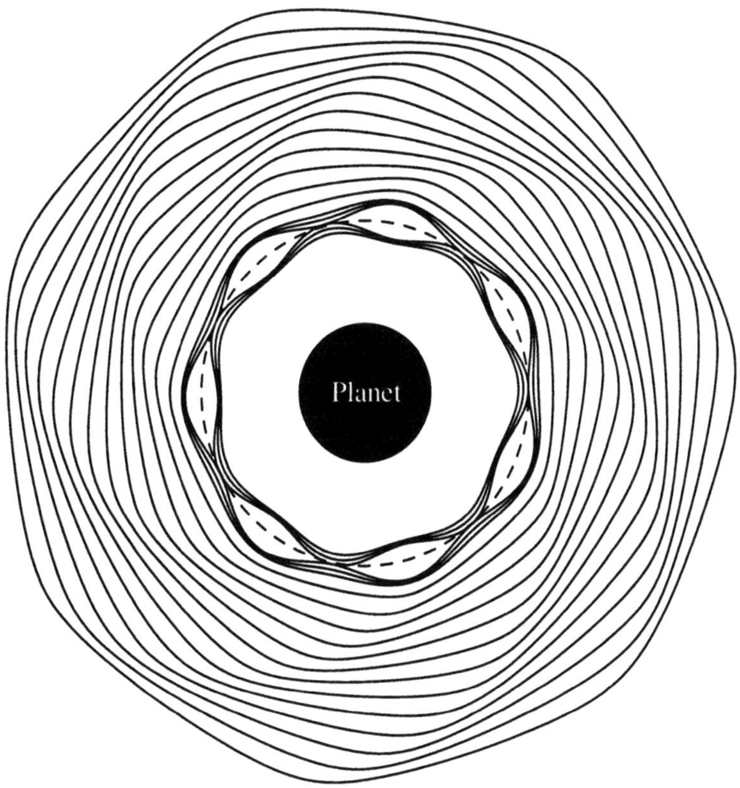

Fig. 29. Schematic representation of the particle streamlines associated with the 7:6 inner Lindblad resonance showing the lobes of the resonance and the resulting seven-armed spiral structure. The dashed line denotes the location of the exact resonance. (Taken from [33].)

of the perturbing object, the location of the *exact* resonance depends on the form of the specific resonant argument.

For planetary ring particles the planet's oblateness usually dominates the motion of a particle's pericentre and node, especially in regions close to the planet. Note from Eqs. (144), (150) and (151) that the contribution of the planet's oblateness to n, $\dot{\varpi}$ and $\dot{\Omega}$ is a function of the semi-major axis. Unfortunately this is a non-linear dependence and so a numerical method is needed in order to find the location of the exact resonance for a given argument.

The results of such calculations for the second-order resonances associated with the 5:3 Mimas commensurability are shown in Table 2. Including terms up to order 2 in the eccentricities and inclinations means that there are six possible resonant arguments. Here we have taken the mean motion, perichrone rate and node rate of Mimas to be $381.9945°\mathrm{d}^{-1}$, $1.0008°\mathrm{d}^{-1}$ and $-0.9995°\mathrm{d}^{-1}$ respectively [35].

Table 2. The 5:3 resonances of the saturnian satellite, Mimas. In the classification column ILR, IVR, CER, and CIR denote inner Lindblad resonance, inner vertical resonance, corotation eccentricity resonance and corotation inclination resonance respectively.

i	φ_i	Type	Class.	j	m	k	p	n (°d^{-1})	a (km)
1	$5\lambda' - 3\lambda - 2\Omega$	I^2	—	5	–	–	–	638.886	131793
2	$5\lambda' - 3\lambda - \Omega - \Omega'$	II'	IVR	5	4	0	1	638.102	131900
3	$5\lambda' - 3\lambda - 2\Omega'$	I'^2	CIR	5	3	0	2	637.324	132007
4	$5\lambda' - 3\lambda - 2\varpi'$	e'^2	CER	5	3	2	0	635.990	132191
5	$5\lambda' - 3\lambda - \varpi - \varpi'$	ee'	ILR	5	4	1	0	635.219	132298
6	$5\lambda' - 3\lambda - 2\varpi$	e^2	—	5	–	–	–	634.454	132404

In Table 2 "type"refers to the terminology of the disturbing function approach while "classification"refers to the terminology of ring dynamics. Note that the effect of the oblateness has caused the resonances to be spread out over more than 600 km in semi-major axis.

7.4 Waves in Rings

We have seen in Fig. 29 how a satellite's inner Lindblad resonance in a ring distorts the particles' streamline shapes around exact resonance. This distortion introduces an azimuthal variation in the gravitational potential. For a $p+1:p$ resonance there are $p+1$ lobes which act to alter the local potential. The effect of this modified potential on surrounding ring material leads to a trailing pattern with $p+1$ spiral arms (see Fig. 29 for the specific case of the 7:6 ILR). This is an example of a *spiral density wave.* In reality the spiral is tightly wound but a radial profile would always show a decrease in wavelength with increasing distance from exact resonance. Vertical resonances lead to the formation of *spiral bending waves.* For an IVR the result is trailing bending waves propagating inwards from the exact resonance.

In Table 2 we showed the exact location of the Mimas 5:3 resonances. Because Mimas has an inclination of 1.5°, its vertical resonances are comparable in strength to its Lindblad resonances. The table shows that Mimas's 5:3 ILR lies 398 km beyond its 5:3 IVR. The effect on the rings is shown in Fig. 30. The spiral density wave due to the 5:3 ILR propagates outwards while the spiral bending wave due to the 5:3 IVR propagates inwards. Note that the bending wave produces vertical motion of the particles that results in localised warping of the ring plane. This explains the different contrast between the two wave features. The vertical distortions from the IVR are more difficult to resolve than the ring-plane density variations resulting from the ILR because of the nature of the geometry.

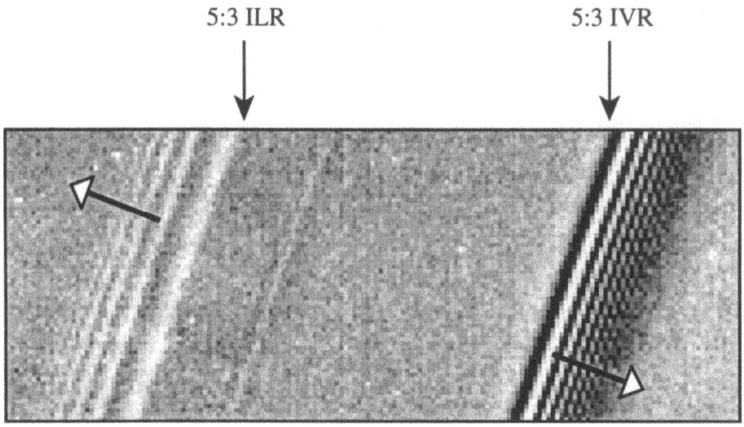

Fig. 30. A Voyager image showing the outward-propagating, trailing spiral density wave resulting from the 5:3 Mimas inner Lindblad resonance and the inward propagating, trailing bending wave resulting from the 5:3 Mimas inner vertical resonance. The superimposed arrows denote the direction of propagation of each wave. The resonances are separated by approximately 400 km (Taken from [33]).

7.5 The Encke Gap and Pan

It can be shown that a satellite of mass m' and semi-major axis a separated a distance Δa from a ring produces a wave on that ring of wavelength $3\pi\Delta a$ and amplitude $2.24a(m'/m_{\mathrm{p}})(a/\Delta a)^2$, where m_{p} is the mass of the planet. Therefore detection of a wave on a ring and measurements of its wavelength and amplitude allow one to deduce the presence, radial location and mass of the satellite that produced it.

Voyager images of the 325 km-wide Encke gap in Saturn's A ring showed the presence of wavy edges [36] (see Fig. 31). The putative satellite, now called Pan, was finally discovered by Showalter [37]. Pan's motion was entirely

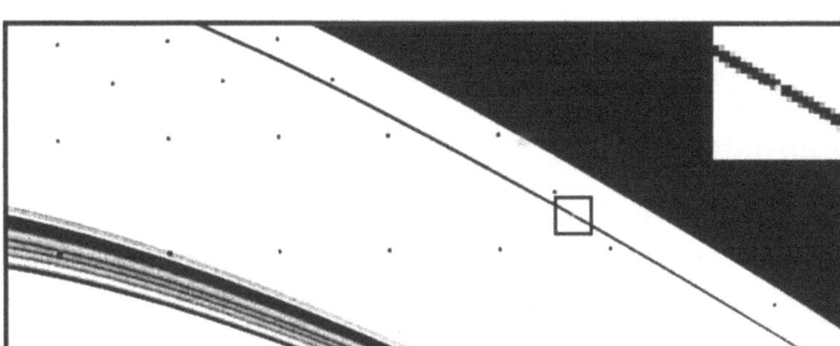

Fig. 31. A "stretched" *Voyager 2* image of Saturn's A ring showing a detection of (see enlarged inset) in the Encke gap in Saturn's A ring. (Taken from [33]).

consistent with that of an object with a semi-major axis of 133582.8±0.8 km, placing it in the centre of the Encke gap [37].

7.6 The Adams Ring of Neptune

Images of the Neptune system returned by the *Voyager 2* spacecraft during the 1989 flyby showed that the planet had an optically thin ring system. The outermost Adams ring contained several arcs of optically thicker (optical depth $\tau \sim 0.04$) material that produced occultation events. In order to have lifetimes that were longer than a few decades, the narrow arcs had to be subjected to some mechanism that would confine them in radius and azimuth.

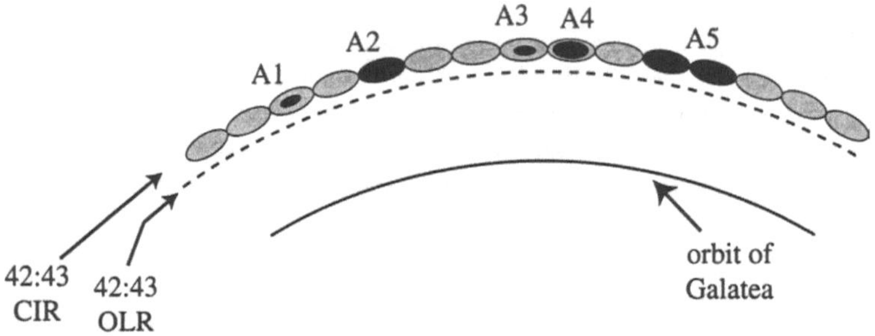

Fig. 32. The resonances thought to be responsible for the azimuthal structure and confinement of Neptune's Adams ring. (Taken from [33].)

Porco [38] suggested that the arcs were maintained by the perturbing effects of a small satellite, Galatea, orbiting $\sim$ 900 km inside the Adams ring. According to Porco's theory Galatea's 42:43 outer corotation inclination resonance provides 84 equilibrium sites (the small ovals in Fig. 32), only some of which are filled, or partially filled by the optically thicker material (the darker ovals in Fig. 32). While the 42:43 outer CIR provides azimuthal confinement, the 42:43 OLR, located $\sim$ 1.5 km interior to the ring (the dashed line in Fig. 32), provides the radial confinement.

8 Summary

Our exploration of the basic properties of the two- and three-body systems, the planetary disturbing function, the dynamics of resonance and chaos has many applications to our understanding of the history and dynamical evolution of our solar system. There is no doubt that spacecraft observations have helped to stimulate numerous aspects of solar system dynamics but it is equally true that the development of new analytical and numerical methods has been crucial too. This article has discussed only some of the intricate dynamical structures that exist in the solar system but hopefully it has given a flavour of what has been found as well as providing a few hints on how to understand it.

References

1. Murray, C.D., Dermott, S.F. (1999) Solar System Dynamics. Cambridge University Press, Cambridge
2. Belton, M.J.S., Chapman, C.R. et al. (1995). Bulk density of asteroid 243-Ida from the orbit of its satellite Dactyl. Nature **374**, 785–788
3. Colwell, P. (1993). Solving Kepler's Equation Over Three Centuries. Willmann-Bell, Richmond
4. Danby, J.M.A. (1988). Fundamentals of Celestial Mechanics, 2nd Edition. Willmann-Bell, Richmond
5. Mikkola, S., Innanen, K.A., Muinonen, K. and Bowell, E. (1994). A preliminary analysis of the orbit of the Mars trojan asteroid (5261) Eureka. Celest. Mech. Dyn. Astron. **58**, 53–64
6. Wiegert, P.A., Innanen, K.A. and Mikkola, S. (1997). An asteroidal companion to the Earth. Nature **387**, 685–686
7. Dermott, S.F. and Murray, C.D. (1981). The dynamics of tadpole and horseshoe orbits. II. The coorbital satellites of Saturn. Icarus **48**, 12–22
8. Nicholson, P.D., Hamilton, D.P., Matthews, K. and Yoder, C.F. (1992). New observations of Saturn's coorbital satellites. Icarus **100**, 464–484
9. Roy, A.E. and Ovenden, M.W. (1954). On the occurrence of commensurable mean motions in the solar system. Mon. Not. R. Astr. Soc. **114**, 232–241
10. Goldreich, P. (1965). An explanation of the frequent occurrence of commensurable mean motions in the solar system. Mon. Not. R. Astr. Soc. **130**, 159–181
11. Malhotra, R. (1993). The origin of Pluto's peculiar orbit. Nature **365**, 819–820
12. Harper, D. and Taylor, D.B. (1993). The orbits of the major satellites of Saturn. Astron. Astrophys. **268**, 326–349
13. Dermott, S.F., Malhotra, R. and Murray, C.D. (1988). Dynamics of the Uranian and Saturnian satellite systems: a chaotic route to melting Miranda? Icarus **76**, 295–334
14. Thomas, P., Veverka, J., Wenkert, D., Danielson, G.E. and Davies, M. (1984). Hyperion: 13-day rotation from Voyager data. Nature **307**, 716–717
15. Wisdom, J., Peale, S.J. and Mignard, F. (1984). The chaotic rotation of Hyperion. Icarus **58**, 137–152
16. Klavetter, J.J. (1989). Rotation of Hyperion. I. Observations. Astron. J. **97**, 570–579
17. Black, G.J., Nicholson, P.D., and Thomas, P.C. (1995). Hyperion: Rotational dynamics. Icarus **117**, 149–171
18. Kirkwood, D. (1867). Meteoric Astronomy (Lippincott, Philadelphia)
19. Wisdom, J. (1982). The origin of the Kirkwood gaps: A mapping technique for asteroidal motion near the 3/1 commensurability. Astron. J. **87**, 577–593
20. Wisdom, J. (1983). Chaotic behavior and the origin of the 3/1 Kirkwood gap. Icarus **56**, 51–74
21. Wisdom, J. (1985a). A perturbative treatment of motion near the 3/1 commensurability. Icarus **63**, 272–289

22. Wisdom, J. (1980). The resonance overlap criterion and the onset of stochastic behavior in the restricted three-body problem. Astron. J. **85**, 1122–1133
23. Murray, C.D. and Fox, K. (1984). Structure of the 3:1 jovian resonance: A comparison of numerical methods. Icarus **59**, 221–233
24. Wisdom, J. (1985b). Meteorites may follow a chaotic route to Earth. Nature **315**, 731–733
25. Murray, C.D. (1986). The structure of the 2:1 and 3:2 Jovian resonances. Icarus **65**, 70–82
26. Wisdom, J. (1987a). Urey Prize Lecture: Chaotic dynamics in the solar system. Icarus **72**, 241–275
27. Kinoshita, H. and Nakai, H. (1984). The motions of the perihelions of Neptune and Pluto. Celest. Mech. **34**, 203–217
28. Applegate, J., Douglas, M.R., Gürsel, Sussman, G.J. and Wisdom, J. (1986). The outer solar system for 200 million years. Astron. J. **92**, 176–194
29. Roy, A.E., Walker, I.W., Macdonald, A.J., Williams, I.P., Fox, K., Murray, C.D., Milani, A., Nobili, A.M., Message, P.J., Sinclair, A.T., Carpino, M. (1988). Project LONGSTOP. Vistas in Astronomy **32**, 95–116
30. Laskar, J. (1988). Secular evolution of the Solar System over 10 million years. Astron. Astrophys. **198**, 341–362
31. Laskar, J. (1994). Large scale chaos in the solar system. Astron. Astrophys. **287**, L9–12
32. Holman, M.J., Touma, J. and Tremaine, S. (1997). Chaotic variations in the eccentricity of the planet orbiting 16 Cygni B. Nature **386**, 254–256
33. Murray, C.D. (1999). The dynamics of planetary rings and small satellites. In *Dynamics of Small Bodies in the Solar System: A Major Key to Solar System Studies*, ed. B. Steves and A.E. Roy (Kluwer, Dordrecht)
34. Porco, C.C. and Nicholson, P.D. (1987). Eccentric features in Saturn's outer C ring. Icarus **72**, 437–467
35. Harper, D. and Taylor, D.B. (1993). The orbits of the major satellites of Saturn. Astron. Astrophys. **268**, 326–349
36. Cuzzi, J.N. and Scargle, J.D. (1985). Wavy edges suggest moonlet in Encke's Gap. Astrophys. J. **292**, 276–290
37. Showalter, M.R. (1991). Visual detection of 1981S13, Saturn's eighteenth satellite, and its role in the Encke gap. Nature **351**, 709–713
38. Porco, C.C. (1991). An explanation for Neptune's ring arcs. Science **253**, 995–1001

Photometry of Resolved Planetary Surfaces

Nicolas Thomas

Max-Planck-Institut für Aeronomie, Max-Planck-Str. 2,
D-37189 Katlenburg-Lindau, Germany

Abstract. The study of reflected light forms the first method of investigating the properties of a planetary surface. The reduction of the observed broad-band radiance into parameters which describe the surface properties can be confusing, however, because of differences in nomenclature and definition used by authors. Here, we present terms and definitions for some of the most commonly used quantities used in broad-band photometry leading to a brief description of Hapke's parameters.

1 Introduction

The investigation of a planetary surface through the observation of the light it reflects usually provides the first knowledge of the physical properties of that surface. While spectroscopy, particularly in the infra–red, can be used to determine composition, photometry of the surface can be used to infer structural properties such as surface roughness.

Imaging of planetary surfaces in a few selected, relatively, broad–band wavelength bands can characterize surface units over an entire object relatively quickly in a well-planned space mission. Spectroscopic experiments tend to be more "data hungry"and therefore a typical approach is one where an imaging device maps a significant fraction of a surface allowing selection of key areas of interest for subsequent detailed investigation by spectroscopy. The ability to reduce observations of reflected sunlight is therefore a powerful and necessary tool in a planetary scientist's analytical arsenal.

In the study of surface photometry, the student is confronted with two major issues. The first is one of nomenclature. The literature on this subject contains many traps and pitfalls resulting from different definitions used by different authors. Secondly, the study of the effects of particle size, roughness, etc. on the observed radiance is not simple and equations allowing a less empirical treatment than previously have only recently been developed (e.g. [3], [4], [5]). This chapter provides an introduction to the subject by presenting the basic definitions and equations which link the observed radiance to a surface's properties.

2 Specific Intensity and Radiance

The radiant energy (dE_ν) transported across an area element (da) in directions confined by a solid angle ($d\omega$) per unit frequency (or wavelength) interval ($d\nu$) is expressed in terms of the specific intensity, I_ν ([1]). This is the fundamental quantity to be measured in planetary photometry (Figure 1) and is defined by

$$dE_\nu = I_\nu \cos\theta d\nu da d\omega. \tag{1}$$

The typical units of I_ν are [W m^{-2} sr^{-1} Hz^{-1}] or [W m^{-2} sr^{-1} nm^{-1}]. It should be noted that interchanging these units is possible by differentiating

$$\nu = \frac{c}{\lambda} \tag{2}$$

(where ν is the frequency, λ is the wavelength, and c is the velocity of light) to obtain

$$d\nu = -\frac{c}{\lambda^2} d\lambda. \tag{3}$$

Here, we recognize that the units are interchangable and express units in terms of wavelength only for convenience.

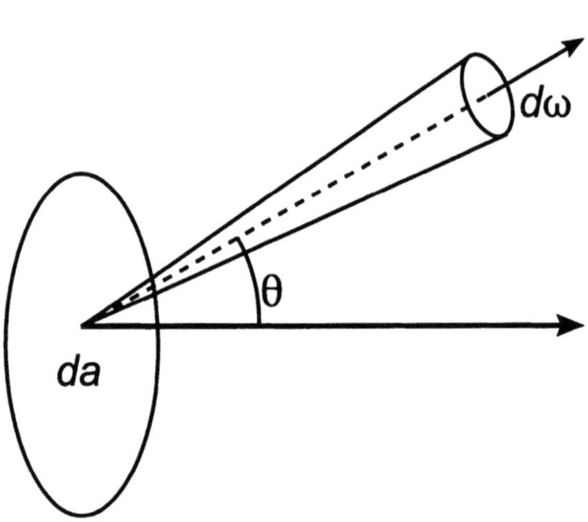

Fig. 1. The definition of specific intensity, I_ν (adapted from Chandrasekhar (1960)).

The radiance, I, is the radiant energy transported across an area element in directions confined to a solid angle element during a specified time. This quantity is, therefore, the specific intensity integrated over a wavelength range. It also called "intensity"or, more rarely, "luminance"and has units of [W m^{-2} sr^{-1}]. The observed radiance is linked to the reflecting properties of a surface by the equation

$$\rho_C(\alpha, \epsilon, \iota) = \frac{\pi I(\alpha, \epsilon, \iota)}{\mu_0 F} \tag{4}$$

where ρ_C is the radiance co-efficient which is a function of the phase angle (Sun-surface-observer), α, the angle of incidence, ι, and the angle of emission,

ϵ (see Figure 2). F is the irradiance which is defined as the radiant flux incident on a surface area and has units of [W m^{-2}]. The solar flux is an irradiance and, in Solar System studies, it is common to define $F_\odot$ as the solar flux at 1 AU so that eq. (4) becomes

$$\rho_C(\alpha, \epsilon, \iota) = \frac{\pi I(\alpha, \epsilon, \iota)}{\mu_0 F_\odot / {R_h}^2} \tag{5}$$

where R_h is the heliocentric distance in [AU]. μ_0 is the cosine of the solar zenith angle at the surface (i.e. $\mu_0 = \cos\iota$).

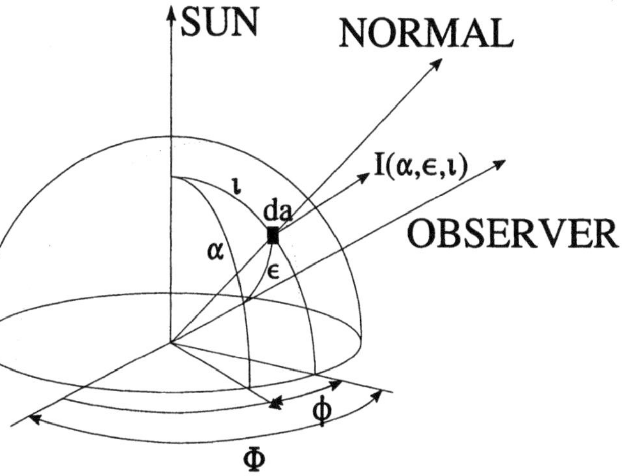

Fig. 2. The definition of angles in surface photometry (adapted from Hanel et al., 1992).

There are two potential sources of confusion here. It is frequently assumed in very broad-band photometry that one measures, I. But all photometric devices have a varying response with wavelength. Therefore, to obtain the correct value of ρ_C, $F_\odot$ must be defined in terms of the same spectral bandpass properties of the instrument used to measure I. Secondly, ρ_C is often referred to as the "bidirectional reflectivity"(e.g. [2] or as a reflectance whereas other authors use these terms for completely different quantities. We use the subscript, C, to denote that this quantity is the radiance co-efficient.

Another pitfall in studying the literature is that many authors (e.g. [8]) define the solar flux to be equal to a flux multiplied by π (i.e. $\pi F = F_\odot$). There is little or no reason for this and it can lead to confusion over definitions when there is a reference to the term "I over ".

The definition of the radiance co-efficient (eq. 5) requires knowledge of the cosine of the angle of incidence ($\mu_0 = \cos\iota$) of the incoming illumination. In many cases, this angle is difficult to define because of irregularities in the surface topography. A good example of this is the nucleus of comet Halley which is crudely approximated by a prolate ellipsoid but with significant local deviations from this shape. A useful quantity in these cases is the radiance factor, ρ_F, defined as

$$\rho_F(\alpha, \epsilon, \iota) = \frac{\pi I(\alpha, \epsilon, \iota)}{F_{\odot}/{R_h}^2}. \tag{6}$$

A related quantity is what we shall define (following Hapke, 1981) as the bidirectional reflectance, ρ_R,

$$\rho_R(\alpha, \epsilon, \iota) = \frac{I(\alpha, \epsilon, \iota)}{F_{\odot}/{R_h}^2}. \tag{7}$$

It is important to note that while ρ_C and ρ_F are unitless, ρ_R has the units of [sr].

3 Reflectance and Albedo

We have just defined the term, bi-directional reflectance, but why is it called "bi-directional"? The definitions of terms related to reflectance were first compiled by [6]. The reflectance is prefixed first by the directionality of the illuminating flux and then by the directionality of reflected radiance (Figure 3). Thus, if an incoming flux from a point source is reflected from a surface and studied over the hemisphere, the reflectance is described as the "directional-hemispherical reflectance". If a reflectance is derived from a directional source and a directional radiance then it is called a "directional-directional reflectance" which is shortened to bi-directional reflectance.

The directional-hemispherical reflectance is an important quantity, which leads to the definition of the hemispherical albedo, A_H. This is the ratio of the total power reflected by a surface element in all directions to the irradiance of light from a collimated source incident from a specific direction,

$$A_H = \frac{1}{\mu_0 F_{\odot}/{R_h}^2} \int_{2\pi} I(\alpha, \epsilon, \iota) \cos \epsilon d\Omega. \tag{8}$$

This equation is exactly equivalent to Hapke's equation in a different coordinate system

$$A_H = \frac{2\pi}{\mu_0 F_{\odot}/{R_h}^2} \int_0^1 I(\alpha, \mu, \mu_0) \mu d\mu. \tag{9}$$

where $\mu = \cos \epsilon$. Since $d\Omega = \sin \epsilon d\epsilon d\phi$,

$$\int_{2\pi} I(\alpha, \epsilon, \iota) \cos \epsilon d\Omega = \int_0^{2\pi} \int_0^{\frac{\pi}{2}} I(\alpha, \epsilon, \iota) \cos \epsilon \sin \epsilon d\epsilon d\phi = \pi I(\alpha, \epsilon, \iota) \tag{10}$$

if $I(\alpha, \epsilon, \iota) = constant$ and, therefore, $A_H = \rho_C$. The condition, $I(\alpha, \epsilon, \iota) = constant$, is an expression of Lambert's law, *i.e.* the radiance of a small

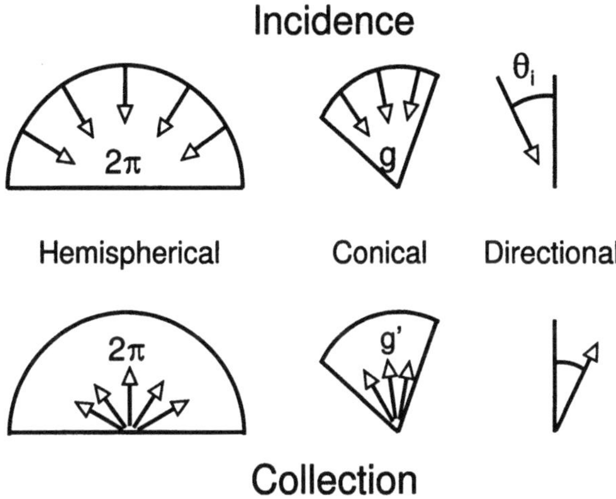

Fig. 3. The definition of terms in reflectometry (adapted from Judd, 1967). To describe a quantity, the incident illumination must first be described by hemispherical, conical, or direction. Then, the collection method must be described (again as hemispherical, conical, or directional) as the second part of the term.

element of a perfectly diffusing sphere in any direction is proportional to the cosine of the angle between the direction and the normal. A surface obeying this law is referred to as a Lambertian surface.

If the surface is Lambertian and one integrates over the entire illuminated surface, then A_H becomes exactly equivalent to the Bond albedo, A_B, which is the total reflected power divided by the total incident power intercepted by an object. Strictly speaking, the Bond albedo is obtained by integrating over all wavelengths. Often I and $F_\odot$ are expressed within a bandpass and hence ρ_C and A_H only apply to that bandpass if determined in this way.

The difficulty in the determination of A_H is that one usually determines the bi-directional reflectance, ρ_R, and thus one has to extrapolate to a hemisphere. The geometric albedo, p, is far more easily measured and is often used to describe the reflecting properties of an entire planet, satellite, or small body. p is defined as the ratio of brightness of a planetary disc observed at zero phase angle to the brightness of a perfectly diffusing disc viewed under the same geometry. By using the spherical cosine law, one can modify the coordinate system so that an integration can be performed in azimuth about the sub-solar point of the planetary body. Thus,

$$\cos\epsilon = \cos\alpha\cos\iota + \sin\alpha\sin\iota\cos\phi \tag{11}$$

where ϕ is the azimuth angle (see Figure 2). Hence,

$$da = r^2 \sin\iota d\iota d\phi \tag{12}$$

where r is the planetary radius and da is the surface element. This leads to the defining equation for geometric albedo,

$$p = \frac{1}{\pi} \int_0^{2\pi} \int_0^{\frac{\pi}{2}} \rho_C(\iota, \phi) \sin \iota \cos^2 \iota d\iota d\phi \tag{13}$$

(e.g. Hanel et al., 1992). It should be noted that the

$$\int_0^{2\pi} \int_0^{\frac{\pi}{2}} \sin \iota \cos^2 \iota d\iota d\phi = \frac{2}{3}\pi \tag{14}$$

and thus, for a Lambertian surface, $A_H = \rho_C = 3p/2$.

Unfortunately, for planetary surfaces, Lambert's law is a very poor approximation. Thus, it is necessary to determine how $I(\alpha, \epsilon, \iota)$ varies over a hemisphere to allow an exact integration of the radiance over the hemisphere. This variation is often expressed in terms a phase function, $\phi(\alpha)$. p is related to A_B by the equation

$$A_B = pq \tag{15}$$

where q is the phase integral determined by

$$q = 2 \int_0^{\pi} \phi(\alpha) \sin \alpha d\alpha. \tag{16}$$

One can see that $q = 3/2$ for a Lambertian surface but is typically of the order of $3/4$ for natural surfaces.

[7] shows how the geometric albedo and the phase function can be related to the integrated photon flux an observer sees from a planetary body.

4 Hapke's Parameters

4.1 Non-isotropic, Multiple Scattering

The ultimate objective of planetary photometry is to relate the measured reflectance to the properties of the surface. Hapke, in a series of papers, has provided a theoretical basis for this type of investigation. He considers dust particle scattering in a planetary regolith. For a single scatter from the surface

$$\frac{I_s \pi}{\mu_o \frac{F_\odot}{R_h^2}} = \frac{\omega}{4} \frac{1}{\mu + \mu_o} P(\alpha) \tag{17}$$

where I_s is the mean radiance from a single scatter, and ω is the single scattering albedo which is equal to S/E where E is the extinction co-efficient and S is the scattering co-efficient. $P(\alpha)$ is the average particle phase function. [3] has generalized the equations for $P(\alpha)$ for a multi-component medium but for the purposes of this demonstration, we can simply assume a single particle size and composition and define $P(\alpha)$ such that

$$\int_{4\pi} P(\alpha)d\Omega = \int_0^{\pi} P(\alpha)\sin\alpha d\alpha = 4\pi. \tag{18}$$

For isotropic scattering $P(\alpha)= 1$ and eq. (17) reduces to the Lommel-Seeliger law which describes the radiance factor as

$$\rho_F(\alpha, \epsilon, \iota) = \frac{\omega}{4}\frac{\mu_0}{\mu + \mu_0} = \rho_C\mu_0 \tag{19}$$

(see also [8]). If one includes multiple scattering in this isotropic scattering approximation then

$$I = I_s + I_m = \frac{F_\odot}{{R_h}^2}\frac{\omega}{4\pi}\frac{\mu_0}{\mu + \mu_0}H(\mu_0)H(\mu) \tag{20}$$

where

$$H(\mu) = \frac{1 + 2\mu}{1 + 2\gamma\mu} \tag{21}$$

and

$$\gamma = \sqrt{1 - \omega} \tag{22}$$

For non-isotropic scatters, the multiply scattered term is relatively insensitive to the phase function and therefore

$$I = I_s + I_m = \frac{F_\odot}{{R_h}^2}\frac{\omega}{4\pi}\frac{\mu_0}{\mu + \mu_0}(P(\alpha) + H(\mu_0)H(\mu) - 1). \tag{23}$$

It is often useful and realistic to define $P(\alpha)$ as a backscattering function of the form

$$P(\alpha) = 1 + \cos(\alpha). \tag{24}$$

An alternative possibility is to use a Lunar-like phase function of the form

$$P(\alpha) = \frac{\pi^2}{5}\left(\frac{\sin\alpha + (\pi - \alpha)\cos\alpha}{\pi} + \frac{(1 - \cos\alpha)^2}{10}\right). \tag{25}$$

4.2 The Opposition Effect

When the phase angle approaches zero (the geometry under which the geometric albedo is measured) the incident beam merely has to be scattered back directly to escape from the surface and reach the observer. If the back scattering is efficient, then a surge in the radiance near opposition ($\alpha = 0$) can be seen called the opposition effect.

Hapke accounted for this in the single scattering, non-isotropic case by

$$I_s = \frac{F_\odot}{{R_h}^2}\frac{\omega}{4\pi}\frac{\mu_0}{\mu + \mu_0}P(\alpha)(1 + B(\alpha)) \tag{26}$$

where

$$B(\alpha) = \frac{B_0}{1 + [\tan(\alpha/2)]/h} \tag{27}$$

where h is a parameter related to the angular half-width of the opposition effect defined through the relation

$$h = -\frac{3}{8}\ln(1 - F_F)Y \tag{28}$$

where F_F is the fraction of the surface volume occupied by particles (typically 0.5 for planetary surfaces) and Y is a complex function strongly dependent upon the particle size distribution. If the size distribution, $n(r)$, can be defined by $n(r) = Kr^{-\beta}$ then for integral values of β analytical solutions exist if the ratio of the size of the largest particle to that of the smallest particle (r_l/r_s) can be defined (Table 1).

Table 1. Y parameter for values of β defining a particle size distribution according to $Kr^{-\beta}$.

β	Y
0	$4/3\sqrt{3}$
1	$3/\sqrt{8ln(r_l/r_s)}$
2	$2\sqrt{r_s/r_l}$
3	$\sqrt{2}(ln(r_l/r_s))^{3/2}(r_s/r_l)$
4	$\sqrt{3}/ln(r_l/r_s)$
5	$1/\sqrt{2}$

Values of h between 0.01 and 0.1 appear appropriate for Solar System bodies ([5]). Also,

$$B_0 = \frac{S(0)}{\omega P_0}. \tag{29}$$

Here $S(0)$ is the fraction of light scattered at zero phase. Including multiple scattering, we arrive at the equation

$$\rho_C(\alpha, \epsilon, \iota) = \frac{\omega}{4}\frac{1}{\mu + \mu_0}([1 + B(\alpha)]P(\alpha) + H(\mu_0)H(\mu) - 1) \tag{30}$$

Figure 4 shows three solutions to this equation for two different phase functions and two different values of h. The strong dependence of the phase curve on the phase function is evident (as might be expected) but the sensitivity of the opposition effect to h is extremely strong. Physically, h is related to the ratio of the mean particle radius to the extinction length and is therefore related to the porosity of the surface.

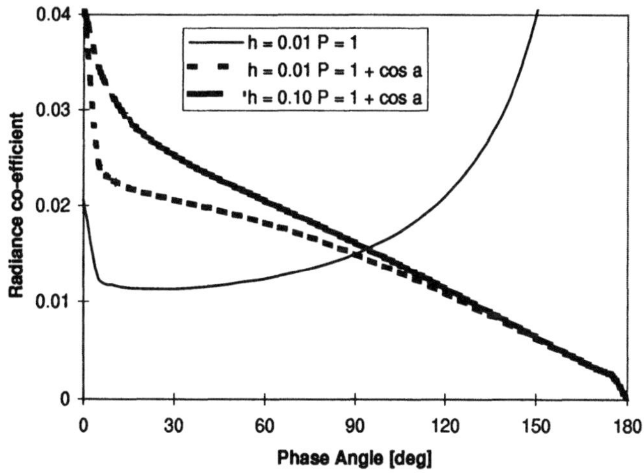

Fig. 4. Three solutions to Hapke's equations for multiple scattering incorporating backscattering (but excluding macroscopic roughness).

4.3 Macroscopic Roughness

In 1984, Hapke developed a series of equations which could be used to correct for the effects of large scale roughness in quantifying the bidirectional reflectance. Although the derivation of these equations is complicated, the resulting formulae are relatively simple involving only a small number of free parameters. The equations can be summarized as follows.

The mean slope angle with respect to the normal of a planar surface, Θ, is defined by

$$\tan\Theta = \frac{2}{\pi}\int_0^{\pi/2} \tan\theta a(\theta) d\theta \tag{31}$$

where $a(\theta)$ is a probability distribution such that

$$\int_0^{\pi/2} a(\theta) d\theta = 1 \tag{32}$$

f is the fraction of the surface in shadow (therefore unilluminated) but also hidden from view (which Hapke refers to, somewhat confusingly, as a visibility shadow). This is assumed to be a function of an azimuth angle, Φ defined from α, ι, and ϵ through

$$\cos\Psi = \frac{\cos\alpha - \cos\iota\cos\epsilon}{\sin\iota\sin\epsilon}. \tag{33}$$

and $0 \leq \Psi \leq \pi/2$.

When $\iota \leq \epsilon$

$$\rho_C(\alpha,\iota,\epsilon) = \tfrac{\omega}{4\pi}\tfrac{\mu_0'}{\mu_0'+\mu'}[[1+B(\alpha)]P(\alpha) - 1 + H(\mu_0')H(\mu')]$$
$$\frac{\mu'\mu_0}{\mu'^0\mu_0'^0\sqrt{1+\pi\tan^2\Theta}[1-f+f(\mu_0/\mu_0'^0\sqrt{1+\pi\tan^2\Theta}]} \tag{34}$$

where

$$f(\Psi) = e^{-2\tan(\Psi/2)} \tag{35}$$

$$\mu_0' = \frac{1}{\sqrt{1+\pi\tan^2\Theta}}(\cos\iota + \sin\iota\tan\Theta\frac{\cos\Psi e^{-\frac{\cot^2\Theta\cot^2\epsilon}{\pi}} + \sin^2\frac{\Psi}{2}e^{-\frac{\cot^2\Theta\cot^2\iota}{\pi}}}{2 - e^{-\frac{2}{\pi}\cot\Theta\cot\epsilon} - \frac{\Psi}{\pi}e^{-\frac{2}{\pi}\cot\Theta\cot\iota}}) \tag{36}$$

$$\mu' = \frac{1}{\sqrt{1+\pi\tan^2\Theta}}(\cos\epsilon + \sin\epsilon\tan\Theta\frac{e^{-\frac{\cot^2\Theta\cot^2\epsilon}{\pi}} + \sin^2\frac{\Psi}{2}e^{-\frac{\cot^2\Theta\cot^2\iota}{\pi}}}{2 - e^{-\frac{2}{\pi}\cot\Theta\cot\epsilon} - \frac{\Psi}{\pi}e^{-\frac{2}{\pi}\cot\Theta\cot\iota}}) \tag{37}$$

$$\mu_0'^0 = \frac{1}{\sqrt{1+\pi\tan^2\Theta}}(\cos\iota + \sin\iota\tan\Theta\frac{e^{-\frac{\cot^2\Theta\cot^2\iota}{\pi}}}{2 - e^{-\frac{2}{\pi}\cot\Theta\cot\iota}}) \tag{38}$$

$$\mu'^0 = \frac{1}{\sqrt{1+\pi\tan^2\Theta}}(\cos\epsilon + \sin\epsilon\tan\Theta\frac{e^{-\frac{\cot^2\Theta\cot^2\epsilon}{\pi}}}{2 - e^{-\frac{2}{\pi}\cot\Theta\cot\epsilon}}) \tag{39}$$

When $\iota \geq \epsilon$

$$\rho_C(\alpha,\iota,\epsilon) = \tfrac{\omega}{4\pi}\tfrac{\mu_0'}{\mu_0'+\mu'}[[1+B(\alpha)]P(\alpha) - 1 + H(\mu_0')H(\mu')]$$
$$\frac{\mu'\mu_0}{\mu'^0\mu_0'^0\sqrt{1+\pi\tan^2\Theta}[1-f+f(\mu/\mu'^0\sqrt{1+\pi\tan^2\Theta}]} \tag{40}$$

where μ_0' and μ' are re-defined by

$$\mu_0' = \frac{1}{\sqrt{1+\pi\tan^2\Theta}}(\cos\iota + \sin\iota\tan\Theta\frac{e^{-\frac{\cot^2\Theta\cot^2\iota}{\pi}} + \sin^2\frac{\Psi}{2}e^{-\frac{\cot^2\Theta\cot^2\epsilon}{\pi}}}{2 - e^{-\frac{2}{\pi}\cot\Theta\cot\iota} - \frac{\Psi}{\pi}e^{-\frac{2}{\pi}\cot\Theta\cot\epsilon}}) \tag{41}$$

$$\mu' = \frac{1}{\sqrt{1+\pi\tan^2\Theta}}(\cos\epsilon + \sin\epsilon\tan\Theta\frac{\cos\Psi e^{-\frac{\cot^2\Theta\cot^2\iota}{\pi}} + \sin^2\frac{\Psi}{2}e^{-\frac{\cot^2\Theta\cot^2\epsilon}{\pi}}}{2 - e^{-\frac{2}{\pi}\cot\Theta\cot\iota} - \frac{\Psi}{\pi}e^{-\frac{2}{\pi}\cot\Theta\cot\epsilon}}) \tag{42}$$

5 Conclusions

The amount of light reflected from a surface in a specified direction depends strongly upon the properties of the surface. However, the derivation of surface parameters from photometric measurements is not trivial and is only now being treated rigorously through the application of Hapke's parameters.

When studying this subject (and also when writing on this subject), it is vitally important to be clear about definitions because authors are not consistent either in their use of terms or in the definition of symbols (particularly with respect to the use of πF to describe the solar flux). On the other hand, recent developments show that useful information about surface properties can be gained by careful application of the relevant formulae.

References

1. Chandrasekhar, S. (1960) Radiative Transfer. Pub. Dover Publications, New York,
2. Hanel
3. Hapke, B., (1981) Bidirectional reflectance spectroscopy 1. Theory, *J. Geophys. Res.*, *86*, 3039–3054
4. Hapke, B., (1984) Bidirectional reflectance spectroscopy 3. Correction for macroscopic roughness, *Icarus*, *59*, 41–59
5. Hapke, B., (1986) Bidirectional reflectance spectroscopy 4. The extinction coefficient and the opposition effect, *Icarus*, *67*, 264–280
6. Judd, D. B., (1967) Terms, definitions, and symbols in reflectometry, *J. Opt. Soc. Am.*, *57*, 445–452
7. Tomasko, M.G., (1976) Photometry and polarimetry of Jupiter In *Jupiter. Pub. Univ. Arizona Press, Tucson. Ed. T. Gehrels*, 486–515
8. Veverka, J., P. Thomas, T.V. Johnson, D. Matson, and K. Housen, (1986), The physical characterisitics of satellite surfaces, In *Satellites. Pub. Univ. Arizona Press, Tucson. Ed. J.A. Burns and M.S. Matthews*, 342–402

Mercury – Goals for a Future Mission

Nicolas Thomas

Max-Planck-Institut für Aeronomie, Max-Planck-Str. 2,
D-37189 Katlenburg-Lindau, Germany

Abstract. Mercury is an end-member in our Solar System being the most dense planet and the closest to the Sun. It is also arguably the least known planet with only 50% of its surface mapped by Mariner 10, limited data on its magnetic field and exosphere, and sparse Earth-based observational data. In the coming years, however, this situation will change as the importance of a detailed investigation of Mercury to complete our initial survey of the Solar System becomes more recognized. The European Space Agency has indicated their interest by making a Mercury Orbiter mission a "cornerstone"of their programme. In this brief article, we shall point out some of the interesting phenomena associated with Mercury and describe some of the goals future missions to Mercury should seek to achive.

1 Introduction

Given the huge attention paid to the Pluto-Charon system in the past 10 years, it is probably fair to say that even though Mercury has been visited by a spacecraft (in fact, three times by the same spacecraft), it is the planet we know least about. The spacecraft in question, Mariner 10, made fly-bys of Mercury in 1974 and 1975 and while remarkable discoveries were made, the following 25 years have seen huge advances in Man's remote sensing capabilities (using instruments such as the Hubble Space Telescope, HST) such that the properties of Pluto appear fairly well-known. Mercury, on the other hand, despite being 30 times closer to the Earth, is an astronomer's nightmare.

At maximum elongation, Mercury is only 28° from the Sun. HST, in common with a whole host of other orbiting observatories (e.g. Extreme Ultraviolet Explorer, Infrared Space Observatory), is not allowed to observe that close to the Sun. Observations in twilight from the ground require telescopes to be pointed at zenith angles of greater than 70° which is often close to or above hardware limits. Even when allowed, the observing time available is just minutes. The surface brightness of Mercury is also so bright that it can exceed the flux limits of some instruments (SOFI at the European Southern Observatory, for example). These difficulties and the pre-occupation of planetary scientists with other objects (e.g. the Jovian system and Mars) have meant that our knowledge of Mercury remains, at best, sketchy.

In this brief article, I will discuss some of the questions that a future mission to Mercury can address. Many of the points were raised in the summary presented by [9]. However, since that book was completed, new observations have revealed further interesting phenomena which also warrant detailed investigation. Particular emphasis will be placed on these new results in the later sections of the article.

Table 1. Properties of Mercury and its orbit.

Property	Value
Mean distance from Sun	0.3871 AU
Periapsis distance	0.3075 AU
Eccentricity	0.20564
Inclination	7.005
Mean orbital velocity	47.89 km s^{-1}
Sidereal period	87.969 days
Rotation period	58.65 days
Inclination of orbit to equator	0°
Mean density	5.42 g cm^{-3}
Mass	0.0558 Earth masses = 0.333 10^{24} kg
Radius	0.382 Earth radii = 2436 km
Escape velocity	4.3 km s^{-1}

2 Mercury's Orbit

The basic parameters of Mercury and its orbit are given in Table 1. An important discovery of Mariner 10 was Mercury's spin-orbit coupling. Mercury rotates on its axis three times during the time it takes to complete two orbits around the Sun. Mariner 10's heliocentric orbit was such that a fly-by was possible every two Mercury orbits. Thus, when Mercury was encountered (three times in total), the same hemisphere was illuminated. As a result, Mariner 10 could only map 50% of the surface of the planet. Full coverage of this hemisphere was acquired at 1-1.5 km resolution, while a small fraction (1%) was obtained at resolutions of 100 to 500 m. A further disadvantage of the fly-by geometry was that most of the data were acquired at phase angles between 80° and 100° which implies that little topographic information can be extracted from the data. The quality and coverage of the imaging data is comparable to Earth-based telescopic observations of the Moon before spaceflight.

The orbit of Mercury is also relatively eccentric for a planet (Figure 1). As a result, the sub-solar point surface temperature on the planet varies by more than 100 K between perihelion and aphelion (Figure 2). However, the low inclination and the orientation of the spin axis means that temperatures in the polar regions can dip well below 273 K.

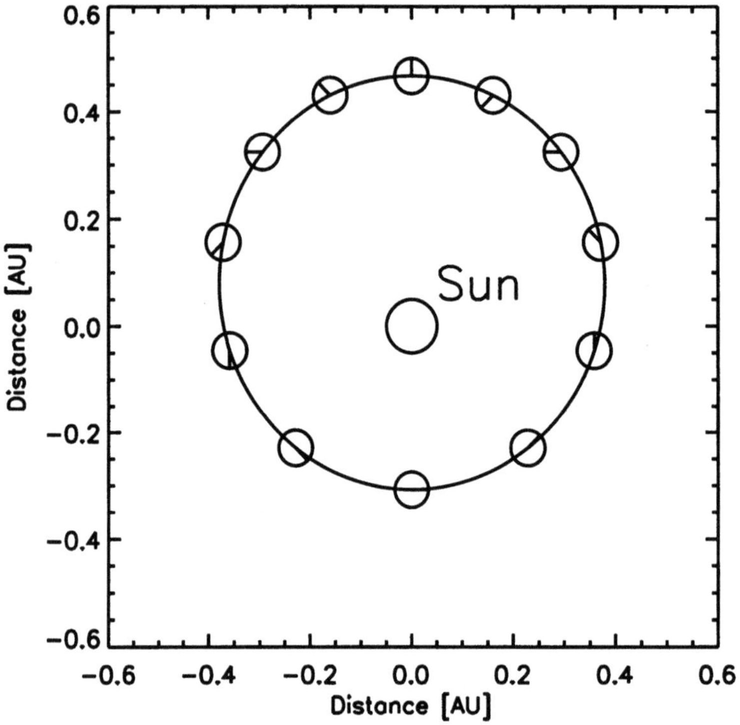

Fig. 1. Mercury's orbit projected onto the ecliptic viewed from the northern ecliptic direction. Mercury moves anti-clockwise. A mark fixed to Mercury is directed towards the anti-solar direction when Mercury is at its maximum y-value. Following Mercury around its orbit, the mark shows how Mercury's rotation brings a surface element to the sub-solar meridian every two orbits. The orbit of Mercury is clearly eccentric.

3 The Structure of the Surface

The surface of Mercurys is heavily cratered but less so than ancient areas on the Moon. This suggest that Mercury may have been re-surfaced at an early stage in its history. The surface shows many inter-crater plains. Crater counts show that these plains have different ages which further indicates a gradual re-surfacing process. However, there is no unambigious evidence of volcanic activity in the past.

Mercury is not tectonically active now. However, its early history would have seen strong forces be exerted on the surface layers. With the planet being close to the early Sun, the heating would have been extreme. However, Mercury then began to cool and tidal forces started to reduce the rotation rate to bring it into the spin-orbit resonance. The de-spinning of the planet led to a reduction in the polar flattening. This induced surface stresses. Evidence of this process, in the form of lineaments in the lithosphere, is seen in the

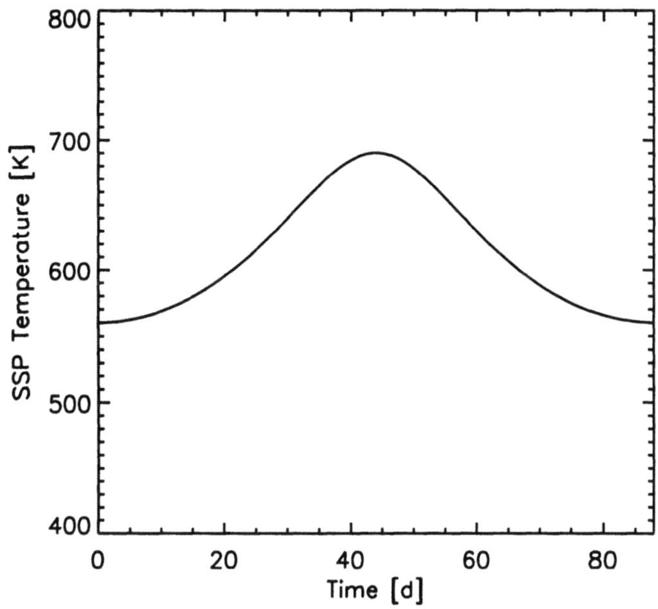

Fig. 2. The sub-solar point temperature on Mercury varies strongly because of the eccentricity of its orbit. Modelled albedo of 0.14 and an infrared emissivity of 0.9.

Mariner 10 data. The directions of the lineaments appear to be strongly correlated with latitude. Cooling of the planet may have reduced the radius by as much as 8 km (0.3%) producing scarps as thrust faulting occurred. Again Mariner 10 images show scarps up to 2 km high and 600 km long over the surface. Subsequently, impact cratering has been responsible for modifying the surface. The analysis of craters on Mercury is affected by the special properties of the planet. For example, crater structure might be affected by the higher temperature which could cause more effective viscous relaxation. The viscosity will vary strongly over the surface because of the high temperature gradients between the sub-solar point and the poles. Hence, the observed crater structure for an identical impact may vary with latitude. A particularly remarkable set of features on Mercury are the "hills"at the exact antipode of Caloris basin. The impact producing Caloris led to shock waves which focussed at the antipode leading to an irregular structure of hills and troughs between 100 and 1800 m high.

Caloris itself, which was half in shadow during the Mariner 10 fly-bys, shows some multiple ring structures. Several multi-ring basins are evident on the surface (e.g. Tolstoj basin) but some are not easily identified and, in any case, their number is limited by comparison with the Moon, for example. The formation mechanism of multi-ring basins is still unknown but the "rock-tsunami"concept remains plausible and one of the easiest mechanisms to imagine. There are four possible reasons for the relative lack of multi-ring basins. Firstly, the lithosphere may have been rather thick at the time of the impacts. Secondly, the rings produced might have subsided. Thirdly, the rings may have undergone viscous relaxation. Finally, the formation of intercrater plains may have been so extensive that rings were "washed out". Future data and studies of the rheology and surface structure of Mercury will undoubtedly help address some of these issues in much more detail.

4 Mineralogy and Volatiles

The global variation of surface chemistry and mineralogy on Mercury is basically unknown. There are limited albedo variations and some recent work on Mariner 10 data has suggested that there are different mineralogical units evident on the surface. Some of the global variations in albedo may be related to radiation darkening although this is speculative. Low resolution spectra show slight differences between the Moon and Mercury with Mercury exhibiting a higher spectral reflectance near 1 micron, the wavelength range associated with orthopyroxene (Fe^{2+}) absorptions.

A major topic of interest is the question of polar volatiles. Some regions at the poles, the bottoms of craters for example, may be in permanent darkness and other areas might also be cool enough to allow condensation of volatile. Sub-surface volatiles are even expected and cold traps may be present.

Delay-Doppler radar images appear to confirm the presence of polar ices [2]. These images show anomalous bright features at the poles which also have unusual polarization properties. The south polar feature is confined to the floor of a crater (Chao Meng-Fu). Unusual polarization and bright reflectivity is a characteristic of the icy Galilean satellites - hence the suggestion that reflections are surface (or near surface) frosts in permanently dark polar regions ([4]).

5 The Exosphere

The presence of an exosphere on Mercury was first established by the Mariner 10 ultraviolet spectrometer ([1]). Emission at 584 $\AA$ from neutral helium was detected above the surface with a rapidly decreasing number density with height. The precise origin of the helium is unknown but it is probably solar and transported to Mercury's surface by the solar wind.

The detection of Na emission from the vicinity of Mercury was the next major discovery. As with the Io neutral clouds and the comet neutral tails (see sections other chapters) Na is a trace species but extremely bright because of the efficiency of resonance scattering of sunlight at optical wavelengths.

There are several candidate mechanisms for the production of the Na gas ([6]). These include

- micro-meteroid impact,
- diffusion from sub-surface reservoirs,
- photon-stimulated desorption,
- energetic charged particle sputtering,
- and chemical sputtering.

Chemical sputtering is an interesting new idea with possible application to other Solar System bodies. In this case, an impactor reacts chemically with the surface to release another species. An example for Mercury ([6]) would be

$$2H + Na_2SiO_3 = 2Na + SiO_2 + H_2O \tag{1}$$

It was assumed previously that energetic particle sputtering of the surface ejecting He was producing the exosphere observed by Mariner 10. Although several other mechanisms for Na have been suggested, these alternatives do not appear to be attractive possibilities for He because of its low atomic weight and inertness.

Clues to the nature of the exosphere are now evident in the observed spatial distribution of Na emission. These measurements seem to show that Na emission is concentrated towards the poles ([7]) although how this observation should be interpreted is rather unclear. The emission is also variable on timescales of hours, which is somewhat surprising considering the slow rotation period of the planet. Potassium emission has also been detected but, rather like the situation at Io, the brightness variations track those of Na but at a much lower intensity.

6 The Magnetic Field

One of the most important discoveries made by Mariner 10 was that of Mercury's internal magnetic field. The dipole moment was found to be around 300 nT R_m^3 ([3]) although there is considerable uncertainty because of the limited data set available. The field strength is not particularly strong but is sufficient to keep the solar wind from the surface under normal conditions. The apex of the field is however within the Roche limit. It remains unclear, however, how often and where the solar wind actually reaches the surface and what effects result. The magnetosphere may have a substantial heavy ion population because of the interaction between energetic particles and the surface. As the solar wind changes, this population must be affected in some way but it is not known how.

Although the magnetosphere is similar in many ways to the Earth's ([8]), the mechanism producing the field may be very different. Mercury's size presents a significant problem because thermal evolution models indicate that the inner core region should have solidified during its early history (although unusual mechanisms to prevent this occurring have been suggested). Thus, generation of the field by a dynamo is not so straightforward. A remnant magnetization is a plausible.

It is important to note that the cusps of the magnetic field are at the poles. This is precisely where the radar anomalies are. Thus, energetic particles spiralling down the field lines may impact the surface or near-surface ices leading to an exosphere comprising ice species or their dissociation products (see e.g. [5]).

7 ESA's Mercury Orbiter Mission – Bepi Colombo

A further spacecraft visit to Mercury is long overdue. Close observations would allow the detailed investigation of an end-member (closest to the Sun, highest bulk density) in our planetary system. The phenomena associated with its surface, exosphere, and magnetic field make it an interesting target for a range of scientists from many different disciplines. The principle objectives can be stated easily and are extremely broad because so little is known.

- What does the unimaged side of Mercury look like and what processes have influenced its surface?
- What is the chemical and mineralogical composition of the surface?
- What is the chemical composition of the atmosphere and how is it generated?
- How is Mercury's magnetic field generated and how does it interact with the solar wind?

The Horizon 2000+ programme of the European Space Agency (ESA) includes a Mercury Orbiter mission as the pre-selected planetary "cornerstone" mission, and as such one of the most prestigious elements of ESA's programme. The mission has been re-named Bepi Colombo and selected for launch in 2004. The mission is currently being defined and studies are being carried out to prepare technologies that will be needed to complete the mission. Of particular concern is the harsh thermal environment with a solar flux 10 times greater than at the Earth and a high thermal emission from the planet itself. However, the mission definition is proceeding rapidly and an Announcement of Opportunity for provision of experiments ist expected early in 2002. The main objectives are

- Investigation of the unimaged hemisphere.
- Study the mineralogical and chemical composition of the surface.
- Investigate the distribution of surface volatiles.
- Study the intrinsic magnetic field and how it is produced.
- Study the internal structure and gravity field.
- Probe the interaction between the magnetic field and the solar wind.

- Study the neutral "atmosphere"and its interaction with the solar wind and the magnetic field.
- Place further constraints on theories of general relativity.

Currently, the mission comprises two interplanetary cruisers powered by solar- electric propulsion. The first cruiser brings a planetary orbiter which will enter low orbit and point continuously towards the nadir for the surface studies. The second cruiser will drop a lander onto the surface. A magnetospheric orbiter will then be release from the propulsion state. This magnetically clean sub-satellite is brought into an elliptical orbit ideal for studies of the energetic particle environment.

Although this concept may change, ESA has made it clear that Europe takes the study of Mercury very seriously and that a mission will be launched in the next decade. Young planetary scientists should take note.

8 Conclusions

Mercury is a remarkably interesting object and deserves considerably more attention than it has received in the past few years. The surface and its magnetic field indicate an unusual history, while its exosphere and magnetosphere show that it is a dynamic planet. Missions to Mercury will be launched in the coming few years and they offer a major opportunity to make fundamental new discoveries about a neighbour and an end-member in our Solar System.

References

1. Broadfoot, A.L., D.E. Shemansky, and S. Kumar, (1976) Mariner 10: Mercury atmosphere, Geophys. Res. Lett., **3**, 577-580.
2. Butler, B.J., Muhleman, D.O. and Slade, M.A., (1993) Mercury - Full-disk radar images and the detection and stability of ice at the North Pole, J. Geophys. Res., **98**, 15003.
3. Connerney, J.E.P. and N.F. Ness, (1988) Mercury's magnetic field and interior, In "Mercury"Ed. F. Vilas, C.R. Chapman, and M.S. Matthews, Pub. Univ. Arizona Press, Tucson, Az, pp. 494-513
4. Harmon, J.K., Slade, M.A., Velez, R.A., Crespo, A., and Dryer, M.J. and Johnson, J.M., (1994) Radar mapping of Mercury's polar anomalies, Nature, **369**, 213-215.
5. Killen, R.M., Benkhoff, J. and Morgan, T.H., (1997) Mercury's Polar Caps and the Generation of an OH Exosphere, Icarus, **125**, 195-211.
6. Morgan, T.H. and Killen, R.M., (1997) A non-stoichiometric model of the composition of the atmospheres of Mercury and the Moon, Planet. Space Sci., **45**, 81-94.
7. Potter, A.E. and T.H. Morgan, (1997) Sodium and potassium atmospheres of Mercury, Planet. Space Sci., **45**, 95-100.

8. Russell, C.T., D.N. Baker, and J.A. Slavin (1988) The magnetosphere of Mercury, In "Mercury" Ed. F. Vilas, C.R. Chapman, and M.S. Matthews, Pub. Univ. Arizona Press, Tucson, Az, pp. 514-561
9. Vilas, F., C.R. Chapman, and M.S. Matthews (Eds.), (1988) Mercury, Pub. Univ. Arizona Press, Tucson, Az.

Physical Processes Associated with Planetary Satellites

Nicolas Thomas

Max-Planck-Institut für Aeronomie, Max-Planck-Str. 2,
D-37189 Katlenburg-Lindau, Germany

Abstract. The study of the formation, diversity, evolution, and present state of planetary satellites is one of the most interesting and challenging subjects in planetary physics. Here we discuss some of the physical processes affecting the shapes, surfaces and atmospheres of planetary satellites and illustrate them with examples from throughout the Solar System. In particular, satellite sphericity, cratering, resurfacing, the effects of surface frosts, the presence of tenuous atmospheres, and satellites as sources of heavy ions in planetary magnetospheres are discussed.

1 Introduction

The satellites of the planets in our Solar System exhibit remarkable diversity. This diversity is evident in their sizes, surface compositions and atmospheric pressures, amongst many other properties. They also differ in how they interact with their "parent"planet and their immediate environment. The extent of background knowledge in physics and chemistry required to treat many of the problems in the study of planetary satellites makes this a challenging and fascinating subject which cannot be covered adequately in a short review. We shall therefore look at a few of the major processes affecting planetary satellites and illustrate them with examples from throughout the Solar System.

2 Satellite Classification and the Exceptions

[3] classified planetary satellites into four categories (Table 1). Regular satellites are large, spherical, and relatively close to the parent. They are in prograde orbits with low inclination and eccentricity and often in synchronous rotation (i.e. the same hemisphere always faces the parent). Their masses are small compared to the parent.

Collisional shards are small, irregularly-shaped objects, extremely close to the planet or co-orbital with a regular satellite. They have prograde orbits with essentially zero inclination and eccentricity. They are thought to be the "debris"left over after planet and regular satellite formation.

Irregular satellites are far from the parent and have substantial inclination, eccentricity, or both. They are small. They are thought to have been

Table 1. Classification of planetary satellites.

	Earth	Mars	Jupiter	Saturn	Uranus	Neptune	Pluto
Regular			Io	Mimas	Miranda		
			Europa	Enceladus	Ariel		
			Ganymede	Tethys	Umbriel		
			Callisto	Dione	Titania		
				Rhea	Oberon		
				Titan			
				Hyperion			
				Iapetus			
Shards		Phobos	Metis	Atlas	Cordelia	Naiad	
		Deimos	Adrastea	Prometheus	Ophelia	Thalassa	
			Amalthea	Pandora	Bianca	Despina	
			Thebe	Epimetheus	Cressida	Galatea	
				Janus	Desdemona	Larissa	
				Telesto	Juliet	Proteus	
				Calypso	Portia		
				Helene	Rosalind		
					Belinda		
					Puck		
Irregular			Leda	Phoebe	Caliban	Nereid	
			Himalia	+IO	Sycorax		
			Lysithea	other			
			Elara	cadidates			
			Ananke				
			Carme				
			Pasiphae				
			Sinope				
Unusual	Moon					Triton	Charon

captured into their present orbits (although it has to be said that a reasonable mechanism for this process remains to be demonstrated).

Unusual satellites do not fall into any of the other categories. Both the Moon and Pluto's satellite, Charon, are excluded from the regular satellite category because they are relatively large compared to the parent and thus their evolution must have been more complicated than would have occurred in a proto-planetary disc structure (as is assumed for the Galilean satellites, for example). The origin of the Moon is unknown but the concept of it being the result of a major impact on the Earth is still the leading hypothesis (e.g. [1]). Because of our lack of knowledge, discussions of Charon's origin remain purely speculative at the present time.

Triton is unusual because physically it resembles a regular satellite but its orbit is retrograde. (It was once thought that the orbit was also unstable but it now appears that Triton will only approach Neptune well after the end of the Sun's main sequence lifetime; [21].) Theories on the origin of Triton are numerous but none are particularly satisfying ([21]). Some of the possibilities which have been considered include,

a. Triton as a perturbed regular satellite but this requires a massive unknown body entering the Neptune-Triton system.

b. Capture of Triton from solar orbit by gas drag.

c. Capture by tidal dissipation but this requires a close pass which might in turn lead to tidal disruption of the satellite and also requires some means of raising the periapsis to the present value.

d. Capture of Triton by collision with a regular Neptunian satellite.

e. 3-body capture using the fortuitous circumstance of a third object in heliocentric orbit in the proximity of Neptune.

f. Mass change of Sun and/or Neptune at the time of capture but this is only achievable during the accretionary epoch because the temporary capture time of Triton is only around 10^4 years.

3 Sphericity

Sphericity is a common property of all satellites larger than 200 km. Mimas is the smallest spherical satellite (radius, $r = 196$ km) while Hyperion (semi-axes of 205 km x 130 km x 110 km) and Amalthea (135 km x 83 km x 75 km) are the two largest irregular satellites. Sphericity is a consequence of gravity overcoming the structural strength of the material which built the satellite. [11] have considered the balance between these two forces in asteroids. The pressure at the centre of a satellite, P, can be estimated by integrating the product of the mass and the gravational acceleration of material in a column above the centre of the satellite, leading to the equation,

$$P = \frac{2}{3}\pi G \rho^2 r^2, \tag{1}$$

where ρ is the bulk density in [kg m^{-3}]. Table 2 gives 4 examples of the internal pressures computed from this formula.

One can estimate crudely the tensile strength from the bulk modulus of rocky material (typically 10^{11} Pa) and compare this to the internal pressure. This comparison suggests that typical tensile strengths of around 10 MPa are exceeded within bodies of $r \geq 150$ km. Bodies of this radius or greater therefore re-structure and equilibrate themselves as the interior fractures and flows under pressure. This process may be influenced by heat generated by radioactive decay or by the gravitational energy released during formation. One can show, however, that the latter is of limited importance for 100 km sized bodies by integrating the energy acquired by a satellite from the gravitational in-fall of material from infinity onto the surface taking into account the increasing gravitational acceleration as the body grows. For uniform density, the analytical solution for the total kinetic energy, E_T gained is

$$E_T = \frac{16}{15}\pi^2 \rho^2 G r^5 \tag{2}$$

(c.f. [25] from which one can determine the temperature rise expected using an estimate for the specific heat capacity, θ. A typical value for θ would be 1.7×10^3 J kg^{-1} K^{-1} (Lewis, 1997) leading to very small temperature rises for bodies of size $r \approx 100$ km. Radioactive decay heating provides 4×10^{-11} W kg^{-1} which suggests that over periods of $> 10^8$ years higher temperatures could be reached. However, the smaller the body, the larger the surface to volume ratio, and hence cooling via conduction and thermal emission should be rather efficient.

Table 2. Internal pressures of selected planetary satellites.

Satellite	Radius [km]	Density [g cm^{-3}]	Mass [10^{19} kg]	Int. Pressure [MPa]
Mimas	196	1.44	4.55	10.9
Enceladus	250	1.13	7.40	12.6
Iapetus	730	1.15	188.	100.
Europa	1569	3.01	4870.	3180.

4 Roche's Limit and Tidal Forces

The sphericity of a satellite is distorted by tidal forces. The equilibrium figure of a synchronously rotating, fluid satellite on a circular orbit distorted by tidal and centrifugal forces is roughly a tri-axial ellipsoid at large planetocentric distances. The long axis of the ellipsoid is always aligned with the object–planet line. However, at Roche's limit (a_R), no closed equipotential surface for the satellite exists. Roche's limit is given by,

$$a_R = 2.456(\rho_p/\rho_s)^{1/3} r_p, \tag{3}$$

where ρ_p and ρ_s are the densities of the planet and satellite, respectively, and r_p is the radius of the planet.

Roche's limit is also important when an object (such as a comet) comes too close to a planet. An example is comet Shoemaker-Levy 9 which is thought to have broken up during a very close pass to Jupiter two years before it actually impacted the planet. It needs to be stressed that the limit does not apply to small moons held together by, for example, Van der Waal's forces rather than self-gravity. Thus, Metis (J16), for example, either has a fairly high density (≥ 3.48 g cm^{-3}), which is somewhat unlikely, or it is held together by chemical forces. Several collisional shards inside the orbits of regular satellites are close to or inside Roche's limit and, hence, the collisional evolution and formations of these objects has also been strongly affected by tidal forces.

The orbits of the shards are sometimes in a resonance with a regular satellite (e.g. Hyperion) which indicates that tidal interaction with the larger satellites is also important. The resonance allows the object to survive. The two co-orbitals of Tethys (Calypso and Telesto) are an extreme example. They must remain close to the leading and trailing Lagrangian points otherwise they would either collide with Tethys or be ejected into another orbit with possible catastrophic results if a close encounter with Saturn were then to arise (see chapter by Murray).

5 Cratering and Re-surfacing

In extreme cases, tidal forces can provide a major source of internal heat dissipation for a satellite. Io is the classic example of this process. A bulge greater than 100 m high is raised on Io by Jupiter. This bulge is distorted by the resonant interaction between Io and Europa (and, to lesser extent, Ganymede). Europa forces Io's orbit to become slightly eccentric. The equilibrium position for the bulge on Io is directed towards the centre of Jupiter. The orbital eccentricity takes the bulge away from the equilibrium position. The bulge tries to re-equilibrate producing huge amounts of heat through friction. The total heat generated by this process is greater than 10^{14} W or 2 W m^{-2}. For comparison, the measured heat flow for the Moon is around 0.017 W m^{-2} ([14]). Prior to the Voyager 1 fly-by, [22] correctly predicted that the magnitude of the internal dissipation on Io would be sufficient to produce active volcanism. The Voyager 1 images (Plate 3) showed a surface totally devoid of impact craters ([13] suggesting re-surfacing rates in excess of 1 mm yr^{-1}.

Re-surfacing rates calculated from the absence of craters critically depend upon our knowledge of the cratering rate and its variation with time. It is generally assumed that after their initial formation, the planets and their satellites experienced a period of heavy bombardment which ended around 3.5 billion years ago. During this time, the existing large bodies swept up small objects left over from planetary formation. The cratering rate subsequently slowed down to today's present rate.

In addition to being time variable, the impact characteristics must be position dependent. In the outer Solar System, the relative velocities between the impactors and the impacted are slower. This reduces the impacting energy and affects the morphology of the resulting crater. However, the velocities are increased by the potential well of a planet. [26] show how impact velocities on the Galilean satellites rise steeply as one moves inward from Callisto to Io. The orbital velocities of the satellites also affect the crater statistics and produce asymmetries between the leading and trailing hemispheres.

The nature of the impacted surface also influences the structure of the craters produced by impacts. The surface properties must therefore be taken into account when interpreting the cratering record. Ganymede and Callisto, for example, show a sharp drop-off in the number of craters with diameter $\geq$ 60 km compared to the terrestrial planets. One possibility is that the large craters on these icy surfaces have relaxed. However, the overall crater distribution on these objects suggests that the entire impactor distribution on these satellites is incompatible with the modern terrestrial planet projectile population ([5]). Thus, entirely different populations of projectiles are required depending upon the position within the Solar System. Furthermore, collisional shards produced during satellite formation which ultimately impact satellites form an additional source of craters which may alter significantly the impactor distribution for the giant planet satellites. Therefore, estimates of absolute ages of satellite surfaces from crater statistics remain extremely imprecise. However, cratering does allow a comparison of the relative ages of surface units.

The cratering on Enceladus clearly indicates a range of different ages. Some surfaces show no craters at all at the limit of the Voyager resolution and geologically-recent tectonic activity driven by tidal heating is suspected. It has even been suggested that Enceladus may be active now with this activity also leading (by an unspecified mechanism) to the production of Saturn's E-ring, the density of which is strongly peaked at the orbit of Enceladus ([28]).

Another candidate for recent activity is Triton. [29] have recalculated the cratering rates based upon the size distribution of objects in the trans-Neptunian region. They suggest that Voyager images show crater densities well below the values expected and that Triton must therefore be active. Because of its orbit, tidal heating leading to surface re-structuring would be a possible mechanism.

Miranda shows evidence of a different type of re-surfacing. Grooved and faulted terrain is evident with a low crater density. These young regions might have been produced by upwelling of lighter material as Miranda began to differentiate. Alternatively, it is thought that the change in the axial tilt of Uranus was produced by an impact. The satellite system which existed prior to the impact was then de-stabilised leading to collisions between the satellites and subsequent formation of the present system. The older surfaces on Miranda may be original surfaces while the grooved and faulted terrain may be connected in some way to the collision and re-equilibration process.

6 Tenuous Atmospheres

Re-surfacing by volcanic activity requires high temperatures and volatiles to drive the ejection of material. In the case of Io, for example, the main gases are thought to be SO_2 and S_2. The gases produced then contribute to the atmosphere of the satellite. Io's SO_2 atmosphere was first detected by the Infrared Imaging Spectrometer (IRIS) on Voyager 1 (Pearl et al., 1979) close to the volcanic vent, Loki. The derived pressure was around 10^{-7} bar. This is remarkably close to the equilibrium vapour pressure (EVP) of SO_2 at the temperature of the sub-solar point on Io ($\approx$ 130 K). A logical conclusion was that the atmosphere was in equilibrium with surface ices over the whole satellite. IRIS measured the temperature of the nightside of Io to be around 87 K. The EVP of SO_2 decreases by 7 orders of magnitude from 130 K to 87 K and therefore huge pressure gradients and substantial transport of SO_2 from the sub-solar point towards the poles were predicted. Recent microwave (Lellouch, 1994) and HST observations, however, suggest that the SO_2 detected by IRIS was actually in Loki's volcanic plume and therefore not representative of the ambient atmosphere. The ambient pressure is much lower than predicted by the EVP model because of cold traps. Regional cold trapping of atmospheric gases can occur in two ways. Firstly, assume we have a naturally dark surface with a slightly brighter surface adjacent to it. The dark surface will be warmer because of its lower albedo. Atmospheric gases will therefore condense preferentially on the lighter surface so that the gas pressure will equilibrate at the temperature of the lighter surface. There is a potential positive feedback mechanism here because ice condensates tend to have very high albedos driving the surface temperature down further. The second type of cold trap is the sub-surface cold trap (e.g. [20]). The sub-surface temperature is controlled by the thermal conductivity. If the surface layers are extremely porous and of low thermal conductivity, the sub-surface layers can be much colder than the illuminated surface. Because of the porosity, gas can flow into the sub-surface layers and condense there so that the equilibrium pressure is at the temperature of the sub-surface layer, not the surface.

The atmosphere of Triton may be somewhat similar to an EVP atmosphere but with a subtle twist. The atmosphere was detected by the phase delay in the Voyager 2 X-band radio signal when the spacecraft was occulted by the satellite. Refraction caused by N_2 (which was subsequently detected in gas phase from the ground at 2.16 μm) gave a tangential column density at the surface of 8.8 10^{26} m^{-2} and scale height equivalent to a surface temperature of 38 K. This is consistent with N_2 in equilibrium with the surface. Haze particles were thought to be present at low optical depths although they have limited significance for dynamics of the atmosphere. Like SO_2 on Io, the EVP of N_2 at Triton surface temperatures is strongly temperature dependent. Therefore, one might expect huge pressure gradients. However, the temperatures on Triton are so low that the latent heat released by condensation on the surface is important for the thermal balance. The balance at the surface can be described by

$$m_s c \frac{dT}{dt} = \frac{S \cos\theta}{{R_h}^2}(1 - A_H) - \epsilon\sigma T^4 + L\frac{dm}{dt} + \kappa\frac{dT}{dz}, \qquad (4)$$

where m_s is the mass of the surface layer, T is the temperature, S is the solar constant at 1 AU, θ is the solar zenith angle, ${R_h}^2$ is the heliocentric distance in [AU], A_H is the directional hemispheric albedo, ϵ is the IR emissivity, σ is Stefan's constant, L is the latent heat of sublimation (typically 2 10^6 J kg^{-1} for H_2O, for example), dm/dt is the condensation rate and κ is the thermal conductivity (Figure 1). The magnitude of the condensation term is such that the N_2 frost on the surface is isothermal independent of position (Yelle et al., 1995).

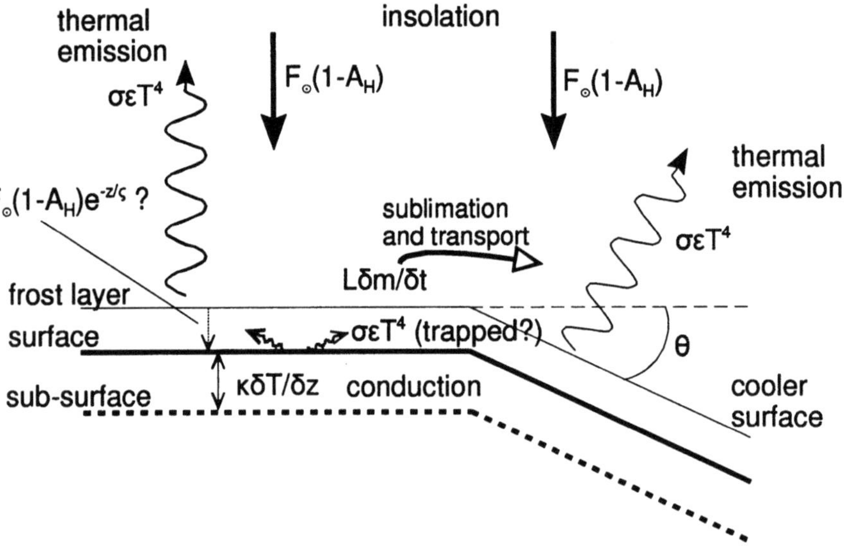

Fig. 1. The physical processes affecting the thermal balance at the surface of a satellite include insolation, thermal emission, and conduction. In addition, surface frosts may sublime or condense (depending upon the temperature) and solid-state greenhouse effects may occur. (Figure modified from [9]).

In EVP systems, there is a net transport of material towards the poles. Triton is particularly interesting here because of its unusual orbit which means that the sub-solar latitude wanders from 50 ° S to 50 ° N over timescales of a few hundred years ([10]). The sub-solar latitude at the present time is very close to its southernmost extreme. Thus, the frost distribution on the surface should be rather mobile. However, the Voyager images of the frost distribution are not consistent with the simple EVP theory. Furthermore, recent evidence has been presented of a doubling of the surface pressure ([7]) which may be caused by the motion of the sub-solar point or by changes in the frost distribution.

Voyager 2 also made the remarkable discovery of eruption plumes on Triton which threw dark material up to an altitude of about 8 km. These columns produced long dark clouds which extended hundreds of kilometres away from the vent. An increase in the number of plumes may result in a darker, more absorbing atmosphere and/or surface leading to higher atmospheric pressures.

7 Solid-State Greenhouse Effect

One theory of the origin of Triton's plumes is the solid-state greenhouse effect. Ice is translucent at visible wavelengths. If an icy surface is also opaque in the thermal IR then a greenhouse is created ([19]) and changes in the vertical temperature distribution of the surface layers can occur. This can result in sub-surface melting of ices and possible phase changes. If the surface ice forms a seal over the sub-surface layers then pressure can build-up which may lead to plume production. Eq. (4) can be adapted to take this into account.

For flow into the interior, the governing equation is,

$$\frac{\delta T(t,z)}{\delta t} = \frac{1}{\rho c}\kappa \frac{\delta^2 T(t,z)}{\delta z^2}. \tag{5}$$

In the case of the solid-state greenhouse, a depth profile for the solar insolation has to be established, e.g.

$$S'(z) = S e^{-z/\xi} \tag{6}$$

where ξ is a scale length. Eq. (5) then becomes,

$$\frac{\delta T(t,z)}{\delta t} = \frac{1}{\rho c}\left(\kappa \frac{\delta^2 T(t,z)}{\delta z^2} - \frac{\delta S'(t,z)}{\delta z}\right). \tag{7}$$

Could this process produce a sub-surface ocean on Europa?

8 Sputtered Exospheres

Triton clearly has a bound atmosphere, albeit extremely tenuous and comparable in pressure to a good vacuum in the laboratory, which is driven by solar insolation. However, tenuous atmospheres can be produced by other means and, in reality, all regular satellites will possess, at the very least, exospheres of neutral atoms. The discoveries of O_2 "atmospheres" on Europa, Ganymede, and most recently, Callisto (Table 3) illustrate this ([8]; [4]).

Table 3. Atmospheric pressures on the Galilean satellites.

	Species	Pressure [bar]
Io	SO_2	10^{-7*}
Europa	O_2	0.4 - 2.5 10^{-11}
Ganymede	O_2	0.2 - 2 10^{10}
Callisto	CO_2	exosphere**

* possible contribution from a volcanic plume

** just discovered by Carlson et al. (1999)

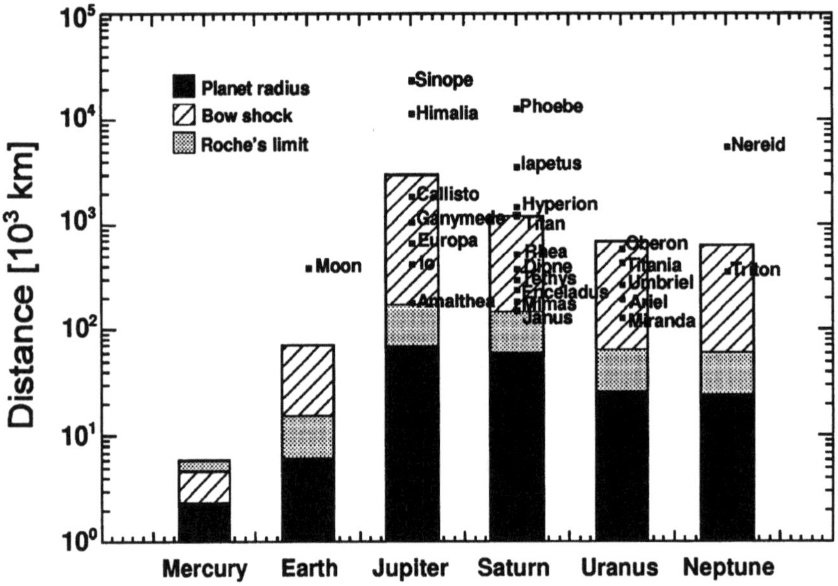

Fig. 2. The semi-major axes of several planetary satellites and their positions relative to Roche's limit and the bow shock of the planet's magnetosphere. Many regular satellites (e.g. Ganymede, Tethys, Ariel, and Triton) are within the magnetosphere of their parent planet. Iapetos, for example, is outside the magnetosphere. Note that Titan is an intermediate case and that Mercury's bow shock position is inside Roche's limit.

All satellites sit in a particle radiation environment whether it be the solar wind or within a planetary magnetosphere (Figure 2). Incident energetic particles can eject atoms from a surface. Yields of $\approx$ 80 ejected atoms per impacting atom have been estimated for surfaces within Jupiter's magnetosphere for example. If the satellite is large, ejected atoms may not have sufficient energy to escape. It is estimated that $<3\%$ of ejected atoms attain the necessary 2.56 km s^{-1} to escape from Io for example. The remainder go to form a bound "atmosphere" (Figure 3). Atmospheres produced by surface sputtering have their exobases at the surface, i.e. the column density is inversely proportional to the collision cross-section and typically equivalent to a surface pressure of $<3\ 10^{-10}$ bar and comparable to the values determined for Europa and Ganymede using HST. It should also be noted that thicker atmospheres formed by other processes (e.g. sublimation of SO_2 at Io) can be scavenged by atmospheric sputtering (Figure 4). Losses from satellite atmospheres in this way can populate the magnetospheres of planets with heavy ions leading to other interesting effects.

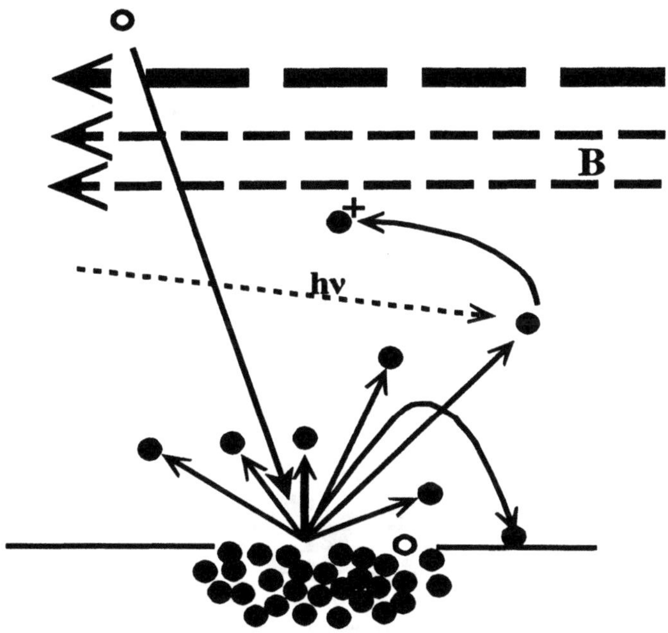

Fig. 3. Schematic diagram of surface sputtering. An external particle (open circle) strikes the satellite and ejects material from the surface. Some of this material falls back to the surface on ballistic trajectories. Other atoms or molecules can undergo photoionization, be picked-up by a passing magnetic field line, and removed entirely from the satellite. The initial impactor will often remain on the satellite.

9 Low Energy Plasma Properties – An Introduction

Charged particles gyrate around field lines with a gyroradius, r_c, of (Figure 5),

$$r_c = \frac{mv_\perp}{qB}, \tag{8}$$

where m is the mass of the particle, $v_\perp$ is the particle's perpendicular velocity, q is its charge and B is the magnetic field strength (see Richardson et al., 1995). The cyclotron frequency, ω, is defined as

$$\omega = \frac{qB}{m}. \tag{9}$$

Particles can also have a velocity, $v_\parallel$, along a magnetic line. The particles therefore execute a spiral motion (called gyromotion) along the field line which has a pitch angle, α defined by,

$$\alpha = \tan^{-1} \frac{v_\perp}{v_\parallel}. \tag{10}$$

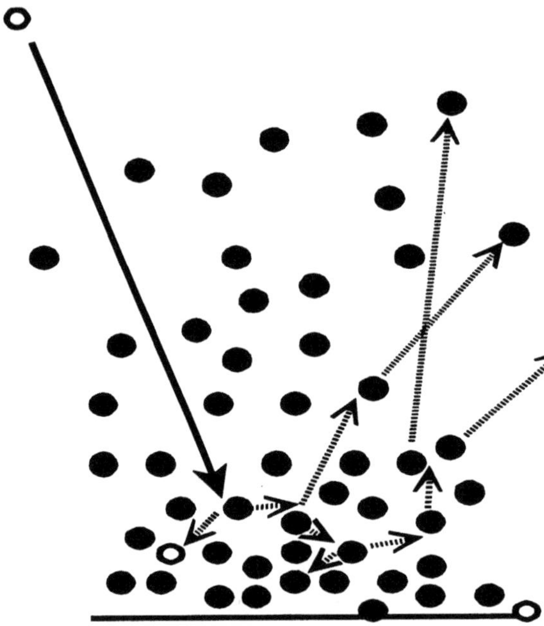

Fig. 4. Schematic diagram of atmospheric sputtering. Here, an external impactor (open circle) enters the atmosphere. Multiple collisions occur near the exobase resulting in the ejection of atmospheric particles. Some of the external impactors remain within the atmosphere after all their original energy has been dissipated.

The speed of the motion allows the definition of a temperature, T. Strictly speaking the definition of T is only possible if the distribution is Maxwellian (Gaussian), i.e.,

$$f(x,v,t) = n(x,t)\frac{m}{2\pi kT}^{3/2} e^{-\frac{mv^2}{2kT}}. \tag{11}$$

However, non-Maxwellians often occur especially when the collision frequency is low (e.g. in the solar wind). Plasma can have different speeds perpendicular and parallel to field lines. If the speeds are the same, the plasma is called isotropic. The guiding centre approximation assumes that the particle does not execute circular motion about the field line but is centred on the field line. This allows us to ignore gyromotion and study only the bulk motion of the plasma.

It is usually assumed that the plasma and the magnetic field move together. This is referred to as the frozen-in condition and is a consequence of Faraday's law,

$$\nabla \times E = -\frac{\delta B}{\delta t}. \tag{12}$$

Plasma therefore co-rotates with the magnetic field at the co-rotation velocity which varies linearly with distance from the planet. For example, Jupiter rotates with an angular velocity of 0.01008 ° s^{-1}. At Europa's distance (9.40

R_J where 1 $R_J = 7.14\ 10^4$ km = 1 Jovian radius), the co-rotation velocity is 118 km s^{-1}.

At typical regular satellite distances from the parent planet, the parent's magnetic field can often be assumed to be a dipole. In this configuration, the equation of a field line is given by,

$$R = R_0 \cos^2 \lambda, \tag{13}$$

where λ is the latitude and R_0 is the distance from the centre of the planet along the magnetic equator (assuming a centred, non-tilted dipole). A convenient number often seen is the McIlwain L-shell defined as,

$$L = \frac{R_0}{R_p}, \tag{14}$$

where R_p is the radius of the planet. In a dipole field, ions are trapped close to the equator by an effect known as the magnetic bottle. As a particle moves to higher latitudes, the field strength increases according to the relation,

$$\sin^2 \alpha = (B/B_0) \sin^2 \alpha_0, \tag{15}$$

until the particle is brought to a halt. This has the consequence that ions injected into the field undergo simple harmonic motion about the magnetic equator with their parallel velocity along the field line being a maximum at the magnetic equator.

Ions can be added or lost. Ions created from neutrals are accelerated to co-rotation velocity gaining an initial thermal speed ($v_\perp$) from the pick-up velocity. Note that the thermal speed depends upon the relative velocity of the neutral with respect to the magnetic line. Thus, although the co-rotation velocity at Io is 74 km s^{-1}, the thermal speed of an ion freshly created from a neutral at Io is only 57 km s^{-1} because Io orbits Jupiter in the same direction as Jupiter's rotation at a speed of 17 km s^{-1}. After creation, plasma can then diffuse or undergo convection and thus distributes itself through a magnetosphere.

10 Satellite Supply of Heavy Ions

The best studied example of a satellite supplying material to a planetary magnetosphere is that of Io. [2] detected emission from neutral sodium near Io. It was quickly recognized that the Na formed a cloud about Io, that its emission was produced by the resonant scattering of sunlight, and that energetic particle sputtering of Io's surface and/or atmosphere was producing the escaping gas. Subsequently, [16] detected emission from ionized sulphur (S^+, also called SII) near Io at 6731 Å and, two years later, O^+ emission was also detected. The Voyager 1 photometry of the surface showed that Io was

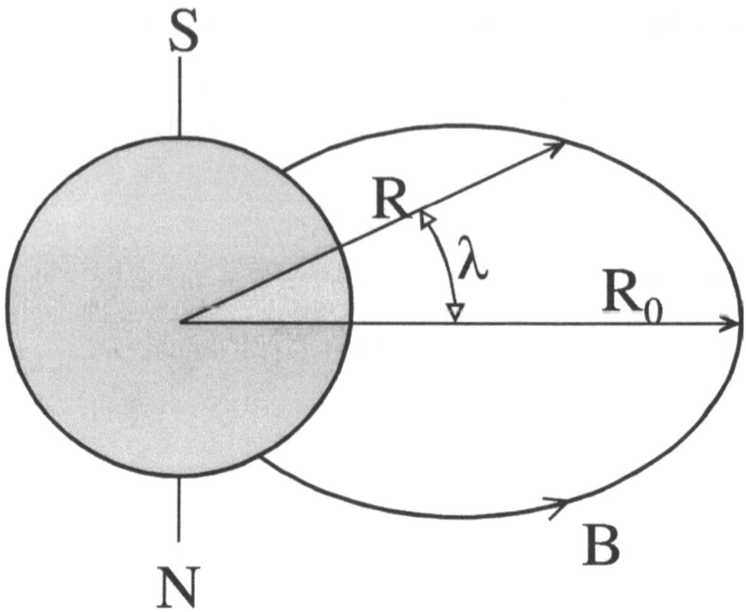

Fig. 5. A magnetic field line in a dipole magnetic field can be expressed in terms of the maximum distance from the dipole centre and a latitude.

covered in sulphur compounds and hence was the source of the material in Jupiter's magnetosphere.

Plasma in Jupiter's magnetosphere sputters material from the surface and/or atmosphere of Io producing a series of neutral clouds. Neutral sodium is the most easily detected from ground-based observation because it is a remarkably efficient scatterer of sunlight. However, it is only a trace element making up only around 5% of the neutral clouds. Sulphur and oxygen neutrals actually dominate the composition. Oxygen atoms have been detected from ground-based observations at 630.0 nm. Electrons in Jupiter's magnetosphere interact with the neutrals to produce ions by electron impact ionization. The ions are then picked-up by Jupiter's co-rotating magnetic field. The ions are thus swept into a ring about Jupiter centred roughly on Io's orbit (5.91 R_J). The ions diffuse both inwards and outwards from this ring to fill a torus-shaped volume in space which is called the Io plasma torus or "IPT".

The density, temperature, and composition of the IPT varies strongly with the radial distance from Jupiter. Inside Io's orbit the plasma is cool and the composition is dominated by singly-ionized species (mostly S^+ and O^+). The electron and ion temperatures, although not in equilibrium, rise rapidly between 5.4 R_J and 6.0 R_J. The higher energy electrons give rise to a more highly ionized plasma and S^{2+} becomes the most numerous sulphur ion around 5.8 R_J. The electron density reaches a maximum of around 3000 cm^{-3} at around 5.7 R_J - just inside the orbit of Io. The estimated composition at

5.9 R_J (at Io's orbit) is estimated to be 14% S^+, 48% O^+, 19% S^{2+}, 2% O^{2+}, 2% S^{3+}, 2% Na^+, $\leq$ 2% SO_2^+. A 10% mixing ratio of protons is usually assumed but this uncertain. The estimated mass of the IPT is around 2 10^6 tonne and the loss rate from Io is thought to be around 4 tonne s^{-1}.

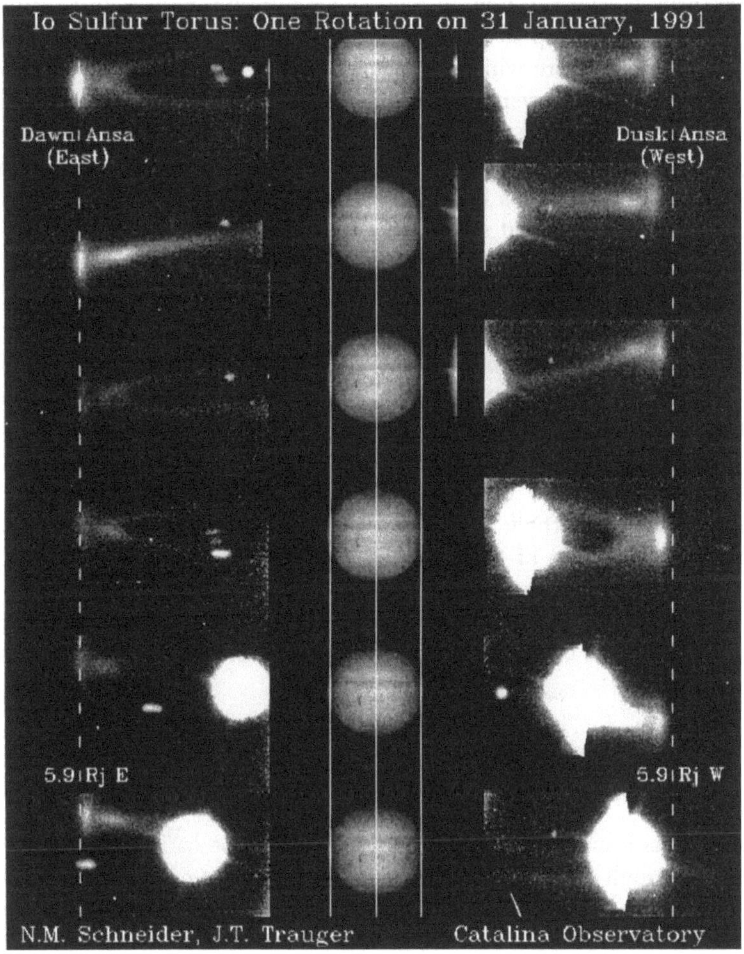

Fig. 6. Images of the Io plasma torus at different phases of Jupiter's rotation. The orientation of the torus changes because the magnetic axis is tilted by 9.8 ° with respect to the rotation axis. Image courtesy of N.M. Schneider and J.T. Trauger.

Figure 6 shows how the IPT appears to wobble in a Jupiter-centred reference frame. Jupiter's magnetic dipole is tilted by 9.8 ° with respect to the rotation axis. The equilibrium position for ions in a dipole magnetic field is the magnetic equator (where the magnetic mirror force becomes zero). However, Jupiter is rotating with high angular velocity and thus the centrifugal force on the ions has to be taken into account. As a result, the equilibrium position for the ions is at the so-called centrifugal equator which is between the rotational equator and the magnetic equator. The centrifugal equator is tilted by about 7 ° with respect to the rotational equator, hence the apparent wobbling motion.

The torus is also variable on long timescales and can vary in brightness from year to year by up to a factor of 4. This may well be connected to the state of the atmosphere which, in turn, possibly depends upon the volcanic activity (although this remains a subject of heated debate) or the extent of the ambient atmosphere. Furthermore, although it is widely assumed that charged particle sputtering and subsequent electron impact ionization is the major source of heavy ions in Jupiter's magnetosphere, this forms a type of positive feedback loop and consequently an alternative possibility has been discussed.

Any bound atmosphere will produce an ionosphere. This can be produced by photo-ionization or by electron impact ionization. It is generally thought that the latter is the dominant process at Io (e.g. [15]) but this remains to be proved. However, if Jupiter's magnetic field can interact with the Io's ionosphere, ions can be scavanged directly from Io's ionosphere. There is considerable ground-based observational evidence for this type of process and it could be the most important loss mechanism.

Titan, Triton, Europa, and Enceladus are also known to influence the heavy ion populations of their parent planets and the processes described above could to a greater or lesser extent be operating there also (e.g. [12]). It is highly probable that Charon and Pluto have an even more complex interaction because of the potential for the flow of the atmosphere of Pluto to Charon.

11 The Unusual Case of Iapetus

Energetic ions impacting a satellite surface might be considered an exogenic process (although often the satellite is itself the source of the impacting ions). However, Iapetus shows evidence of a completely different exogenic process. The trailing hemisphere of Iapetus has an albedo of around 0.5 and looks like a typical icy satellite surface. The leading hemisphere, however, is extremely dark with an albedo of 0.05. Voyager images suggest that the dark material is overlying the icy surface. Iapetus is in synchronous rotation and the dark material lies exactly symmetrically about the apex of the leading hemisphere. The most plausible theory of the origin of this material is that the outer Saturnian satellite, Phoebe, was, at one point in its history, emitting dust (i.e. some form of cometary emission). This material spiralled inwards towards Saturn under the effect of Poynting-Robertson drag and was swept up by the leading hemisphere of Iapetus. Unpublished numerical simulations have demonstrated the feasibility of this concept (D. Hamilton, personal communication). Several of the outer satellites of Jupiter and Uranus might also be captured comet-like objects and hence they may also have influenced the surfaces of regular satellites in a similar but less profound manner. Hopefully, the Cassini mission will provide a better understanding of the surface of Iapetus.

12 Summary

Most planetary satellites have had complex histories involving initial formation, equilibration, differentiation, subsequent break-up and reaccumulation, cratering, and tidal evolution and heating. While we are aware of many of the physical processes that might have occurred during satellite evolution, the details are obscure because we lack sufficient information. The study of the present state of planetary satellites and the physical processes that are associated with them also shows that the objects are dynamic. Tidal heating continues. Surface frosts cover surfaces, migrate, and produce tenuous atmospheres. Energetic particles sputter surfaces and atmospheres. All of these processes can provide information on how a satellite has evolved. But they can also mask other evidence of a satellite's history. This makes the study of the formation, diversity, evolution, and present state of planetary satellites one of the most interesting and challenging subjects in planetary physics.

References

1. Bennett, J., Donahue, M., Schneider, N., and M. Voit (1999) The Cosmic Perspective, Pub. by Addison Wesley Longman.
2. Brown, R.A. (1974) in Exploration of the Planetary System, eds. Woszczyk, A., Iwaniszewska, C., D. Reidel, Dordrecht, 527
3. Burns, J.A. (1986) Some background about satellites, in Satellites, Ed. Burns, J.A. and Matthews, M.S., Pub. by Univ. Arizona Press, Tucson, Az.
4. Carlson, R.W. and the Galileo NIMS team (1999) NASA JPL Press Release 99-007, Feb. 1st, 1999.
5. Chapman, C.R. and W.B. McKinnon (1986) Cratering of planetary satellites, in Satellites, Ed. Burns, J.A. and Matthews, M.S., Pub. by Univ. Arizona Press, Tucson, Az.
6. Chyba, C.F., Jankowski, D.G., and P.D. Nicholson (1989) Tidal evolution in the Neptune-Triton system, Astron. Astrophys., **219**, L23-26.
7. Elliot, J.L., H.B. Hammel, L.H. Wasserman, O.G. Franz, S.W. McDonald, M.J. Person, C.B. Olkin, E.W. Dunham, J.R. Spencer, J.A. Stansberry, M.W. Buie, J.M. Pasachoff, B.A. Babcock, T.H. McConnochie (1998) Global warming on Triton, Nature, **393**, 765–767
8. Hall, D.T. and P.D. Feldman and M.A. MCGrath, D.F. Strobel (1998) The Far-Ultraviolet Oxygen Airglow of Europa and Ganymede, Astron. Phys. J., **499**, 475
9. Hansen, C.J. and D.A. Paige (1992) A thermal model for the seasonal nitrogen cycle on Triton, Icarus, **99**, 273-288

10. Harris, A.W. (1984) Physical properties of Neptune and Triton inferred from the orbit of Triton, In Uranus and Neptune, Ed. J.T. Bergstralh, NASA CP-2330.
11. Hughes, D.W. and G.H.A. Cole (1995) The asteroidal sphericity limit, MNRAS, **277**, 99-105
12. Ip, W.-H., D.J. Williams, R.W. McEntire, and B.H. Mauk (1998) Ion sputtering and surface erosion at Europa, Geophys. Res. Lett., **25**, 829-832
13. Johnson, T.V., and L.A. Soderblom (1982) Volcanic eruptions on Io: Implications for surface evolution and mass loss, In Satellites of Jupiter, Ed. D. Morrison, Pub. Univ. Arizona Press, Tucson, Az.
14. Kaula, W.M., Drake, M.J., and J.W. Head (1986) The Moon, in Satellites, Ed. Burns, J.A. and Matthews, M.S., Pub. by Univ. Arizona Press, Tucson, Az.
15. Kumar, S., (1984) J. Geophys. Res., **89**, 7399
16. Kupo, I., Mekler, Y., and Eviatar, A. (1976) Astrophys. J., **205**, L51-L54
17. Lellouch, E. (1996) Urey Prize Lecture. Io's Atmosphere: Not Yet Understood, Icarus, **124**, 1-21
18. Lewis, J.S., (1997) Physics and Chemistry of the Solar System, Pub. Academic Press, San Diego, Ca.
19. Matson, D.L. and R.H. Brown (1989) Solid-state greenhouses and their implications for icy satellites, Icarus, **77**, 67-81
20. Matson, D.L. and D. Nash (1983) Io's atmosphere: Pressure control by regolith cold trapping and surface venting, J. Geophys. Res., **88**, 4771-4783.
21. McKinnon, W.B., Lunine, J.I., and D. Banfield (1995) Origin and evolution of Triton, in Neptune and Triton, Ed. D. Cruikshank, Pub. Univ. Arizona Press, Tucson, Az.
22. Peale, S., Cassen, P., and R. Reynolds (1979) Melting of Io by tidal dissipation, Science, **203**, 892-894.
23. Pearl, J., Hanel, R., Kunde, V., Maguire, W., Fox, K., Gupta, S., Ponnamperuma, C., and F. Raulin (1979) Identification of gaseous SO_2 and new upper limits for other gases on Io, Nature, **280**, 757-758.
24. Richardson, J.D., Belcher, J.W, Szabo, A. and R.L. McNutt, Jr. (1995) The plasma environment of Neptune, in Neptune and Triton, Ed. D. Cruikshank, Pub. Univ. Arizona Press, Tucson, Az.
25. Schubert, G., Spohn, T., and R.T. Reynolds (1986) Thermal histories, compositions and internal structures of the moons of the Solar System, in Satellites, Ed. Burns, J.A. and Matthews, M.S., Pub. by Univ. Arizona Press, Tucson, Az.
26. Shoemaker, E.M. and R.F. Wolfe (1982) Cratering time scales for the Galilean satellites, In Satellites of Jupiter, Ed. D. Morrison, Pub. Univ. Arizona Press, Tucson, Az.
27. Spencer, J.R. and N.M. Schneider (1996) Io on the Eve of the Galileo Mission, Ann. Rev. Earth Plan. Sci., **24**, 63-88
28. Squyres, S.W. and S.K. Croft (1986) The tectonics of icy satellites, in Satellites, Ed. Burns, J.A. and Matthews, M.S., Pub. by Univ. Arizona Press, Tucson, Az.
29. Stern, S.A. and W.B. McKinnon (1999) Science, submitted
30. Yelle, R.V., Lunine, J.I., Pollack, J.B., and R.H. Brown (1995) Lower atmospheric structure and surface-atmosphere interaction on Triton in Neptune and Triton, Ed. D. Cruikshank, Pub. Univ. Arizona Press, Tucson, Az.

Light Scattering in the Martian Atmosphere: Effects on Surface Photometry

Nicolas Thomas

Max-Planck-Institut für Aeronomie, Max-Planck-Str. 2,
D-37189 Katlenburg-Lindau, Germany

Abstract. The coming decade will see the beginning of the detailed exploration of Mars. Our knowledge of phenomena associated with the planet is already extensive and increasing rapidly. Each individual aspect can only be addressed adequately in a relatively large review. Here, we concentrate on the atmosphere and in particular on the dust content. The dust influences our interpretation of remote sensing data in subtle ways which require extensive modelling effort. We illustrate this using data from the Imager for Mars Pathfinder. We show how the dust produces a diffuse, reddened illumination which alters the perceived colour of the surface. We then describe the means by which models can be constructed to study these effects.

1 Introduction

Of the planets, Mars is the one that has consistently sparked Man's imagination. Although orbiter and lander missions to Mars in the past 25 years have clearly disproved the more fanciful theories of the early observers, the images returned have only added to our fascination. They have shown us

- the largest volcano in the Solar System (Olympus Mons, 600 km in diameter and 20 km high),
- a huge rift valley (Vallis Marineris, 4000 km long, up to 600 km wide and 8 km deep) created by massive uplifting of the mantle (Tharsis bulge),
- a remarkable global dichotomy where the northern hemisphere is, on average, 3 km below the average height of the southern hemisphere,
- evidence for liquid water in Mars' past in the form of run-off channels, valleys, and stream-lined islands (e.g. Ares Vallis),
- and a dry, rocky surface covered with dust, dominated by iron oxides.

These missions have also shown that Mars is not particularly hospitable towards Man. Maximum surface temperatures are below 270 K even with the Sun directly overhead. Nighttime temperatures drop below 190 K at equatorial latitudes. The thin (7 mbar) atmosphere of mostly CO_2 (Table 1) provides little or no protection from solar UV radiation. Water ice is present at the polar caps but probably only in the form of a permafrost layer well under the surface over the rest of the planet. However, the fact that liquid water was once present on the surface (albeit 3 billion years ago) has again triggered imaginations ([2]) and experiments designed to search for evidence of past life are now being constructed for launch in the coming decade.

Our knowledge of Mars is now so extensive that research books on the subject can run to more than 1400 pages (e.g. [7]). Therefore, in this lecture, we choose to restrict ourselves to a discussion of the Martian atmosphere and particularly the dust content. Dust plays an important role in the dynamics of the atmosphere and in the illumination of the surface. Dust acts as a heat source for atmospheric gases, it can act as a nucleation site for condensates, it coats the surface via sedimentation, and it scatters incoming radiation to provide a diffuse illumination at the surface. Its atmospheric number density is also highly time-variable. As we shall see, the physical processes involving the dust provide a series of interesting problems with far-reaching implications.

2 The Mars Pathfinder Mission

We will frequently be referring to the Mars Pathfinder spacecraft (MPF) throughout the lecture and hence, we start with an introduction to this mission.

Mars Pathfinder was originally designed to be a technology demonstration mission in preparation for the construction of a network of stations on Mars. Although NASA eventually cancelled this programme, MPF was allowed to go ahead as a demonstration of how new technology could be used to land on Mars for a fraction of the cost of the Viking landers, 20 years before. MPF was launched on 4 Dec 1996 and landed on the surface of Mars on 4 Jul 1997. The spacecraft was shaped like a tetrahedron and bounced onto the surface using a set of inflatable airbags. After coming to rest, the airbags were deflated and pulled under the spacecraft. The tetrahedron then unfolded to reveal the main body of the spacecraft and expose its solar cells to sunlight.

The landing site was selected with three points in mind. The site had to be below the level of the mean surface level in order to guarantee that the parachutes would have sufficient atmospheric column density to brake the spacecraft before impacting the surface. Secondly, the site had to be equatorial in order to provide sufficient power through the solar cells. Thirdly, part of the payload complement, comprising a camera (the Imager for Mars Pathfinder - IMP) with 24 interference filters ([16]) and a rover (Sojourner) with an alpha-proton-X-ray spectrometer (APXS), was optimum for the investigation of different rock types on the surface. The site selected was 32.8° W, 19.0° N at the common mouth of two outflow channels, Tiu Vallis and Ares Vallis. The northen latitude was suitable because the landing occurred during northern summer and, being in the northern hemisphere, the site was below the global mean surface level. A large variety of rocks from many geologically different areas were expected to have been brought down to the landing site by the floods which created the two valleys. Hence, the selected site, which was around 800 km SW of the Viking 1 landing site appeared to be ideal ([4]).

In addition to observing the surface through 15 geological filters, the IMP was also capable of taking images of the sky and could observe the Sun through a series of neutral density filters in order to measure the optical depth at the landing site. Furthermore, MPF carried an atmospheric sciences package designed to measure temperature, pressure, and wind speed at the surface.

3 Atmospheric Composition and Surface Pressure

The composition of the atmosphere of Mars is summarized in Table 1. It is dominated by carbon dioxide. The tilt of the rotation axis of Mars with respect to the ecliptic (the obliquity) is similar to that of the Earth (Table 2). Hence, the polar regions on Mars experience long periods when no solar heat input occurs. When the atmospheric temperature drops below about 148 K during these periods, CO_2 condenses to form a polar cap. This process provides heat to the cap through the latent heat released so that the condensation rate depends upon a balance between the thermal emission from the cap and the latent heat supplied by the condensation. When the Sun rises again in spring, the process is reversed so that the sublimation rate depends upon a balance between the solar insolation and the latent heat required to sublime the CO_2.

Table 1. The atmospheric composition of Mars relative to a total pressure of 7.5 mbar (after [7]).

Species	CO_2	N_2	Ar	O_2	CO	H_2O	Ne	Kr	Xe	O_3
Abundance [%]	95.3	2.7	1.6	0.13	0.07	0.003	2.5 ppm	0.3 ppm	0.08 ppm	0.10 ppm

The permanent polar caps on Mars are important because they buffer the atmospheric pressure (this is sometimes referred to as the Leighton-Murray model - [9]) so that the average pressure is the result of a balance between solar insolation and thermal emission over the annual cycle. Mars is not, however, uniform in surface properties (e.g. albedo) and has appreciable orbital eccentricity and therefore the atmospheric pressure at any given time can vary about the average pressure depending upon the season ([6]). Furthermore, there is a marked difference between the North and South poles in the amounts of CO_2 which condense on them. Southern winter is significantly longer than northern winter and hence more condensation occurs (Table 2). The size of the permanent southern polar cap is much smaller than the northern cap but the area that becomes covered by condensates during southern winter is huge and extends up to -40° latitude. The end of southern winter corresponds to a minimum in the Martian global atmospheric pressure which was beautifully illustrated by the pressure sensor on the MPF. Figure 1 shows the pressure variation with Sol (Martian day) with Sol 0 as the date of the Pathfinder landing.

The areocentric longitude of the Sun (L_s) is an expression of Mars's position in its orbit about the Sun. $L_s = 0°$ marks the vernal equinox. L_s at the time of the MPF landing was 142°. The minimum in the atmospheric pressure occurred at $L_s = 151°$. Ground-based observations show that once this point is reached the southern cap (which is roughly symmetric about the geographic pole) begins to sublime and the visible cap radius drops from 40° to 10° as L_s increases from 180° to 260°.

Table 2. Mars orbital data and the variability of the polar caps ([7]).

Property	Value
Semi–major axis	1.52366 AU
Eccentricity	0.0934
Inclination	1.85°
L_s of perihelion	250.87°
Length of day	24h 37m 22.663s
Mean orbital period	686.98 Earth days
	669.60 Sols (Mars days)
Obliquity	25.19°
Mass	6.418 10^{23} kg
Mean radius	3389.92 km (3397 km eq.)
Surface gravity	3.73 m s^{-1}
Escape velocity	5.027 km s^{-1}
Area of perennial polar caps	88,000 km^2 (south)
	837,00 km^2 (north)
Area of polar layered terrain	1,395,000 km^2 (south)
	395,000 km^2 (north)

Water vapour is also present in the atmosphere but in relatively small amounts. MPF results indicate that if all the atmospheric water were condensed as liquid onto the surface, the resulting layer covering the planet would be only 6 microns thick (e.g. [17]; [20]). There is an indication in the Pathfinder data that the water may be concentrated in the lower 3 km of the atmosphere rather than uniformly mixed with the CO_2. A possible explanation is that the regolith forms a reservoir and that the partial pressure of water vapour varies with the diurnal temperature variations. However, seasonal variations in the water vapour content are much stronger which suggests that the permanent polar caps form the main reservoir for H_2O in the atmosphere.

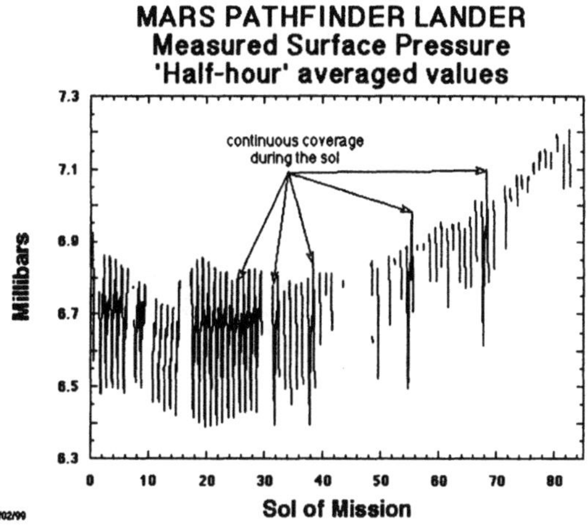

Fig. 1. The variation in the atmospheric pressure at the Mars Pathfinder landing site starting on 5 July 1997. Note the minimum at around sol 16. This date corresponds to the end of southern winter and the maximum extent of the southern polar cap. (Image courtesy of Jim Murphy and Tim Schofield.)

4 Atmospheric Opacity

The gas in the atmosphere contributes very little to the optical thickness of the atmosphere. However, as the Viking landers demonstrated, dust in the atmosphere produces optical depths, τ, consistently greater than 0.2 and, during dust storms, optical depths in excess of 3 were found. The amount of dust in the atmosphere is important for several reasons. Dust absorbs light and re-radiates isotropically at longer wavelengths and thus, heats the atmosphere. It also reduces the energy incident at the surface and re-distributes energy by scattering light into shadowed areas, for example. The latter effect has far reaching implications. In particular, photometric and mineralogical studies can be misleading unless both the attenuation of the direct flux and the scattered flux at the surface are taken into account.

The atmospheric opacity is easily measured using a method well-known to ground-based astronomers. The direct solar flux at a planetary surface, F, is related to the flux at the top of the atmosphere, $F_{\odot}$, through Beer's law,

$$F = F_{\odot} e^{-\tau A_m} \tag{1}$$

where A_m is referred to as the airmass. When the Sun is at the zenith ($z = 0°$), $A_m = 1$. For solar zenith angles less than about 75°, $A_m = \sec z$. For larger values of z, A_m is a function of the radius of the planet and the scale height, H. Thus, by measuring the direct solar flux at the surface at different airmasses, one can derive τ. Furthermore, by measuring precisely how F varies at large airmasses, the scale height of the dust can be determined.

Figure 2 shows the optical depth at the MPF landing site during the mission ([15]). The scale height was found to be 13 km ([16]; [18]).

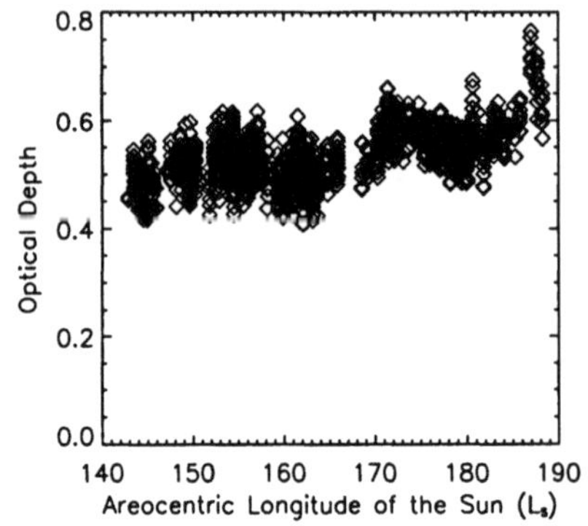

Fig. 2. The optical depth at the MPF landing site over the mission at 885 nm.

These values of optical depth indicate that even with the Sun at zenith only 60% of the solar flux at the top of the atmosphere reaches the surface without being scattered.

The short time scale fluctuations in the curve are not "noise" but represent real variations with time. The atmosphere is dynamic on several time scales. For example, the optical depth at blue wavelengths relative to other wavelengths showed a clear minimum at 14:00 local solar time which has been interpreted as evidence for the sublimation of ice-coated particles during the morning as the atmosphere heats. Weather systems passing over the spacecraft also produced changes in the optical depth. Very short time scale phenomena, such as dust devils, were also seen in the pressure and wind data and on at least two occasions were actually imaged by the IMP. These too produces changes in the local opacity.

5 Colour of the Sky

The dust in the atmosphere gives the sky a pronounced red colour (Figure 3). The sky is also very bright. On a clear day on the Earth, the sky is blue and almost uniform in brightness. (It was not possible to demonstrate this at the Summer School in Ireland for one very obvious reason!) This is because scattering in the Earth's atmosphere is dominated by molecules (Rayleigh scattering). The blue colour is derived from the fact that the scattering efficiency is proportional to λ^{-4} ([21]). On Mars, the situation is completely different because gas plays little or no role and the dust particles are large compared to visible wavelengths. For large particles, the scattering efficiency

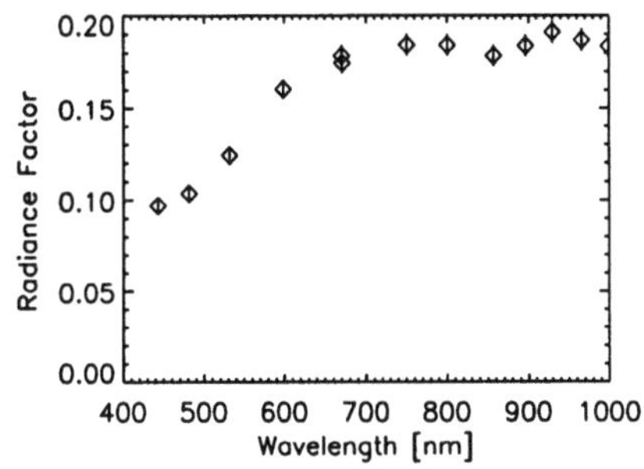

Fig. 3. A spectrum of the sky showing a strong increase in radiance factor from short wavelength to long wavelength. Note also the absorption band at 850 nm ([19]).

strongly depends upon the phase angle and this is clearly evident in a highly non-uniform sky brightness distribution on Mars.

6 Sky Brightness Models

To quantify the sky brightness, the IMP made measurements in a series of well-defined geometries around midday with the Sun close to the zenith. These measurements have been modelled using Mie theory but adapted for irregular particles using a semi-empirical approach ([12]).

Light striking a dust particle can either be scattered or absorbed. The scattering and absorption cross-sections (C_{sca} and C_{abs}) of the particle are often normalised by the geometric cross-section (e.g. $C = \pi a^2$ where a is the radius of a spherical particle) to produce efficiency factors (Q_{sca} and Q_{abs}). Extinction is defined to be the sum of scattering and absorption, i.e.,

$$Q_{ext} = Q_{sca} + Q_{abs}. \tag{2}$$

Mie theory can be used to determine Q_{sca} and Q_{abs} when the particle size is comparable to the wavelength of the light interacting with the particle. A dimensionless parameter called the size parameter is used as a basis for Mie calculations. The size parameter, x, is defined as the ratio of a sphere's circumference to the wavelength of the scattered light, λ, i.e.,

$$x = \frac{2\pi a}{\lambda}. \tag{3}$$

[1] describe the algorithms necessary (including a FORTRAN code) to compute phase functions using Mie theory. Four examples are shown in Figure 4 determined using the input size parameters and complex indices of refraction given in Table 3.

Table 3. Parameters used for the Mie theory calculations shown in Figure 4.

line	x	Refr. Index	Q_{sca}	Q_{ext}	Q_{back}
a	1.000	(1.55,0.0)	0.259197	0.259197	0.221648
b	5.213	(1.55,0.0)	3.10500	3.10500	2.92421
c	1.000	(1.55,0.2)	0.257269	0.776138	0.211055
d	1.000	(3.00,0.0)	8.14673	8.14673	5.27721

Note that the scattering angle is NOT the phase angle but 180° minus the phase angle. Bohren and Huffman also give a code for ice-coated spheres which has some applications (e.g. [11]).

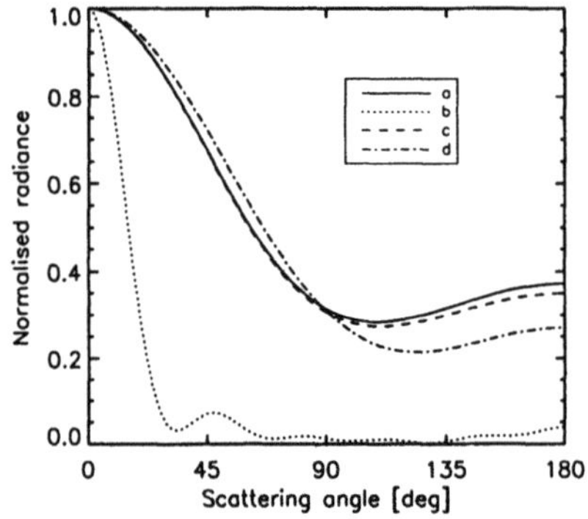

Fig. 4. Phase functions derived from Mie theory using the input parameters given in Table 3.

Mie theory has many restrictions, the principle one being that it is applicable only to spherical particles. [12] introduced a semi-empirical theory of scattering and absorption of non-spherical particles. Below a certain size parameter, x_0 (typically around 5), the particles were considered to follow Mie theory. Above this size, three effects were examined: diffraction, reflection, and transmission. A factor, S, was introduced which describes the ratio of the total surface area of the irregular particles to the surface area of the ensemble of equal volume spheres. An effective particle radius, $\overline{a} = a\sqrt{r}$ was then defined. The diffracted component was then determined by

$$I_D = \int_{x_0}^{\inf} d(\overline{x})\pi\overline{x}^2 n(\overline{x})d\overline{x} \tag{4}$$

where $n(\overline{x})$ is the size distribution and,

$$d(\overline{x}) \approx \frac{2C_D sin^2(\overline{x}\sin\theta - \pi/4)k}{\pi^2\overline{x}\sin^3\theta} \tag{5}$$

where $k = (1 + \cos^2 \theta)/2$ and θ is the scattering angle. The diffraction cross-section, C_D is obtained by normalising I_D so that

$$\int \frac{I_D}{4\pi} d\Omega = 1. \tag{6}$$

The reflected component describes the effect of reflection from the surfaces of large particles. This component is described by the equation

$$I_R = \frac{1}{2} C_R [\frac{\sin(\theta/2) - [|\overline{m}|^2 - 1 + sin^2(\theta/2)]^{1/2}}{\sin(\theta/2) + [|\overline{m}|^2 - 1 + sin^2(\theta/2)]^{1/2}}]^2$$
$$+ \frac{1}{2} C_R [\frac{|\overline{m}|^2 \sin(\theta/2) - [|m|^2 - 1 + sin^2(\theta/2)]^{1/2}}{|m|^2 \sin(\theta/2) + [|\overline{m}|^2 - 1 + sin^2(\theta/2)]^{1/2}}]^2 \tag{7}$$

where

$$|\overline{m}|^2 = m_r^2 + m_i^2 \tag{8}$$

and m_r and m_i are the real and imaginary parts of the refractive indices of the particles and C_R is found by normalising I_R in the same way that I_D was normalised.

The transmitted component is the main reason why the scattering behaviour of irregular particles differs from spherical particles of similar size. [12] addressed this by using an empirical constant, b, to define the phase function as

$$I_T = C_T e^{(1+b\theta)}. \tag{9}$$

By again normalising I_T and defining a constant, G, as

$$G \equiv \frac{\int_0^{\pi/2} I_T d\theta}{\int_{\pi/2}^{\pi} I_T d\theta}, \tag{10}$$

b can be related to G which is a quantity similar to (but not the same) as the ratio of energy scattered into the forward and backward directions.

The full phase function of the irregular particles is then determined by summing the individual components weighted by their contribution to the total scattering. Thus, Pollack and Cuzzi have adjusted Mie theory to take into account irregular particles (assuming random orientation) by using three free parameters (S, G, and x_0).

This approach (with a subsequent modification by [14]) has been used to construct the scattering properties of dust in the atmosphere of Mars by several authors (e.g. [10]). A plane-parallel approximation for the atmosphere is usually used. One can also solve for the optical depth although, in the case of Mars Pathfinder, direct measurements of τ were obtained. For high values of τ, multiple scattering must be taken into account. One approach is to use the "doubling and addin"method which produces an exact solution for multiple scattering ([5]). This method separates the atmosphere into identical, discrete, thin layers. If the transmission and reflection functions of two discrete layers of very low optical depth (say 2^{-20}) can be determined then the reflection and transmission from the combined layer can be computed by determining the successive reflections between the layers ([8]). If the atmosphere is homogeneous, then repeated adding of identical layers can be used to produce the transmission and reflection functions of an atmosphere of any optical depth.

The observations made by the IMP have been used to constrain models of the sky brightness. The models have then been extrapolated to the whole hemisphere to produce maps of the sky brightness (Figure 5).

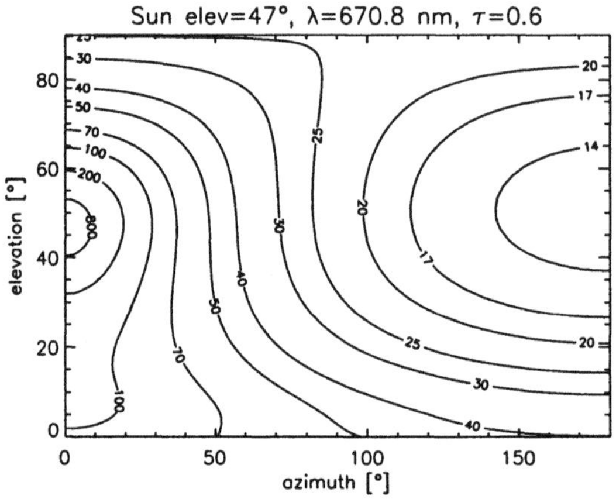

Fig. 5. A model of the brightness over the entire sky on Sol 56 of the MPF mission. The Sun was at an elevation of 47° and $\tau = 0.55$. The sky brightness is strongly wavelength dependent. This model is for $\lambda = 670$ nm ([19]).

There are other approaches to solving the problem of the scattering properties of irregular particles. Of particular note is the discrete dipole approximation (DDA) which models a particle as a series of discrete dipoles and solves Maxwell's equations to determine the scattering properties ([13]; [3]). DDA is an exact method but is computationally time consuming (a disadvantage which is no longer as serious as it was). Furthermore, an advantage of Pollack and Cuzzi's method is that with a limited number of parameters, albeit ones that are somewhat difficult to interpret physically, the scattering properties of an atmosphere can be defined. With DDA, however, one computes scattering properties for one particle and a specific orientation. It is not a method easily adapted to solving the inverse problem of extracting parameters from an observed brightness distribution. However, developments in this field are progressing rapidly.

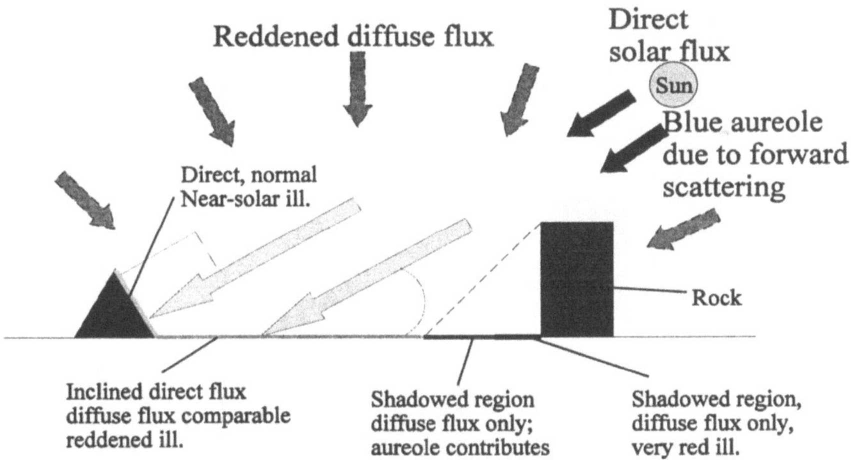

Fig. 6. Schematic diagram showing how the changing orientation of rocks on the surface of Mars leads to differences in the colour of the illumination. Direct sunlight aligned with the surface normal provide the highest direct/diffuse flux ratio. However, if the surface normal is not parallel to the sunlight, the magnitude of the direct flux decreases by the cosine of the inclination angle and hence the direct/diffuse flux ratio decreases. As the diffuse flux is red, the illumination becomes increasingly reddened as the inclination angle of the surface increases. Once the surface goes into shadow, it is not unilluminated. The diffuse flux ensures, however, that the illumination is extremely red and hence shadow areas appear remarkably red in the scene.

7 Effects of Dust on Surface Photometry

The sky brightness is sufficiently strong that the photon flux onto the surface is dramatically affected. Plate 1 shows a colour view of the MPF site. In this colour enhanced view, the rock Yogi appears to be relatively blue on the right side and red on the left side. This was initially interpreted as an effect of wind abrasion because the right side is towards the north-east which is the direction of the prevailing wind at the site. However, further images revealed that the colour of Yogi's surface changed with time of day. It then became clear that the illumination was responsible for the effect. The principle is illustrated in Figure 6 ([19]).

The sky brightness model was therefore used as input to a simple model of the geometry of Yogi to determine the magnitude of the diffuse illumination of the surface ([19]). A result of this process is shown in Figure 7. Two areas on Yogi (A and C) were selected, calibrated, ratioed and normalised at 670 nm. The open diamonds with the error bars show that region A is more strongly blue than region C. The filled diamonds show the result of the model incorporating both the direct and diffuse fluxes and assuming that A and C under identical illumination are the same colour. The model fits the data extremely well. Although it is unlikely that Yogi is completely uniform in

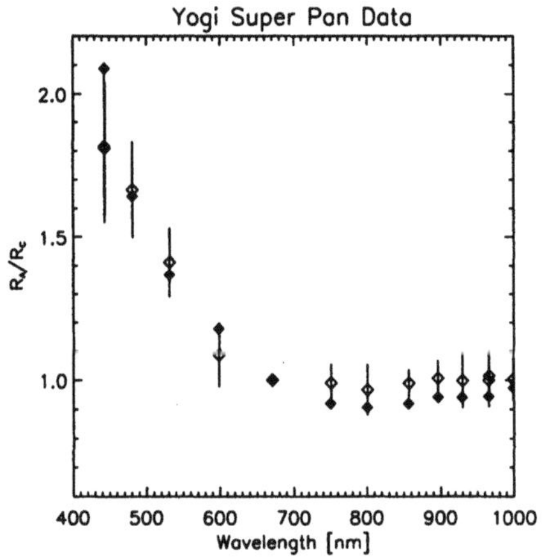

Fig. 7. Open diamonds: The ratio of the observed intensity on two faces of the rock, Yogi (normalised to 1 at 600 nm). Notice that one face is much bluer than the other. Filled diamonds: The ratio of the integrated direct plus diffuse flux onto the two faces computed using the sky brightness model for the diffuse flux under the assumptions that under similar illumination conditions, the faces would be the same colour. Notice that the model fits the data fairly well.

colour, any differences are small compared to the effects of the illumination. Hence, any attempt to interpret spectra of the surface must take into account the diffuse illumination either by modelling or by ensuring that adequate calibration "standards"are present, in the form of calibration targets, for example. One difficulty with the latter approach is that surfaces on Mars collect dust at a fairly rapid rate. MPF data suggest that sedimentation of dust covers a surface area equivalent to 0.25% per day. Calibration targets were used on MPF. Analysis of the data has shown that the model describes the radiance from the calibration targets well if dust sedimentation is taken into account ([19]).

The remaining difficulty in interpreting spectra of the surface is the presence of the absorption band in the spectrum of the sky. Figure 8 shows a spectrum of a region referred to as "Photometry Flats"because it was devoid of rocks with a smooth darkish surface. The spectrum shows a shallow absorption band at 850 nm. However, it might be argued that this feature has been produced by the diffuse flux affecting the illumination of the surface and without detailed modelling this hypothesis cannot be ruled out. However, the dust on the surface of Mars is obviously the source of the dust in the atmosphere. The dust is picked-up by winds. The IMP observed several examples of this process in the form of small dust devils. Therefore, it would be somewhat surprising if dust surfaces did not show a shallow absorption.

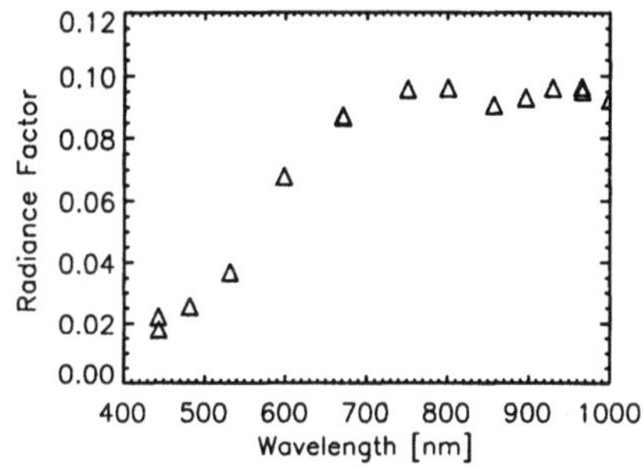

Fig. 8. A spectrum of a region of dark soil at the MPF landing site called Photometry Flats. Note the strong red colour and the relatively weak absorption band at 850 nm. The solar elevation angle was 46° and the optical depth around 0.5.

8 Conclusions

After the Viking missions and the latest series of missions to Mars (Mars Pathfinder and Mars Global Surveyor), our knowledge of the red planet is now quite extensive. If the current political momentum is maintained (with two launches every 25 months), Mars will be really explored in the coming decade culminating in a sample return mission. This implies that studies of the properties of Mars will become increasingly sophisticated. The investigation of dust in the atmosphere and its effect on the interpretation of surface photometric measurements is an excellent example. The huge volume of data expected from these missions also means that there are sure to be significant opportunities for young people to participate and work on many different aspects of the Martian environment including the investigation of its interior, its mineralogy, its atmosphere, its magnetosphere, and, of course, its evolution. In the coming decade, Mars will undoubtedly be a "hot"topic!

Acknowlegdments. I would like to thank my colleague, W.J. Markiewicz, for discussions and assistance in clarifying the descriptions of the models.

References

1. Bohren, C.F. and D.R. Huffman (1983) Absorption and scattering of light by small particles, Pub. by Wiley Interscience, New York.
2. Carr, M.H. (1996) Water on Mars, Pub. by Oxford University Press, New York.
3. Draine, B.T. (1988) The discrete-dipole approximation and its application to interstellar graphite grains, Astrophys. J., **333**, 848-872
4. Golombek, M.P. and Cook, R.A. and Moore, H.J. and Parker, T.J. (1997) Selection of the Mars Pathfinder landing site, J. Geophys. Res., **102**, 3967-3988
5. Hansen, J.E., and L.D. Travis, Light scattering in planetary atmospheres, Space Science Reviews, **16**, 527–610, 1974.

6. James, P.B., Kieffer, H.H., and D.A. Paige (1992) in Mars, Eds Kieffer, H.H., Jakosky, B.M., Snyder, C.W., and Matthews, M.S. (1992) Mars, Pub. by University of Arizona Press, Tucson, USA.
7. Kieffer, H.H., Jakosky, B.M., Snyder, C.W., and Matthews, M.S. (1992) Mars, Pub. by University of Arizona Press, Tucson, USA.
8. Lenoble, J. (1985) in Radiative Transfer in scattering and absorbing atmospheres: Standard computational procedures, Pub. by A. Deepak Publishing, Hampton, Virginia, USA.
9. Leighton, R.B. and Murray, B.C. (1966) Behavior of carbon dioxide and other volatiles on Mars, Science, 153, 136-144.
10. Markiewicz, W.J., Sablotny, R.M., Keller, H.U., Thomas, N., Titov, D., and Smith, P.H. (1999) Optical properties of the Martian aerosols as derived from Imager for Mars Pathfinder midday sky brightness data. J. Geophys. Res., in press.
11. Petrova, E., H.U. Keller, W.J. Markiewicz, N. Thomas, and M.W. Wuttke (1996) Ice hazes and clouds in the Martian atmosphere as derived from the Phobos/KRFM data, Planet. Space Sci., **44**, 1163–1176
12. Pollack, J.B., and J.N. Cuzzi (1980) Scattering by nonspherical particles of size comparable to wavelength - A new semi-empirical theory and its application to tropospheric aerosols, Journal of Atmospheric Sciences, **37**, 868–881
13. Purcell, E.M. and C.R. Pennypacker (1973) Scattering and absorption of light by nonspherical dielectric grains, Astrophys. J. **186**, 705-714.
14. Showalter, M.R., J.B. Pollack, M.E. Ockert, L.R. Doyle, and J.B. Dalton (1992) A photometric study of Saturn's F Ring, Icarus, **100**, 394–411
15. Smith, P.H. and Lemmon, M. (1999) Opacity of the Martian atmosphere measured by the Imager for Mars Pathfinder, J. Geophys. Res. in press.
16. Smith, P.H., Tomasko, M.G., Britt, D., Crowe, D., Reid, R., Keller, H.U., Thomas, N., Gliem, F., Rueffer, P., Sullivan, R., Greeley, R., Knudsen, J.M., Madsen, B.M., Gunnlaugsson, H.P., Hviid, S., Goetz, W., Soderblom, L., Gaddis, L., Kirk, R. (1997a) The Imager for Mars Pathfinder (IMP) experiment. J. Geophys. Res., **102**, 4003-4025
17. Smith, P.H., Bell, J.F., Bridges, N.T., Britt, D.T., Gaddis, L., Greeley, R., Keller, H.U., Herkenhoff, K.E., Jaumann, R., Johnson, J.R., Kirk, R.L., Lemmon, M., Maki, J.N., Malin, M.C., Murchie, S.L., Oberst, J., Parker, T.J., Reid, R.J., Sablotny, R., Soderblom, L.A., Stoker, C., Sullivan, R., Thomas, N., Tomasko, M.G., Ward, W., Wegryn, E. (1997b) Results from the Mars Pathfinder camera, Science, **278**, 1758-1765
18. Thomas, N. Britt, D.T., Herkenhoff, K.E., Murchie, S.L., Keller, H.U., and Smith, P.H. (1999a) Observations of Phobos, Deimos, and bright stars with the Imager for Mars Pathfinder. J. Geophys. Res., in press
19. Thomas, N., Markiewicz, W.J., Sablotny, R.M., Wuttke, M.W., Keller, H.U., Johnson, J.R., Reid, R.J., and Smith, P.H. (1999b) The color of the Martian sky and its influence on the illumination of the Martian surface. J. Geophys. Res., in press
20. Titov, D.V., Markiewicz, W.J., Thomas, N., Keller, H.U., Sablotny, R., Tomasko, M.G., Lemmon, M., and Smith, P.H. (1999) Measurements of the atmospheric water vapor on Mars by the Imager for Mars Pathfinder. J. Geophys. Res., in press
21. van de Hulst, H.C. (1957) Light scattering by small particles, Pub. by Dover Publications, New York

The Small Bodies of the Solar System

Iwan P. Williams and Alan Fitzsimmons

Astronomy Unit, Queen Mary, London E1 4NS, UK
Dept of Pure and Applied Physics, Queens University, Belfast BT7 1NN, Northern Ireland

Abstract. Comets have been seen since time immemorial but asteroids were only discovered at the begining of the nineteenth century while the first members of the family of bodies that orbit in the region beyond Neptune were only discovered within the last ten years. Here, we wish to review our current state of knowledge of these three classes of bodies and discuss possible interrelations.

1 Introduction

Comets have fascinated the human race since antiquity. Records of their appearances as well as illustrations and descriptions can be found in many sources. A very early account of what was seen comes from Seneca(~4BC–65AD), who described a comet seen in 147 BC in the following terms. "It was as large as the Sun, reddish like fire and bright enough to dissipate the darkness". Seneca obviously could not have seen this comet. The comet of 44BC was visible during the games that Octavian was holding in honour of the assassinated Julius Caesar and is depicted on a Roman coin. [21] contains interesting accounts of many early comets. Most of the reliable accounts of cometary appearances through history are found in Chinese, Korean, and Japanese records and have been collected by [7], while comets were, and sometimes still are, regarded as messengers of doom and were not thus part of the natural order of things, any scientific discussion of their origin and nature was meaningless. Indeed any sensible progress had to wait until Halley demonstrated that the comets of 1531, 1607 and 1682 were one and the same comet and so in effect demonstrating that comets moved on periodic orbits within the Solar System

In contrast to comets, asteroids have only been discovered within recent history. The study of asteroids started on the first day of the 19th century when Piazzi, observing from Palermo, discovered a new object in the Solar System. Though the specific discovery by Piazzi was serendipidous, a campaigne had been initiated by von Zach, whereby the zodiac was divided into 24 zones and a different astronomer was assigned to search each zone for a suspected planet. Piazzi was not one of these 24 astronomers. The expectation of finding a planet was based on the belief that the Titius-Bode law that predicted planetary distances was correct. We must bear in mind that only twenty years earlier in 1781, its correctness had apparently been demonstrated through the discovery of the planet Uranus by Herschel. The only unexpected element in the discovery by Piazzi was that the new planet, named Ceres, was rather faint, much fainter than expected, indicating that the body was somewhat smaller than the "predicted" planet. Within the next four years, three similar objects were discovered, Pallas, Juno and Vesta (two by Olbers and Juno by Harding). No further objects of this class were discovered for 40 years and it was during this period that the group were called, minor planets - for they clearly were not proper planets. It was also during this time interval that the hypothesis was first put forward that these minor planets were remnants of a proper planet that had been broken up by some mechanism.

The story repeats itself nearly two centuries later as the twentieth century draws to a close. In the 1940's and 50's, theories of the formation of the planets by [3],edg49 and [9] had pointed out that there was no reason to suppose that the Solar System finished with the known planets. Indeed, they went further and suggested that a swarm of smaller objects should exist beyond Pluto. In the mid-1980s serious searching for these trans-neptunian objects began and in 1992, [6] discovered an object now given the number 15760, but better known by its temporary designation $1992QB_1$. Five further objects were discovered within twelve months, and ever since, discoveries have been made at an increasing rate, the current total being of order 200.

Returning to the story of the asteroids, as already mentioned, the initial pace was much slower with no asteroids beyond the first four being discovered for forty years. At the end of this period Astraea was discovered, and the floodgates opened so that by 1900 there were 450 known, 1000 by 1923, 5000 by 1991 and now well over 10000.

The aim of this chapter is to describe the general characteristics of each of these classes of objects in terms of dynamics and broad physical properties and also to discuss possible inter-relationships.

Before considering each class individually, it is meaningful to comment on the orbital motion of bodies in the Solar System. As a first approximation, the gravitational field of the Sun is totally dominant so that all bodies can be assumed to satisfy Kepler's three laws.

The first states that all bodies move on an elliptic orbit about the Sun.

The second states that the radius vector from the Sun to the body sweeps out equal areas in equal times.

The third states that the square of the orbital period is proportional to the cube of the mean heliocentric distance.

To specify the position of a body assumed to be moving on an elliptical orbit, six parameters are required. The first two specify the dimensions of the ellipse, and are a, the *semi-major-axis* and e the *eccentricity*. The nearest point to the Sun is called the *perihelion point* and is at a distance of $a(1-e)$ from the Sun, often denoted by q. The maximum distance is $a(1+e)$ and is denoted by Q. The next three parameters specify the orientation of orbit in relation to the plane of the Earth's orbit, *the ecliptic.* The first of these is the inclination i, being the angle between the two planes. $i = 0$ implies that the orbit is in the ecliptic, $i = \pi/2$ implies that the orbit is perpendicular to the ecliptic while values of i between $\pi/2$ and π denote that the motion about the orbit is retrograde. The *longitude of the ascending node Ω* denotes the angle between the crossing point of the orbit through the ecliptic and the vernal equinox, while ω is the angle from the ascending node to the perihelion, called *the argument of perihelion.* In order to specify the position of the body on the orbit, one more parameter is required, either the angle from perihelion to the body at the given time, f and called the *true anomaly* or alternatively the time at which the body passed through perihelion, T.

The equation of an ellipse, where r is the heliocentric distance, is

$$\frac{q(1-e)}{r} = 1 - e\cos f. \tag{1}$$

It is also useful to define two other angles, the *eccentric anomaly,* E and the *mean anomaly,* M. The eccentric anomaly is related to the true anomaly by

$$r\cos f = a(\cos E - e) \tag{2}$$

while the eccentric anomaly and the mean anomaly are related by Kepler's equation,

$$M = E - e\sin E \tag{3}$$

If P denotes the orbital period and t is the time since the last perihelion passage, then

$$\frac{t}{P} = \frac{M}{2\pi} \tag{4}$$

The second of Kepler's three laws is a statement that angular momentum about the Sun is conserved, or

$$h^2 = GM_{\odot}q(1+e) = Constant \tag{5}$$

The third law is perhaps the best known and gives

$$P^2 \propto a^3 \tag{6}$$

when P is measured in *years* and a in *astronomical units* (1.496×10^{11} m).

If all bodies were only affected by the gravitational field of the Sun then their motion would be exactly given by the solution of the above equations. Unfortunately, this is not so. Other forces, though small, can have an effect. The most important are gravitational perturbations by the planets and radiation pressure. Both of these are discussed in the chapter on meteor streams. The subject of orbital dynamics and gravitational perturbations is taken further in the chapter by Murray. The important parameter when discussing radiation pressure is the ratio of this force to gravity, which is independent of heliocentric distance since both are proportional to the inverse square of this distance. This ratio is usually characterized by β and is given by

$$\frac{F_{rad}}{F_{grav}} = \frac{5.7\times10^{-5}Q_{pr}}{b\rho} \tag{7}$$

where b and ρ are the radius and density of the grain. For comets there is also a force that might exist due to the outgassing process since this is unlikely to be isotropic in relation to the nucleus. This is usually accounted for by the inclusion of an arbitrary *nongravitational force* term in the equations of motion, the values of the parameters being obtained by fitting the orbit d. Seen in past apparitions of the comet.

For an inert body, the apparent magnitude at heliocentric distance r and geocentric distance Δ is given by

$$m = H_0 + 5log\Delta + 5logr, \tag{8}$$

where H_0 is the absolute magnitude, taken to be the magnitude when the body is 1 AU away from both Sun and Earth and at zero phase angle (an impossible configuration).

Let us then consider individually the three classes of objects, Asteroids, Comets and Edgeworth-Kuiper, or Trans-Neptunian Objects, taking them in alphabetical order.

2 Asteroids

As already mentioned, the first asteroid, Ceres, was discovered two centuries ago, orbiting between Mars and Jupiter, in a location very close to where the "missing" planet was being sought. With a diameter of 933km, Ceres remains the largest known asteroid. At a mean distance of 2.77 AU, its period is 4.6 years. Most of the known asteroids lie in what is called the *main belt*, the region between about 2 and 3.5 AU from the Sun. Within, or adjacent to the main belt, families of asteroids moving on very similar orbits can be recognized. These are generally thought to be the product of a collision between two parent asteroids. Examples of well known families are Flora and the Hildas.

Resonances with Jupiter have also been recognized for a long time as being dynamically important. In some cases, such as the 3:1 mean motion resonance, there is an absence of asteroids, while others such as the 1:1 produces an overabundance. This latter resonance is very important, producing the Trojan group of asteroids. We will now take a brief look at each of the main groupings.

2.1 The Main-Belt Asteroids

The main belt, as its name implies, contains most of the known asteroids. It lies roughly between heliocentric distances of 2.0 and 3.5 AU. The orbits of these asteroids are near circular, with eccentricities generally less than 0.3 and inclinations less than about 25°, with the majority being less than 10°. Kirkwood recognized gaps within this main belt which were later identified as being at the location of mean motion resonances with Jupiter. This means that the orbital periods are small integer ratios of each other such as 2:1 or 3:2. Most of these do correspond to gaps in the asteroidal distribution and the reason for this was first demonstrated by [20], who showed that the effect of Jupiter was to increase eccentricity until the asteroid became a Mars crosser which led to its loss from the belt. However, for two of the resonances, the 1:1 and the 3:2, the opposite seems to be the case and the Trojan and Hilda groups are found at these locations. The Trojans will be discussed later. The reason for the existence of the Hildas is not fully understood, but one school of thought has them as the remnant of a fairly recent collision. For a given large asteroid, the mean collision time is of the order of 10^{10} years and thus, with an estimated 10^7 asteroids in the belt, the mean asteroid/asteroid collision time is only about 1000 years. It is not thus surprising that we see collisional signatures as the asteroid families. A further consequence of this collisional history is that many asteroids will be irregularly shaped, a result borne out by the few spacecraft images currently available.

One significant benefit of an irregular shape is that the light curve has a periodicity which is easily observed so that rotation rates can be determined. The majority lie between about 6 and 24 hours, though there are a few very slow rotators and a handful rotating as fast as 2.5 hours.

There are also clear compositional differences between asteroids, with upwards of ten different classes having been defined by various authors. We shall confine ourselves to discussing four of the main classes.

S-type. The S stands for stony and these are composed of common rock, or silicates. The spectra shows strong absorption bands due to olivine and pyroxene.

C-type. These are thought to be similar to Carbonaceous Chondrite meteorites, i.e., normal minerals heavily enriched with volatile elements such as hydrogen, carbon and nitrogen.

M-type The M stands for Metal, and these are very rich in metals, particulary iron-nickel.

D-type. D for dark. There is no known meteoritic analogue. Possibly contain large amounts of ices (H_2O and CO) held in a rocky matrix.

2.2 The Trojan Asteroids

Lagrange proved that there were five equilibrium positions within the restricted three body problem. Three of these lie on the line joining the primary to the secondary while the other two, called L_4 and L_5 form an equilateral triangle with the primary and secondary. They are thus on the same orbit as the secondary and may be regarded as being in a 1:1 resonance with the secondary. Asteroid 588 Achilles was discovered by Max Wolf in February 1906 and was found to be librating about L_4 of Jupiter. In October of the same year, Kopff discovered 617 Patroclus librating about L_5. By now there are several hundred such asteroids discovered, and the tradition has continued from the first two of naming them after heroes of the Trojan wars, hence their group name of Trojan Asteroids. At present, about 250 bodies either have permanent designation or are awaiting one and occur in roughly equal numbers around the L_4 and L_5 points of Jupiter. It is a debate that flares up from time to time as to whether or not there is a bias for one point rather than the other and, if so, why. Almost all the known "Trojan" type asteroids are associated with Jupiter, though 5261 Eureka, discovered in 1990, is a Mars "Trojan".

Trojan orbits can be destablized both through collisions and dynamical chaos and the loss rate is such that the current Trojans must either be the remnants of a much larger population or else the population is being replenished from some other, as yet unidentified, source. The origin of the Trojans is also a matter of some debate. Some claim that they were captured at an early stage in the Solar System formation process possibly during the rapid growth stage of Jupiter. Others claim that they represent the capture of asteroidal fragments.

2.3 Near Earth Asteroids

The first asteroid to be discovered outside the main belt was 833 Eros,by Witt in 1898 and spends its life between 1.133 and 1.783 AU. It does not thus cross the orbit of the Earth. This asteroid has recently been the subject of detailed investigation from the NEAR-Shoemaker spacecraft. The first real Earth crosser was 1862 Apollo discovered in 1832 by Reinmuth, which has given its name to the Apollo group of Asteroids. Two other groups of asteroids can approach the Earth namely the Amors and the Atens. These asteroids have currently become very news-worthy in view of the recognition that they may pose a threat to life on Earth. A very good review of these aspects is given in *Report of the Task Force on potentially Hazardous Near Earth Objects* submitted to the UK Government.

3 Comets

Since the early days, comets were recognized by the fact that they appeared *fuzzy* on an image. As comets approach the Sun, they develop a tail. As studies progressed, it was recognized that two tails existed, an ion or plasma tail and a dust tail. Finally, about half a century ago it was postulated that a nucleus must exist, where the bulk of the comet mass actually resided, but observational confirmation of the existence of this had to wait for the space age and the Giotto mission. The standard model for a comet at present follows that proposed by [18], where the nucleus is an icy matrix within which dust grains are embedded. The ice may well be more akin to snow that to the solid lumps that come out of our fridge, but the basic model is the same. It follows that as the nucleus approaches the Sun, it will heats up causing sublimation of the ice and forming an expanding gas and dust coma. Solar radiation pressure and the solar wind then drives material away from the Sun forming the tails. This of course has an effect on the appearance of a comet.

For a comet the apparent size increases due to the formation of the coma and in consequence there is a stronger dependence of apparent magnitude on r than was assumed in the earlier discussion, it being usual to take the apparent magnitude as

$$m = H_0 + 5log\Delta + 2.5nlogr. \quad (9)$$

Here, n is called the activity index, a value of 2 indicating no activity. For most comets n has a value between 3 and 4 when the nucleus is within 3AU of the Sun. $m - 5log\Delta$ is often called the heliocentric magnitude because the geocentric correction has already been applied.

3.1 The Comet Nucleus

The onset of activity, characterized by n exceeding 2, is obviously governed by the beginning of sublimation. The fact that activity generally becomes obvious at about 3AU suggests that the dominant component of the ice is water ice. However, other ices may be present, in particular CO and CO_2 which sublimate at much greater distances.

To produce detailed models of the nucleus and its evolution, three fundamental physical properties are required-size, albedo and rotation rate. Only one comet nucleus has been optically resolved, that of comet Halley by the Giotto spacecraft which obtained dimensions of $16 \times 7 \times 7$ km. However, reasonable estimates of the dimensions of a cometary nucleus can be obtained by comparing the point distribution function for stars and the given comet on a CCD image or by subtracting the coma image from the comet image. A discussion of these methods can be found in [12] and [10]. By now radii have been estimated for several tens of comets. Most values lie in the range of 1-100km with a prepondence towards lower values.

The albedo is a measure of how much radiation is reflected rather than absorbed (and re-radiated in thermal infrared). Accurate determination requires a measurement of both the reflected and re-radiated energy. For Halley the albedo was 0.04. A small number of comets have been observed at the required wavelengths and Halley does appear to be fairly typical.

For most bodies, a rotation rate is determined by observing variations in the light curve. For comets, the nucleus is not generally visible and so the rotation of the nucleus has to be inferred from variations in the coma brightness or the location of jets. It is thus not surprising that the number of comets with known rotation rates is small. As might be expected the best determined are Halley and Hale-Bopp, though in the case of Halley there is a definitional argument as to what is precession and what is rotation. In either event Halley has a very long period, 178 hours by the traditional definition of rotation, while Hale-Bopp has a period of 11.33 hours. The other measured values lie in the range of 5-18 hours [5].

3.2 Cometary Tails

Once in the coma, the material ejected from the comet becomes subject to the effects of solar radiation. It has three effects that are important. First, it causes dissociation of moleclules. The primary consequence of this is that the molecules detected from observations may not be those that are present in the nucleus but rather some daughter products. The most obvious example is perhaps OH as a daughter product of H_2O. The problem, of course, is that water is not the only possible parent for OH so that the detection of OH is not conclusive unambiguous proof of the existence of water. The same is true for all daughter products and indeed to make matters worse we do not know from the observations alone whether we are observing parent or daughter molecules. In comet Hale-Bopp, the following molecules were identified, OH, NH, NH_2, CH, CO, H_2O, CN, C_2, C_3, Na, K, O, HCN, CO_2, CH_4, C_2H_6, C_2H_2, CH_3OH, HDO, OCS, CS, H_2S, HNCO, NH_3, H_2CO, CH_3CN, CH_3OH, DCN, SO, SO_2, H_2CN, HC_3N, NH_2CHO, HCO, H_3O, HCO_2H, and CH_3OCHO.

The second possible effect is to ionize some molecules. This results in them acquiring a charge and thus forces them to move along magnetic field lines which are essentially radially outwards from the Sun. This forms the well known ion tail. In the very early days this was used as a probe of the solar wind properties, and indeed the existence of a solar wind was first inferred from observations of a plasma tail.

The third effect is that of radiation pressure driving grains outwards. Very small grains will be driven out of the Solar System while others will remain on a much enlarged Keplerian orbit. This gives rise to the observed curved fan shaped dust tail. In passing, even larger grains are hardly affected and these form meteoroid streams discussed in another chapter.

3.3 Comet Dynamics

All comets move essentially on Keplerian orbits about the Sun, though they are subject to perturbations by the planets and also effects arising from the ejection of material (the rocket effect). Though the actual orbital changes due to this non-gravitational effect may be small, it can have a large effect on orbital evolution because of the way the interaction geometry with Jupiter changes. For this reason, long-term integrations of the orbits of individual comets are impossible (as opposed to a statistical investigation of general trends) and the longest possible integrations to date have only been for a few thousand years.

Because most cometary orbits have eccentricities that are very different from zero, their orbits cross those of the major planets, in particular Jupiter. As a consequence, their survival time within the inner Solar System is generally of the order of tens of millions of years - considerably less than the age of the Solar System. Hence, all the comets that we observe today are recent visitors to the inner solar system. Consequently, comets have to be stored somewhere where they can not ordinarily be observed but, from time to time get disturbed and fall into the inner Solar System. [14] postulated that such comets resided in a vast cloud at distances significantly larger than any of the known planets and that the necessary random perturbations came from passing stars. Such a notion was very consistent with the *snowball* model of comets proposed by [18] since such nucleii would be totally inactive at such large heliocentric distances and hence could survive there for the required length of time. This cloud of comets became known as the Oort cloud. It has still not been directly observed, but a large amount of indirect evidence seems to point to its existence.

This theory also explained the two types of cometary orbits, *new* comets would be those entering the inner Solar System after being perturbed from the Oort cloud. They would have a long period, high eccentricity, and random inclinations, as is observed. In contrast, *old* comets would have been perturbed by the major planets, principally Jupiter, and would have much shorter periods, of the order of the Jovian period, low eccentricity and low inclination, again as observed.

However, in the late 1980's it became apparent that a problem existed with the Oort cloud model as described above. Numerical simulations of the evolution of the Oort cloud under the action of planetary perturbations were unable to reproduce the main observed characteristics of the orbits of the short period comets. Particular difficulties were encountered in reproducing, from an initial isotropic distribution of Oort cloud comets, the observed near coplanar distribution of orbits found in the short-period comet population (see for example [2], [16], [15]). In order to produce the observed low inclination population of short period comets, and initial population also lying close to the plane of the ecliptic is required.

[3,4] and [9] had independently produced theories for the origin of the Solar System in which the material in the disk or nebula out of which the planets formed did not suddenly cease at around the distance of Neptune. However, since accumulation into planets at such distances is difficult, a swarm of smaller bodies might form. In the late 1980's ground-based observing technology reached a level where it was feasible to search for bodies lying within the Edgeworth-Kuiper belt, culminating in its discovery.

4 The Trans-Neptunian Region

Discovery does not usually come with the first attempt and this was also the case for Trans-Neptunian Objects. Some early, but unsuccessful searches were by [13], [11], [1] and [17]. Another early search was by [8] who surveyed $6400deg^2$ of sky down to a V magnitude of 20. He failed to find any Trans-Neptunian objects, but the survey discovered the first member of a new class of Solar System objects which has been given the collective name of *Centaurs*, namely 2060 *Chiron*.

4.1 The Centaurs

The orbit of 2060 *Chiron* has a perihelion distance of 8.46AU and an aphelion distance of 18.96AU. It is thus a Saturn crosser and is also nearly an Uranian crosser. Since Chiron's discovery, 60 other similar bodies have been discovered, including 5145 *Pholus*, 7066 *Nessus*, and 10199 *Chariklo.* Some of these, including 5145 *Pholus* and 7066 *Nessus* have their aphelia beyond Neptune/ and are generally thought to be closely related to the Trans-Neptunian population.

4.2 The Cubiwanos

As already mentioned, the first detection of a potential member of the Edgeworth-Kuiper belt, 1992 QB_1, was made at the end of August 1992 [6]. A second object was found early in 1993, called 1993 FW. Subsequent observations confirmed the orbits of these two as being near circular with a radius at around 43 AU, thus allowing QB_1 to be given a permanent number (15760). If the albedo of these objects is about 0.04 (similar to comets) then their diameters are in the $200km$ range, considerably larger than cometary nuclei and comparable to some of the larger asteroids. There are now several hundred such objects knownand the group is called after the first, namely *Cubiwanos*, characterized by near circular orbits between about 40 and 50 AU, and essentially being the type of bodies and orbits envisaged by both Edgeworth and Kuiper. Based on current discovery statistics this is significantly the most populus of the groups.

4.3 The Plutinos

Following the impetus given by these discoveries, it was natural that further searches would be conducted. Four objects were discovered in 1993 and were given the temporary designations 1993 RO, 1993 RP, 1993 SB and 1993 SC. Assuming circular orbits, the heliocentric distances for these four new objects were respectively 32.3, 35.4, 33.1 and $34.5AU$, much smaller than for the initial two objects and indeed, placing all four well within the orbit of Pluto and only slightly larger than the orbit of Neptune so that close encounters with Neptune would occur at frequent intervals, leading inevitably to major perturbations of the orbits. 1993 SC was considerably brighter in apparent magnitude than the others, and is still one of the brightest known Trans-Neptunian objects. This brightness allowed it to be observed by other smaller telescopes so that a number of astrometric positions became available for it in the months following discovery. One possible orbit, consistent with the observational constraints, was that 1993 SC moved on a Pluto-type orbit very close to the 2 : 3 mean motion resonance with Neptune. Such an orbit would have a period of order 248 years, and a semi-major axis of around $39.5AU$, the eccentricity being around 0.13, a perfectly sensible value but significantly different from zero [19]. Both 1993 SB and 1993 SC have now been observed over many oppositions and have been given permanent numbers 15788 and 15789. by now a large number of bodies on similar orbits have been discovered and are popularly called the Plutinos because of the similarity of their orbits to that of Pluto.

4.4 The Scattered Objects

In 1996, objects were discovered that did not fit into either of the above two categories. The best known was given the temporary designation 1996 TL_{66} and subsequently the permanent number 15874. It appears to be on a very different type of orbit. It has a semi-major axis of $84.5AU$ and an eccentricity of 0.585. At aphelion, it is thus more that $130AU$ from the Sun while the perihelion distance is only $35AU$, similar to the Plutinos. It also has a high inclination of 24°.

5 Inter-relationships

It is now a days commonly believed that originally, that is at the formation epoch, there were two populations of minor bodies - what might be termed the *asteroidal* population that resided roughly where the current main-belt asteroids reside and a *Trans-Neptunian* population that may have originally formed in the Uranus-Neptune region. Both of these populations have undergone considerable evolution, through both inter-body collisions and gravitational perturbations (both by the planets and other stars) to reach the current state.

In the case of the asteroids, a considerable fraction have been lost entirely from the system, some have been captured as satellites of the planets, some are observed as Near-Earth Objects and some are present as Trojan asteroids. The trans-Neptunian population has also evolved, some being affected by other stars and evolved into the *Oort cloud*, some were captured into resonance and became the *plutinos*, some remained as the *cubiwanos*, some are now seen either as *scattered objects* or as *Centaurs* and some, possibly via some of the other stages as comets which may in turn be currently observable as Near-Earth objects. Based on where they are currently located, it is thus a difficult task to decide on the origin of a particular object, hence the importance of physical studies. An icy body is more likely to be an original *Trans-Neptunian* while a rocky one an original *asteroid*. Going into details of the determination of physical characteristics is beyond the scope of this broad overview and some aspects will be considered elsewhere.

References

1. Cochran A.L. Cochran W.B. & Torbett M.V., 1991, A deep image search for the Kuiper disk of comets, *Bull. Amer. Astron. Soc.*, **23**, 1314
2. Duncan M. Quinn T. & Tremaine S., 1988, The origin of short period comets *Astrophys. J.*, **328**, L69-73.
3. Edgeworth K.E., 1943, The evolution of our Planetary System, *J. of the British Astron. Assoc.*, **53**, 181-188
4. Edgeworth K.E., 1949, The origin and evolution of the solar system, *Mon. Not. R. Astr. Soc.*, **109**, 600-609
5. Jewitt D.C.,1997, Cometary Rotation: An overview, *Earth, Moon and Planets*, **79**, 35-53
6. Jewitt D.C. & Luu J.X.,1993, Discovery of the Candidate Kuiper Belt Object 1992QB_1, *Nature*, **362**, 730-732
7. Ho Peng Yoke, 1964, Ancient and medieval observations of comets and novae in Chinese sources, *Vistas in Astronomy*, **5**, 127-225
8. Kowal C., 1989, A Solar System Survey, *Icarus*, **77**, 118-123
9. Kuiper G.P., 1951, On the origin of the Solar System, In *Astrophysics, A Topical Symposium*, (J.A. Hynek ed.), 357-424, McGraw-Hill, New-York
10. Lamy P.L. Toth I. Grun E. Keller H.U. Sekanina Z. & West R., 1996, observations of Comet P/Faye, *Icarus*, **119**, 370-384
11. Levison H.F. & Duncan M.J., 1990, A search for Protocomets in the outer Regions of the Solar System, *Astron. J.*, **100**, 1669-1675.
12. Lowry S.C. Fitzsimmons A. Cartwright I.M. & Williams I.P., 1999, CCD photometry of distant comets, *Astron.Astrophys.*, **349**, 649-659
13. Luu J.X. & Jewitt D.C., 1988, *Astron. J.*, **95**, 1256
14. Oort P., 1950, The structure of the cloud of comets surrounding the Solar System and a hypothesis concerning its origin, *Bull. Astron. Inst. Netherlands*, **11**, 91-110
15. Quinn T. Tremaine S. & Duncan M., 1990, Planetary Perturbations and the origin of short period comets, *Astrophys. J.*, **355**, 667-679
16. Stagg C.R. & Bailey M.E., 1989, Stochastic capture of short period comets, *Mon. Not. Roy. Astr. Soc.*, **421**, 507-541
17. Tyson J.A. Guhathakurta P. Bernstein G. & Hut P., 1992, Limits of the surface density of faint Kuiper Belt objects, *Bull. Amer. Astron. Soc.*, **24**, 1127
18. Whipple F.L., 1950, A comet model 1, The acceleration of comet Encke, *Astrophys. J.*, **111**, 375-394
19. Williams I.P. Fitzsimmons A. O'Ceallaigh D. & Marsden B.G., 1995, The slow-moving objects 1993SB and 1993SC, *Icarus*, **116**, 180-185

20. Wisdom J., 1982, The origin of Kirkwood gaps: A mapping for asteroidal motion near the 3:1 commensurability, *Astron.J.*, **85**, 1122-1133
21. Yeomans D.K., 1991, Comets: a Chronological History of Observations, Science, Myth and Folklore, J. Wiley and Son, New York

Dust in the Solar System and in Other Planetary Systems

Ingrid Mann

Max-Planck-Institut für Aeronomie, 37191 Katlenburg-Lindau, Germany

1 Introduction

Dust particles play a key role in the evolution of the solar system and in other circumstellar systems. The formation of the Sun and stars is initiated by the collapse of a cloud of dust and gas. An accretion disk is formed around the central body and the planetary system eventually evolves. Some of the original dust particles are blown out-off the newly forming solar system by solar radiation and solar wind pressure, possibly being deflected in the gravity field of the newly formed large planetesimals. Once a planetary system is formed the small bodies in the solar system are a source of dust production. The activity of comets and the collisional evolution of asteroids, meteorites and possibly Kuiper belt objects lead to the formation of a second generation dust cloud, called the zodiacal cloud in our solar system. Similar dust debris clouds, although denser than the zodiacal cloud are now observed around other stars. The zodiacal dust cloud and other debris clouds show the evolution of small bodies in the solar system and in planetary systems in general. The dust also exhibits physical effects caused by the interaction with the surrounding plasma and radiation which are typical for dust in a cosmic environment.

We first introduce the experimental basis of dust studies: brightness observations and in-situ detection of dust particles. The basic forces acting on dust in the solar system and the overall structure of the dust cloud are then described. While this gives the "classical"understanding of the solar system dust cloud, a more complex picture has evolved during the past decade. The dust cloud near the Sun gives an example of dust dynamics in a cosmic plasma and radiation environment. In situ measurements have shown the existence of interstellar dust and we can now attempt to sample interstellar material by measurements from spacecraft. Moreover, the discovery of Kuiper belt objects has initiated the study of possible dust production in the outer solar system. We discuss the different regions of the solar system dust cloud, the population of interstellar dust in the solar system and finally the dust in the Kuiper belt region and its comparison to dust shells around main sequence stars.

2 Brightness Observations and in Situ Detection of Dust

Observational evidence for the existence of dust in interplanetary space is given by the zodiacal light: a diffuse brightness of the night sky that appears to be distributed along the constellations of the zodiac. It is produced by sunlight scattered at, and thermal radiation emitted from, dust particles in the solar system concentrated on the ecliptic plane. Especially in the solar system, brightness observations which reveal the overall structure and average properties of the dust can be compared to local information, which the in situ measurements provide.

2.1 Zodiacal Light and Zodiacal Emission Observations

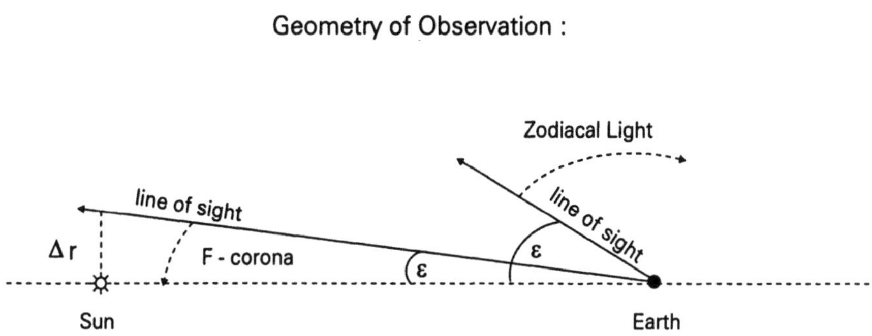

Fig. 1. Geometry of zodiacal light observations. The zodiacal light is seen from Earth as function of the elongation ϵ of the line of sight (within the ecliptic plane), for small elongations the zodiacal light continues into the solar corona. For a given elongation, the line of sight crosses a minimum distance Δr from the Sun.

The visible zodiacal light is measured from the Earth at elongations > 30° from the Sun as indicated in Fig. 1. Some measurements in the visible have been made from spacecraft, namely from Helios between 1 AU and 0.3 AU and Pioneer 10 and 11 at distances beyond 1 AU. The thermal emission brightness was measured by the IRAS and COBE satellites as well as from sounding rockets. More recent rocket observations of the zodiacal light also cover the near-infrared regime and demonstrate the transition from the scattered light to the thermal emission brightness. A detailed description of zodiacal light observations compared to the other components of the diffuse night sky brightness is given in [14]. The spectral slope of the zodiacal light mainly follows the solar spectrum in the visible regime and the brightness increases again to the mid-infrared showing a second maximum caused by the thermal emission of dust particles. Although the zodiacal light stems from the integrated signal along the line of sight, the majority of the thermal brightness seen at large elongations comes from regions near the Earth orbit. Particles with black body temperature of 280 K at 1 AU show maximum thermal emission at 11 μm wavelength. The maximum of the zodiacal emission brightness is observed between 10 and 20 μm. The brightness at smaller solar elongations stems, to a large extent, from regions of high dust number density and high dust bulk temperature near the Sun. A blackbody at 0.1 AU attains a temperature of 880 K and its maximum thermal emission is at 3 μm wavelength. Consequently the maximum of the thermal emission brightness shifts to shorter wavelengths and scattered light and thermal emission brightness overlap near the Sun. Based on the brightness data, the interplanetary dust properties can be compared to models of light scattering and thermal emission from single dust particles.

2.2 Light Scattering at Dust Particles

The scattering efficiency of single grains as function of the scattering angle is described by Mie theory which gives an analytical description of the interaction of electromagnetic waves with a particle with defined boundaries. Exact solutions exist only for some special cases such as spheres, spheroids, and cylinders, with size parameters $\alpha = 2\pi s/\lambda$ close to unity [23], [4] where s is the radius of the particle and λ the wavelength of scattered light. For $\alpha \gg 1$, which is the case for particles in the 1 to 100 μm size range that produce the zodiacal light, analytical descriptions are not yet available.

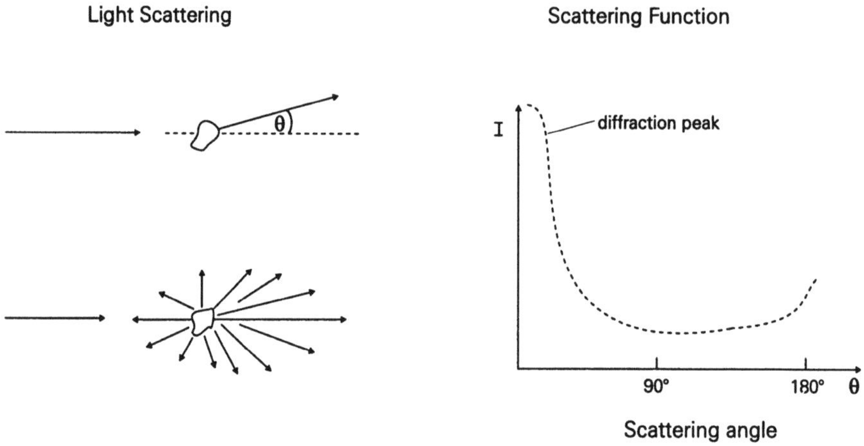

Fig. 2. The light scattering at a small obstacle shown on the left hand side and the relative slope of the scattered intensity of light at particles large compared with the wavelength as a function of scattering angle, Θ.

Empirical scattering functions as shown in Fig. 2 describe a diffraction peak at small scattering angles and a flat slope for scattering angles around

90°. The diffraction peak for small angles is determined by the size of the obstacle, the intensity of scattered light at medium angles depends on the material composition and the surface structure of the particle. The multiple reflection at a grain with an irregular surface reduces the scattering efficiency compared to a perfect sphere at medium scattering angles. The enhanced back scattering around 180° scattering angle is diagnostic of the geometric effect of the irregular surface.

The albedo is a measure for the reflectivity of objects [8] and is often used to compare the properties of dust in different environments. The average albedo of dust particles is either inferred from the comparison of thermal emission and scattered light brightness, or it can be derived from the brightness based on a priori assumptions of dust geometric cross section and dust number density distribution. The albedo of dust particles is given as the generalized geometric albedo for a scattering angle of 90 ° [8]. It has been found to be between 8% and 15% at maximum for particles at 1 AU applying different thermal emission data as well as other assumptions. Laboratory experiments with irregular particles of meteoritic as well as terrestrial material yield albedo values between 5% and 9% and are also comparable to brightness observations [24]. The derived average albedo is typically greater than the albedo of dust observed in the vicinity of comets. This means that either a significant component of the zodiacal dust originates from other sources, that is asteroids, or that dust particles are processed in the interplanetary medium, so that their structure and optical properties change [21]. Further information about dust properties can be obtained by comparing the thermal emission brightness to model calculations of the dust temperature.

2.3 Temperature of Dust Particles

Dust particles in interplanetary space usually attain an equilibrium temperature which is a balance between the absorbed radiation integrated over the solar spectrum and the emitted radiation. Both, absorption and emission are determined by the optical properties of the particles. The conditions for the equilibrium temperature are described by:

$$\int_{\lambda_1}^{\lambda_2} F_o(r,\lambda)\pi s^2 Q_{abs}(s,\lambda)d\lambda = \int_{\lambda_1}^{\lambda_2} 4\pi s^2 \pi B(\lambda,T) Q_{abs}(s,\lambda)d\lambda \tag{1}$$

where F_o is the solar flux at a distance r, s is the radius of the grain, Q_{abs} the absorption coefficient, λ the wavelength of absorbed and emitted radiation and B denotes the Planck function. Note that the dust grains are assumed to attain a homogeneous temperature and, as opposed to larger bodies in the solar system, emission takes place over the total surface area $4\pi s^2$. Further smaller contributions to the energy budget are provided by the sublimation energy and kinetic energy from the impact of plasma particles.

Since the absorption coefficient varies with wavelength and reflects the composition of the material, small particles show typical emission features such as the features around 10 μm attributed to the emission from silicates. Even if these features are not observed, the low absorptivity of silicates in the visible and their higher absorptivity in the near infrared lead to temperature profiles that are different from those of a blackbody. Silicate particles near the Sun tend to be colder than a blackbody, since thermal emission of the hot dust particles occurs in the near infrared, while silicate particles near the Earth orbit are warmer than a blackbody. Large dust particles, especially when they consist of different materials, show only a weak wavelength dependence of the absorptivity. With Q_{abs} being approximately unity over wavelength, they reach blackbody temperature $T_{bb}(r) = T_o(r/r_o)^{-0.5}$ where, r is the distance from the Sun and $T_o = 280$ K, is the temperature at $r_o = 1$ AU. Comparison of model calculations and brightness data has shown that interplanetary dust can be described as dirty silicates, but exact properties can not be inferred from the data. While the analysis of brightness data yields information about the average properties of grains, in-situ measurements from spacecraft provide local information.

2.4 In Situ Measurements

As indicated in Fig. 3, dust particles over a size range of 0.01 to 1000 μm can be detected with different methods. The flux of small solid bodies into the

Earth atmosphere has been known for a long time from the existence of meteors (so called "shooting stars"). The particle and surrounding atmosphere evaporate and produce the observed brightness in the night sky (i.e., photographic visual meteors). Particles below about 1000 μm in size are too small to be observed in the visible, but the ionized gas that they produce reflects radar signals. Depending on their speed, particles as small as several 10 μm can be detected as radar meteors. If the relative velocity of the particles is sufficiently small, they survive the entry and can be collected in the upper atmosphere from aircraft. Although this allows one to study at least some cosmic dust particles in the laboratory, these have to be separated from terrestrial dust that is collected at the same time. The collected particles consist mainly of elements comparable to primitive meteorites but have a higher content of carbon and volatiles. As opposed to the methods that are limited to studies from Earth, similar to brightness observations, measurements from spacecraft describe the distribution and the properties of dust in interplanetary space. While thermal emission and scattered light brightness describe the size range of particles that have the maximum geometric cross-section area (that is the range from 1 to 100 μm for particles near 1 AU), in situ measurements best describe particles which yield a large flux rate, i.e., particles with sizes below 1 μm. The surfaces of many bodies in the solar system provides natural areas for the detection of dust. If the impacting particles are not decelerated by the presence of an atmosphere they produce a crater on the surface which depends on the size and density as well as on the impact speed of the particle. Analysis of micro-craters on samples of the

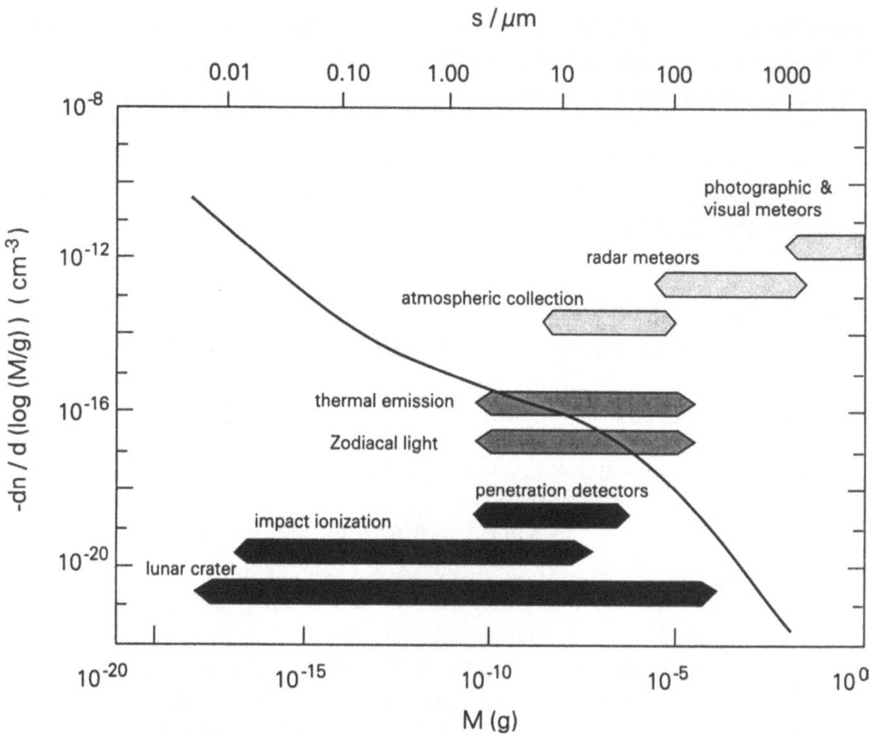

Fig. 3. The differential number density of dust at 1 AU derived (from [7]) as a function of mass and size (based on the assumption of 3 kg m^{-3} bulk density). The size ranges of different detection methods are marked.

lunar surface that were brought back to Earth with the Apollo flights allowed detailed studies of the dust flux. In a similar way, in situ experiments use the large speed of impacting particles. Some of early experiments relied on the fact that thin foils are destroyed by the dust impact. The dust detectors onboard the Pioneer 10 and 11 spacecraft, for instance, consisted of an array of pressurized cells, which were destroyed by dust impact. More advanced instruments detect the material from the dust particle and of the target that is evaporated and ionized upon impact (i.e. impact ionization detectors).

In situ measurements of interplanetary dust have been made from Earth orbiting satellites and the Helios spacecraft covering from 1 AU to a distance as close as 0.3 AU from the Sun. Pioneer 10/11 impact experiments have measured the dust flux between 1 - 18 AU. Recent analysis of data from the plasma wave experiments on Voyager 1 and 2 revealed signals that are consistent with impacts of dust particles. The masses of the particles are estimated to be greater than 1.2×10^{-14} kg and the mass threshold of the detection is given with an uncertainty of one order of magnitude. Data obtained in the inner solar system are summarized in [7]. For a discussion of the data obtained beyond 1 AU see [17]. Present measurements are mainly restricted to the mass and speed of dust particles but improved detector capabilities allow the derivation of information about the dust material composition, as was done for the first time in the vicinity of comet Halley [11]. The results of impact experiments are in good agreement with brightness observations so that we have a good understanding of the overall structure of solar system dust and the effects that determine it.

3 Predominant Forces and Overall Structure of the Dust Cloud

3.1 Acting Forces

The motion of dust particles in the solar system is primarily determined by solar gravity:

$$\mathbf{F}_g = -G\frac{M_\odot m}{r^2}\mathbf{r}^0, \tag{2}$$

where G is the gravitational constant, $M_\odot$ is the solar mass, m is the mass of the dust particle, r is the heliocentric distance of the dust particle, and $\mathbf{r}^o$ is the unit vector in radial direction. The particles move in Keplerian orbits described by the orbital semimajor axis a, eccentricity e, inclination i, the argument of the perihelion ω, and the longitude of the ascending nodes Ω. The position of a particle along the orbit is determined by the time of the perihelion passage.

Further effects stem from other forces, the predominant being the solar radiation pressure force:

$$\mathbf{F}_{rad} = \int_0^\infty B_\odot(\lambda)\pi(R_\odot/r)^2(c^{-1}\pi s^2)Q_{pr}(\lambda, N(\lambda), s)d\lambda, \tag{3}$$

where $B_\odot$ is the solar radiance, $R_\odot$ is the solar radius, c is the velocity of light, and Q_{pr} is the efficiency factor for the radiation pressure. The efficiency factor depends on the structure of the particle, the wavelength λ of scattered light, the index of refraction N which varies with wavelength, and the size s of the particle.

The radial (main) component of the radiation pressure force counteracts the solar gravity and moreover has the same variation with solar distance r. Defining β as the ratio of radiation pressure force to gravity we can consider particles to move in a reduced gravitation field F = (1-β) F_g.

As long as the gravity exceeds radiation pressure the particles are still moving in bound Keplerian orbits. If radiation pressure exceeds gravity, particles leave the solar system in hyperbolic orbits. A further consequence of radiation pressure is the Poynting-Robertson (P-R) drag force, $\mathbf{F}_{PR}$, acting on particles in bound orbit about the Sun. Seen from the moving frame of the dust particle the infalling radiation has a tangential component decelerating the particle in its orbital motion. Since the tangential component of the radiation pressure force depends on the velocity of the particle it gives rise to dissipation. The semimajor axis, and eccentricity of orbits are reduced and the particles drift to the Sun on time scales of 1000 to 100000 years.

3.2 Electric Grain Charge and Lorentz Force

Dust particles in space are electrically charged by photoelectron emission, sticking and recombination of plasma particles, secondary electron emission, thermionic emission, and field emission [6]. The grain charge depends on the size of the particles, their velocity relative to the plasma and its temperature which defines the velocity distribution of infalling plasma particles. Since photoelectron emission, secondary electron emission, and thermionic emission vary with the material, the grain charge also depends on the dust composition. As a result of the dominating photoelectron emission caused by the solar radiation, dust particles in interplanetary space are usually positively charged, as opposed to dust in denser plasmas, such as in planetary magnetospheres. The charge corresponds to surface potentials relative to infinity of between 5 and 10 V. The equilibrium surface charge of grains in the solar system is attained on timescales of less than a day. Temporal variations of the solar wind parameters yield fluctuations of the surface charge of 20 % and less [10].

The Lorentz force acting on a dust particle in the solar magnetic field is (in CGS system):

$$\mathbf{F}_L = \frac{Q}{c}\mathbf{V} \times \mathbf{B}, \tag{4}$$

where Q is the electric surface charge, $\mathbf{V} = \mathbf{v} - \mathbf{v}_{sw}$ is the velocity of the dust relative to the solar wind, and $\mathbf{B}$ is the strength of the magnetic field carried with the solar wind.

The Lorentz force changes the orbital motion of particles and, as dust grains move through the sectored magnetic field of the Sun with alternative polarities, the Lorentz force changes its direction. As long as the orbital period is long compared to the time span between polarity changes, the Lorentz force causes random variations mainly in the inclination, i, but also in a and e. For submicrometre particles with a large charge to mass ratio, the Lorentz force has some influence on the dust dynamics. At great distances from the Sun, particles are moving through the sectored magnetic field and alternately are carried to higher and lower latitudes. This broadens the initial distribution of inclinations of a given dust population. The Lorentz force is more significant in the vicinity of the Sun. The particles are further influenced by the direct solar wind drag from infalling solar wind particles and the indirect (Coulomb) solar wind drag. The direct solar wind drag can amount to up to 20 % of the radiation pressure force. The Coulomb drag is about three orders of magnitude smaller.

3.3 The Overall Structure

Aside from the listed forces, mutual collisions determine the overall structure of the dust cloud and the mass distribution of particles. The relative velocities of the particles are of the order of several 1 - 10 km/s so that the collisions are catastrophic and produce a number of smaller fragments. The collisions of larger, meteoritic bodies are a continuous source of dust particles in the solar system. The increase in number density from the large end of the size distribution down to about 100 μm in size can be approximated with a power law which is explained by the distribution of collision fragments. The lifetime of smaller particles is limited by the Poynting-Robertson effect which causes their drift toward the Sun, in turn, which changes the slope in the size distribution.

The presence of planets and their gravity can be approximated as a perturbation term expanded as a central force in inverse powers of solar distance r. As long as this approximation is valid, the angular momentum of particles is conserved and the inclinations of orbits are constant. The perturbations change the orbital elements with the effect that the argument of the perihelion and the longitude of the ascending node are randomized. Starting from an initial distribution of dust in similar orbits (such as is produced by collision of larger meteorites) the perturbations lead to a rotational symmetric cloud. This is usually assumed to explain the rotational symmetry of the zodiacal dust cloud in its overall shape.

The formation of the dust cloud is illustrated in Fig. 4. As a result of the Poynting-Robertson drift, the initial dust torus extends to smaller distances from the Sun. For particles in circular orbits the resulting slope of the number density is proportional to r^{-1}. Indeed the analysis of zodiacal light observations shows that the increase in dust number density towards the Sun follows approximately this slope and that the particles are predominantly in orbits with low eccentricity. The main component of the dust cloud is concentrated to the ecliptic plane, which means that the inclinations of orbits are less than about 30°. While we have a good understanding of the interplanetary dust cloud at 1 AU, collisions change the cloud at shorter distances from the Sun. Moreover the properties of the dust change and the dynamics of the dust becomes more complex in the solar corona.

4 The Solar Corona: Dust Inward from 0.3 AU

4.1 The Solar Corona

The scattering of solar radiation by electrons, ions and dust particles produces the brightness of the solar corona seen during solar eclipses around the lunar limb. While interplanetary dust particles are slowly "falling"into the Sun, a hot ionized gas of electrons and ions (plasma) is expanding from the Sun and forming the solar wind. The most visible component of the solar corona

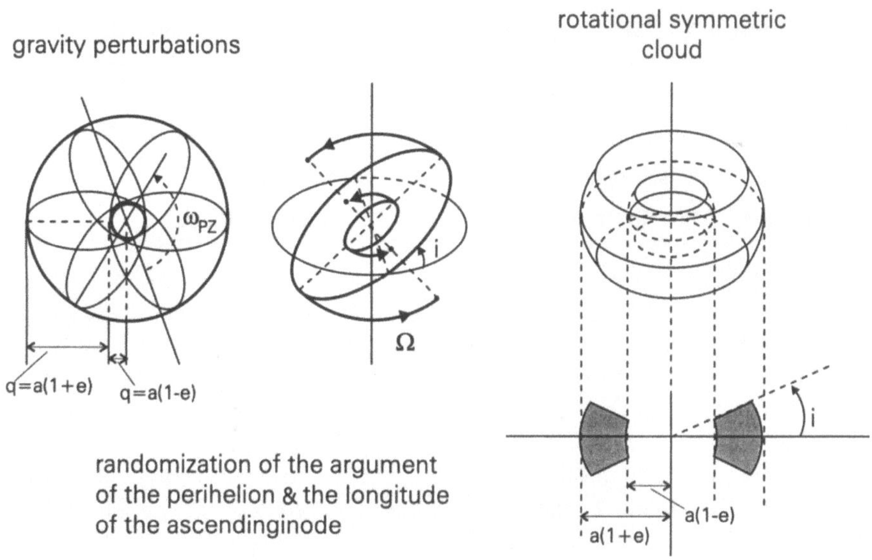

Fig. 4. The formation of a rotational symmetric dust cloud from particles with similar initial orbits under gravitational perturbations. The randomization of the argument of the perihelion distributes the particles within the plane of the original orbit. The random distribution of the longitudes of the ascending node describes the rotation of the orbital plane.

is the K-corona produced from the hot electrons in the solar environment. Although the K-corona mainly follows the solar spectrum, the Fraunhofer lines are smeared out by the Doppler shift induced with the scattering by the hot electrons in random motion. The "K"stems from the German word "Kontinuierlich", (i.e., continuous). The ions in the plasma can be seen due to induced emission of lines ("L-corona") characteristic for the element and the charge state of the ion. The particles are ionized by mutual collisions in the plasma and their charge states depend on the random motion of particles, that is the temperature of the plasma. The scattering of sunlight by the dust particles that are moving around the Sun, causes a small spectral shift but the Fraunhofer lines of the solar spectrum are still observed in the coronal brightness, leading to the name Fraunhofer-corona. Both, F-corona and zodiacal light are based on the same effect and the brightness of the zodiacal light extends smoothly into the solar corona.

4.2 Solar Corona Observations

In 1947 van de Hulst showed that the solar corona brightness is influenced by diffraction of light by dust particles near the Earth. The diffraction part in the forward scattered light at dust particles (see Fig. 2 again) is very effective such that the light scattered (with small scattering angles) from

Table 1. The solar corona is produced by scattering of sunlight

	Effect	Observation
K - corona	scattering by free hot electrons	spectrum ≈ "continuous"
L - corona	line scattering by ions	line emission
F - corona	Mie scattering by dust	≈ solar spectrum

obstacles near the observer is very intense. Corona observations, as well as all other observations of faint objects that are located close to a very bright object, are sensitive to this straylight produced by scattering from objects near the Earth or in the Earth atmosphere, or at impurities in the telescope itself. The slope of the zodiacal light and the F-corona brightness compared to different straylight levels is shown in Fig. 5. To avoid straylight, corona observations are made during solar eclipses: The geometry of the solar eclipse is depicted in the upper part of Fig. 5 The moon shades the direct sunlight with the effect that the corona is visible out to about 6 solar radii from the centre of the Sun. For a further reduction of the atmospheric straylight, observations are often performed from high mountain sites. Lyot, for the first time constructed a so called "coronagraph" for observations from Pic du Midi in France. A coronagraph simulates the conditions of a solar eclipse with an occulting disk that shades the direct solar brightness. But still the viewing conditions are limited by atmospheric straylight. Better conditions can be achieved with spaceborne coronagraphs. A white light coronagraph (Large Angle Spectrometric Coronagraph) has been working for nearly 5 years on the SOHO satellite.

4.3 Dynamics of Dust Near the Sun

The influence of solar radiation pressure and Lorentz force becomes particularly important in the vicinity of the Sun and changes the dynamics of dust grains. Early observation of features in the radial slope of the equatorial near infrared F-corona brightness initiated studies of near-solar dust dynamics in order to discuss the formation of dust rings (i.e., regions of enhanced dust density) around the Sun. The left hand side of Fig. 6 demonstrates the dynamics of dust near the Sun that leads to the formation of a dust ring. Particles drift towards the Sun where their size is diminished by sublimation. Since, for smaller particles, radiation pressure becomes more important, particles may be carried outward again as shown for carbon particles. Particles are either carried away in hyperbolic orbits or drift in and out as a result of direct radiation pressure force that carries them outward and Poynting-Robertson effect that carries them inward again. Depending on the parameters this interplay can lead to the formation of a dust-ring. For silicate particles, in contrast, the radiation pressure force is not sufficient to carry particles away

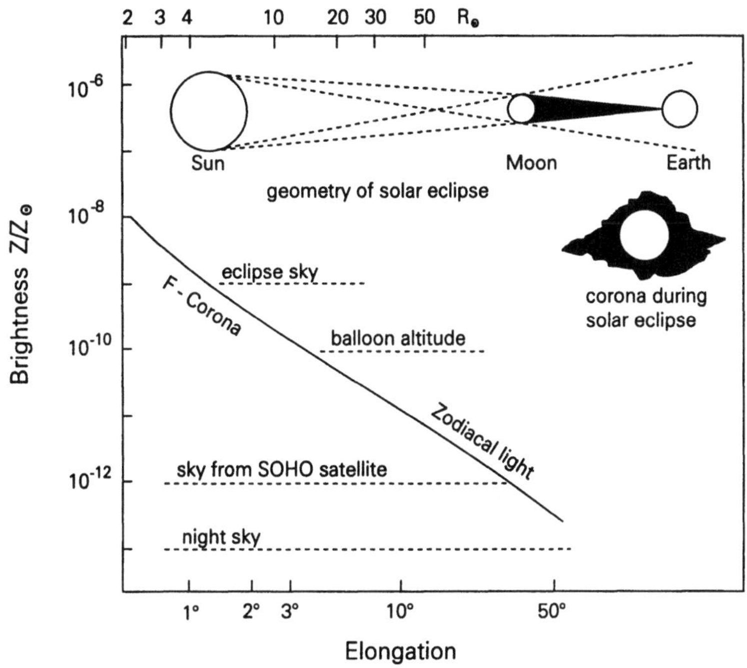

Fig. 5. The brightness as seen with the geometry depicted in Figure 1, i.e., the scattered light from dust seen in the ecliptic plane. The corresponding minimum distance from the Sun that the line of sight crosses is given in the upper scale in units of solar radii. The horizontal lines show the stray light levels of the night sky, the eclipse sky from ground and in balloon altitude and the straylight level that is reached with the LASCO coronagraph on SOHO.

from the vicinity of the Sun. The right hand side of Fig. 6 shows the near infrared F-coronal brightness calculated for this given dust orbit. Silicate particles produce a smooth slope of F-coronal brightness determined by scattered light. Carbon particles produce a brightness profile that is dominated by thermal emission of the absorbing grains. The brightness produced by carbon particles shows a pronounced drop off in the brightness at the point where the line of sight crosses the dust free-zone. This feature is enhanced by the presence of a dust concentration, (i.e., dust ring) which is produced by the above mentioned dynamical effect. Hence the the strength of the feature in the near infrared brightness allows one to draw some conclusions about the properties of dust near the Sun. Although detector capabilities have improved during the last two decades, most of the recent observations do not show any infrared features in the solar corona. This indicates that the amount of very absorbing grains in the vicinity of the Sun is probably small.

The change of the radial distribution is not the only change in the appearance of the dust cloud near the Sun. As a result of the increasing solar magnetic field small grains are strongly influenced by the Lorentz force. As-

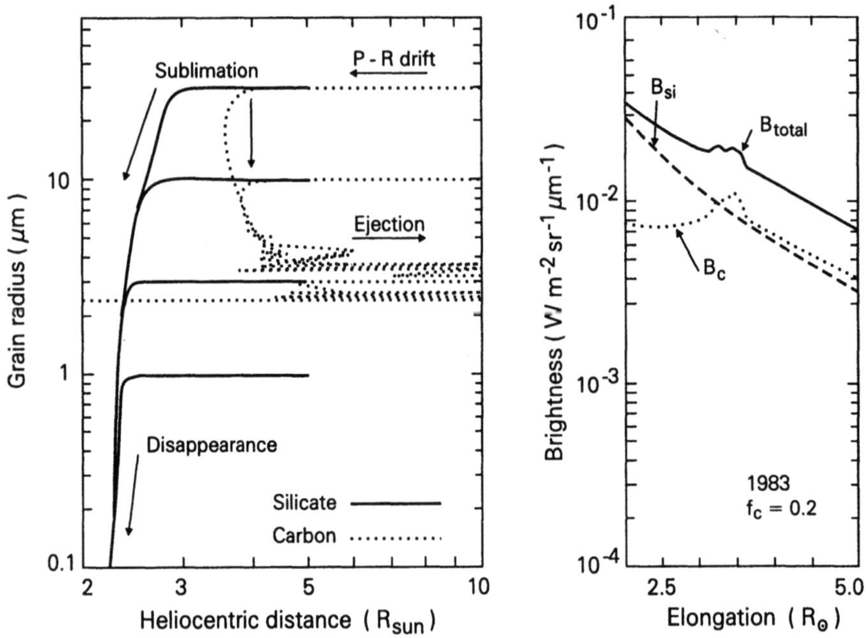

Fig. 6. Formation of dust rings and resulting features in the coronal brightness. The brightness model is based on the assumption of 20% carbon grains. A lower amount of carbon grains reduces the excess of the brightness feature. Further explanations are given in the text.

suming that the interplanetary dust cloud is mainly formed by particles in low inclination orbits close to the ecliptic plane, the influence of the Lorentz force becomes so important for dust grains below 0.1 μm in size, that they are carried into orbits with random inclination. Depending on the solar cycle and the magnetic field strength, particles up to sizes of 1 μm can form an isotropic dust cloud around the Sun at distances smaller than 0.1 AU. Since the particles which are not strongly absorbing can reach very close to the Sun, we expect a complex picture of time variable dust fluxes in the solar magnetic field.

5 Interstellar Dust in the Solar System

5.1 Dust in the Interstellar Medium

The dust cloud beyond the asteroid belt contains a significant amount of interstellar dust particles. The existence of dust particles in interstellar space was first deduced from the attenuation of starlight through the interstellar medium (ISM). Optical extinction measurements revealed the existence of gas and dust in the ISM and showed that their relative abundances on large scales are correlated. As derived from studies of the ISM, the number density of interstellar particles surrounding the solar system is 0.1 - 0.15 cm^{-3} for neutrals, 0.04 - 0.1 cm^{-3} for ions and electrons, and 10^{-14} cm^{-3} for dust particles. Although the number densities of grains in the ISM are small, heavier elements, such as Al, Ca, Mg and Fe are predominantly condensed into dust grains. The study of interstellar grains is important for understanding elemental and isotopic abundances of the ISM as well as for the physics of grain formation and grain evolution. Some of its properties are derived from ground-based optical and near-IR observations as well as far-IR and ultraviolet observations from satellites. A better understanding is expected from the in-situ detection of interstellar grains, within the solar system. Interplanetary dust and interstellar dust can be separated by their speed and flux direction. But also a part of the interstellar dust is deflected from entering the solar system, so that measurements within the solar system do not always represent ISM conditions. Therefore the conditions under which grains enter the solar system need to be studied.

5.2 The Heliosphere and the Local Interstellar Medium

The formation of the heliosphere results from the interaction between the outward streaming solar wind and the plasma of the local interstellar medium. The heliosphere is the region around the Sun which is inflated with the solar wind plasma, the motion of which determines the electric and magnetic field. While the majority of the electrically charged interstellar particles are deflected from entering the heliosphere, high-energy cosmic rays and interstellar neutral gas penetrate the heliosphere. When the charge of dust particles is sufficiently small compared to their mass, dust particles can also enter the heliosphere. While the dust and gas are expected to be coupled in the interstellar medium at least on large scales, close to the heliopause, which is the boundary to the heliosphere, the plasma flow is stopped by the solar wind. Assuming that the magnetic field of the interstellar medium is coupled to the plasma, relative motion between the dust particles and the surrounding plasma imposes a Lorentz force on electrically charged grains. The component of the magnetic field which is perpendicular to the motion of the dust particles will deflect them. Particles with large electric surface charge are either accelerated along the heliopause or reflected back into interstellar space, while sufficiently "neutral" interstellar dust particles will stream into the solar system.

5.3 The Flux of Interstellar Dust into the Solar System

First evidence for the in situ detection of interstellar dust was discussed when an Earth orbiting satellite detected a variation in the dust flux that was explained by the focussing effect of interstellar dust. More recent measurements onboard the Ulysses spacecraft have detected interstellar dust particles and neutral interstellar helium. The antapex direction of the motion of the Sun relative to the local interstellar medium as derived from the flux of interstellar helium is at 73.9°± 0.8° ecliptic longitude and - 5.6°±0.4° ecliptic latitude [25]. This gives the orientation of the velocity vector of the flux of interstellar matter into the solar system–if simply determined by relative velocities. The flux direction is almost parallel to the ecliptic plane and the Sun and Earth form a line nearly parallel to the flow direction of interstellar matter around December 5 with the Earth being behind the Sun, as seen from the interstellar downwind direction. The Lorentz force associated with the solar magnetic field, the solar radiation pressure force, and the solar gravity, influence particles that enter the solar system. Neglecting their surface charge, interstellar dust particles would form a mono-directional flux from the interstellar upstream direction relative to the Sun that is modified by solar gravity and solar radiation pressure forces. If gravity is the dominant force, particles are in hyperbolic orbits, with the focus of orbits behind the Sun: the flux will be collimated in the interstellar downwind direction (i.e., gravitational focussing). If the radiation pressure force exceeds gravity then particles are exposed to a repulsive force and are on hyperbolic orbits with their focus in the interstellar upwind direction (i.e., radiation pressure repulsion). For small particles the charge to mass ratio of particles is large and orbits are further influenced by the solar magnetic field. Depending on the solar cycle, electrically charged particles are either deflected to higher or to lower helio-ecliptic latitudes.

5.4 Experimental Results

The in situ measurements of interstellar dust within the solar system (Fig. 7) provide data for particles with masses between 10^{-19} kg and 10^{-13} kg. Assuming average densities of 2 - 4 $\times$ 10^3 kg m^{-3} and particles of spherical shape, the detected interstellar dust particles are in an interval $0.015\mu m < s < 4.1\ \mu m$. These particles are large compared to the particles that are observed in the interstellar medium (ISM). The distribution for the ISM shows a sharp cut off to larger masses at 10^{-16} kg and for instance the distribution for classical models of the interstellar extinction is described as $n(s) \propto s^{-3.5}$. Interstellar grains that exceed the mass of the interstellar grains expected from the conventional astronomical models are also identified in meteorite samples, as shown in Fig. 7, where diamonds, graphite, SiC and oxide grains of interstellar origin with masses 10^{-18} kg $< m < 10^{-12}$ kg have been separated.

The interstellar dust that is detected with smaller masses may not represent ISM conditions for two reasons: Small particles for one can be deflected at the heliopause from entering the solar system. Furthermore the efficiency of the detector decreases with the size of grains. The measured mass distribution of interstellar dust in the solar system varies with the distance of

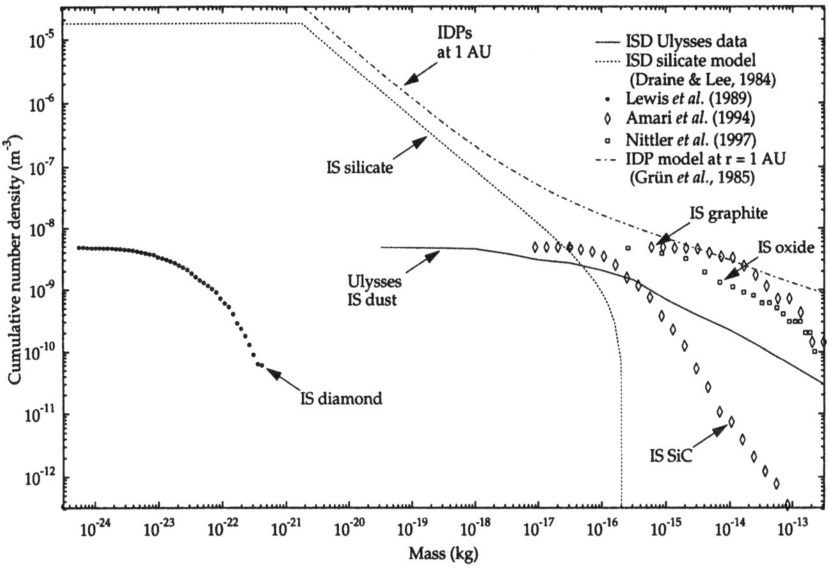

Fig. 7. In situ detection of interstellar dust compared to astronomical models [17]. The cumulative number density distribution of interstellar grains detected by Ulysses from 1990 until end of 1995. The solid line shows the mass distribution of dust that explains interstellar extinction data with scattering and absorption by silicate particles (cf. [6]). The other data points show inclusion of interstellar dust found in meteorite samples: diamonds [15], graphite [1], SiC [1] and oxide grains [22].

observation. This is explained through the influence of the radiation pressure force. Depending on their size and scattering properties, particles are repelled by radiation pressure from further approaching the Sun. Since the orbit of the spacecraft is located in the interstellar upstream direction approximately perpendicular to the interstellar flux into the solar system, the data represent the flux interstellar upstream direction. The distribution of interplanetary dust, shown for comparison exceeds the interstellar component within about 5 AU around the Sun. Nevertheless, those results provide us with the opportunity of analysing the properties and material composition of dust from interstellar space by measurements within the solar system and even to bring samples of interstellar dust to laboratories on Earth.

6 Dust in the Kuiper Belt Region and in Interstellar Debris Shells

As can be seen from the general sketch of small bodies in the solar system in Fig. 8, the discussed dust cloud covers only innermost part. Another dust component- so far not well-understood - is the dust produced in the outer solar system beyond Neptune. Estimates of the outer solar system dust component have been made in connection with studies of Edgeworth-Kuiper objects at solar distances > 40 AU. Dust particles may be produced by mutual collisions of Kuiper belt objects as well as by impact erosion from interstellar grains hitting the surface of the objects [26]. About 20 % of the particles with sizes greater than 3 μm which are produced in the Kuiper belt region are expected to enter the inner solar system [16]. However, even if the Kuiper belt dust is not a significant component in the inner solar system, it would be detected if the solar system were observed from larger distances outside of the solar system. The Kuiper belt dust cloud is probably more comparable to the dust clouds which are observed around other stars, than the zodiacal dust cloud is.

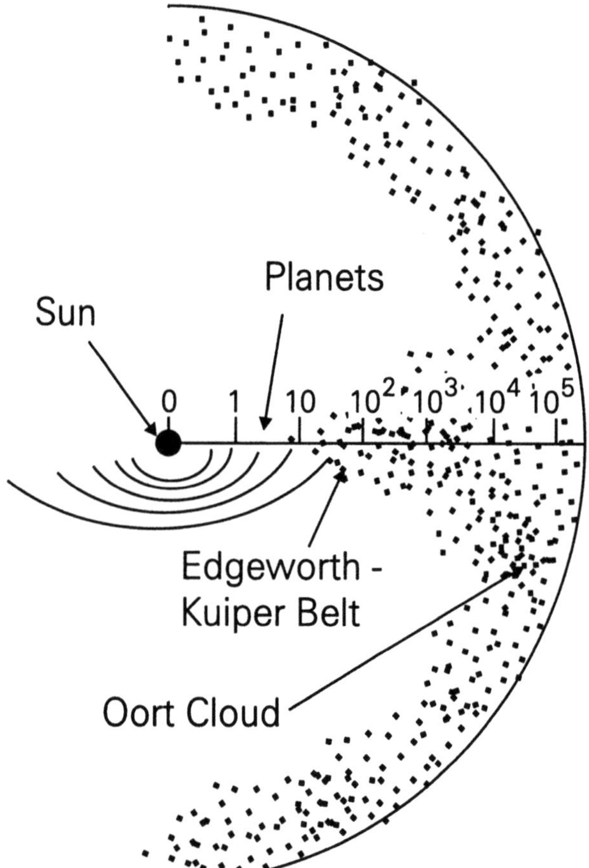

Fig. 8. The large scale structure of the solar system: planets, Kuiper belt objects and Oort cloud. The zodiacal dust cloud is concentrated to the most inner region within the asteroid belt, note that distances from the Sun are on a logarithmic scale.

6.1 New Observations of Circumstellar Dust: The Vega Phenomenon

Shells of dust particles have been observed recently for a number of main-sequence stars. IRAS provided first observations of the so-called Vega phenomenon: the IR excess in the spectral slope of the stellar brightness discovered when Vega (α Lyr) was observed for calibration purposes. Whereas younger stars show a spectral energy distribution which is determined by the hot dust particles surrounding the star, stars in a later stage of evolution show a spectral energy distribution which reflects the temperature of their stellar photospheres. However, the above mentioned observations show a small peak in the infrared brightness which is several orders of magnitude fainter than the brightness peak at shorter wavelength produced by the photospheric emission (in the visual or UV spectral range). Spatially resolved observations of some of the systems show that this second IR peak stems from a disk-like brightness structure around the star.

The observed excess brightness is assumed to stem from the thermal emission of dust particles in orbital motion about the stars. Although the exact ages of the single stars is estimated with great deal of uncertainty, it is assumed that a significant number of these observed objects have already reached the main sequence stage. Moreover, the color temperature shows that the dust is much cooler than the dust in young stellar envelopes and the dust shells are optically thin and the amount of gas is small. Hence, most of these observed dust shells are not part of a proto-planetary nebula which cools over a time period of the order of 10^4 to 10^6 yr. Compared to the stages of the formation of a solar system depicted in Fig. 9, we can assume that these dust shells are mainly attributed to the last stage of formation. Far-IR emitting dust shells occur around at least 15% of the nearby normal field stars [2], [12] of spectral type A, F, G, and K with stellar surface temperatures be-

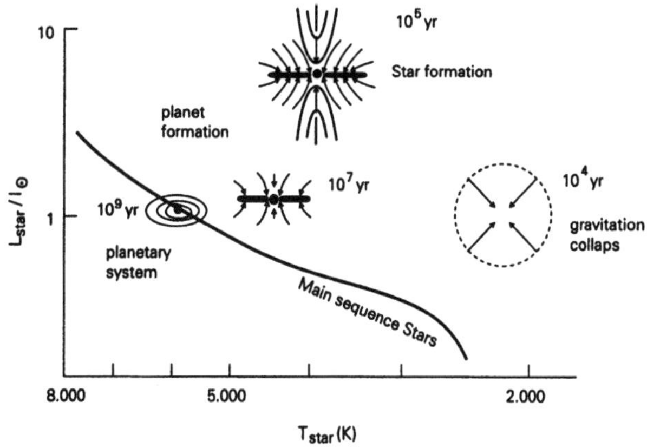

Fig. 9. The stages of the formation of a solar system (from [3]).

tween about 9000 K and 4000 K. It is argued that not that many extremely young stars would be observed in the vicinity of the Sun. The dust is not a remnant of the star formation process but has to be produced by larger parent-bodies after the star is formed. Observations also show the presence of small submicron-sized grains in the dust shells. Since these are blown away by radiation pressure and have a short lifetime in the dust shell, larger bodies embedded in the disks must continuously replenish the dust. The vertical extension of the dust shells requires gravitational perturbations from bodies of size s $\approx$ 1000 km.

Table 2. Circumstellar debris disks observed spatially resolved with IRAS, respectively ISO. Listed are the characteristic radii, r_c, derived from the color temperature, the Poynting Robertson lifetime, τ_{PR}, the collision lifetime τ_{coll} at r_c and the estimated age T_{star} of the star (from [2])

star	r_c	τ_{PR}	τ_{coll}	T_{star}
α Lyr	150 AU	10^7 yrs	10^7 yrs	$4 \cdot 10^8$ yrs
α PsA	150 AU	10^7 yrs	10^6 yrs	$2 \cdot 10^8$ yrs
β Pic	75 AU	10^6 yrs	10^4 yrs	$1 \cdot 10^8$ - $1 \cdot 10^8$ yrs
ϵ Eri	39 AU	10^8 yrs	10^6 yrs	$8 \cdot 10^8$ yrs

Vega or β Pic - like systems are characterized [12] on the basis of the luminosity and the mass of the shell compared to the star, and by the gas to dust ratio within the shell. As opposed to young stellar objects, the luminosity of these shells is significantly lower than the luminosity of the star. The mass of dust and gas is significantly less than the mass of the star and moreover is significantly less than the mass of the minimum proto-solar nebula, which is assumed to be 0.01 $M_\odot$. This distinguishes the shells from planetary systems in their early stage of formation. Finally the dust mass is at least about 10 times larger than the gas mass, consequently the dynamics of dust is not controlled by gas and this again distinguishes these shells from planetary systems in an earlier stage of formation. The listed criteria also apply to the solar system dust cloud, although the dust density in the circumstellar debris disks that were observed so far, is of the order of $\tau \sim 0.001$ to $\sim 10^{-6}$ and therefore greater than in the zodiacal cloud. Debris shells similar to the zodiacal cloud are too faint to be identified with present observational techniques.

6.2 Spatial Distribution of Dust in Vega-Type Systems

Observations show that the spectral brightness from the dust shells peaks at wavelengths corresponding to temperatures below about 200 K. The characteristic radii, r_c, listed in table 2 are inferred from the peak in the spectral

Table 3. Parameters of observed circumstellar debris shells: distance from the Sun, d, dust colour temperature, T_{dust}, estimated size of grains, s_{dust}, and luminosity L_{dust}/L_{star} of the shell compared to the star

star	d [pc]	$T_{dust}[K]$	$s_{dust}[\mu m]$	L_{dust}/L_{star}
α Lyr	8	85	10 - 100	$2 \cdot 10^{-5}$
α PsA	7	60	ca 10	$8 \cdot 10^{-5}$
β Pic	19	110	ca 1	$2 \cdot 10^{-3}$
ϵ Eri	3	50	ca 10	$7 \cdot 10^{-5}$
zodiacal dust	-	280	1-100	$1 \cdot 10^{-7}$
KBO dust	-	<40	> 1	$< 10^{-7}$

slope of the brightness: the characteristic radius of the dust shell is derived at the distance of the star at which a blackbody would attain the temperature that corresponds to the derived wavelength of peak thermal emission. Assuming that the number density of dust increases towards the star, the integrated thermal emission brightness would represent the emission of dust particles very close to the star with the greatest dust number density and temperature. Hence, it is inferred that there is a depletion of dust in the inner regions around the stars. Spatially - resolved images, both in the visual and the IR, have been obtained for β Pic as well as some other systems. They demonstrate that the slope of the radial brightness distribution changes in the inner part of the disks. Observations further show that the material is concentrated in a disk-like structure, which extends to distances of 100 to 1000 AU around the star. Studies of the β Pic system show that the southwest extension of the disk appears smaller and fainter than the northeast wing. The southwest wing is also less flattened than the northeast wing. The spatial structure and the observed asymmetry of the β Pic dust shell have been studied in greater detail. The non-axisymmetric distribution of orbiting dust particles between 150 and 800 AU may result from different types of gravitational perturbations over the last 10^3 to 10^4 years [9]. However, the great extent of this asymmetry, makes it unlikely that it results from a single perturbing planet. It was suggested that particles are trapped in resonance with a large planet and consequently stop the inward motion of the dust particles with most of them being ejected from the system. Whether these zones in every case are really a consequence of the presence of planets has to be studied in more detail. It should also be noted that the observations point to a lack of dust in the inner regions, but not to a dust-free zone.

6.3 Optical Properties and Size Distribution of Eust in Vega - Type Systems

Estimates of the dominant sizes in the debris disks result from a comparison of the dominant grain temperature (derived from the maximum of the spectral energy distribution) against the angular scale of the emitting region. More information can be obtained for the case of β-Pic. The gray color of the disk indicates that particles are typically larger than a few microns, while the observation of silicate features points to the existence of particles with sizes 1 μm < s < 10 μm, most probably particles in the 1 - 3 μm size range. Larger grains (s > 5 - 10 μm) are required to explain far infrared and mm observations. We can assume that, similar to the solar system dust cloud, the dust in the debris shells covers a broad range of sizes. Compared to the solar system dust cloud the derived average albedo of grains is a little higher. The observation of IR features indicates the existence of silicate particles. The scattering phase function and the polarization function of zodiacal dust can fit the β Pic observational data.

6.4 The Gas Component in Vega Type Systems

High-resolution spectroscopy in the visible and UV provides information about the gas component surrounding the star. Spectroscopic observations of β Pic show strong narrow absorption features superimposed on the cores of some broad photospheric lines. While the photospheric absorption of the rotating stars produces the broad absorption features, the narrow lines are clearly separated from the stellar rotation indicating the existence of a gas component in the surrounding of the star. A stable gas component of neutrals and low ionized metals is seen as close as < 1 AU from the star with a relative velocity < 2 km/s. Transient absorption components are observed lasting from hours to days. The observed absorption lines are red-shifted by 10 to a few 100 km/s and some high excitation lines demonstrate collisional excitation of the ions. Atomic lifetimes and time scales for radiation pressure ejection indicate both gas components to be "second generation" populations. The first transient absorption component may result from falling evaporating bodies such as comets falling onto the star. The observed gas production will also supply the latter, persistent gas component. Although other systems show evidence for a gas component surrounding the star, there is no simple relation between the detection of gas and the presence of dust in the systems [12]. The observed circumstellar debris shells may still not be well understood, but nevertheless can be seen as candidate systems for the existence of extra-solar planets in them.

7 Future Studies: Dust and Extrasolar Planets

The age of the dust shells discussed above is of the same order of magnitude as the time scales expected for the formation of planets and we can assume

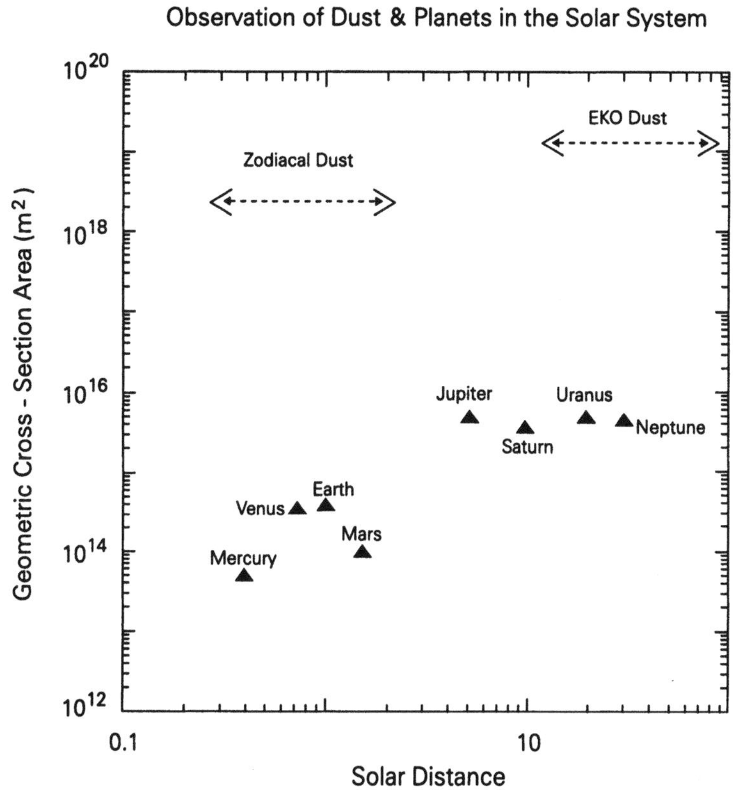

Fig. 10. The geometric cross-sectional area of the zodiacal dust cloud and the estimated geometric cross sectional area of the Kuiper belt dust cloud ("EKO Dust") compared to the geometric cross-section of planets in the solar system.

that the observed systems are solar systems at an earlier evolutionary stage. It is most likely that the dust density in these systems decreases in time, and that the dust shells at a later stage are similar to the solar system dust cloud. This links the study of these circum-stellar dust shells to the search for and the study of extra-solar planets.

The largest fraction of successful detections of a planet around a star applies the Doppler - planet detection technique ([18], [19]). It is based on the fact that the motion of the planet around its central star imposes a small relative motion on the star. The resulting periodic motion of the star relative to the Earth can be detected by the Doppler shift of stellar spectral lines. This method allows one to estimate the mass (to be more specific the product M *sini*, where M is the mass of the companion and its the inclination of the orbit relative to the line of sight of observation), and the size and the period of the orbit. The direct imaging of an extra-solar planet requires telescopes with improved spatial resolution which are expected to be available in the near future.

Since the first detection of a planetary companion around a sun-like star was reported in 1995 by [19]; more than 40 such extra-solar planet candidates have been detected (at the time of writing). Still the possible detection is limited to very large planets that orbit the star at small distances. Surveys of main-sequence stars show that 5 % harbour companions of (0.5 - 8) the mass of Jupiter and that their orbits are within 3 AU of the star ([18]). Their orbits are either within 0.2 AU from the star and/or eccentric. These results differ from the solar system conditions, where the massive planets are located further away from the Sun (Jupiter, the biggest planet is at 5.2 AU) are on almost circular orbits. Although some observational bias may still be considered, since, for instance Earth-like planets can not be detected yet, the observations have changed our picture of the structure of a planetary system. Consequently, the results have stimulated new theoretical studies of the formation of planetary systems. We may finally come to the result that the formation of a system such as our solar system with planets in stable circular orbits about a central star may be a special case depending on special initial conditions.

The dust debris shells that are discussed above are not necessarily all associated with a planet. However, at least large bodies, such as planetesimals, must exist in these systems and some of the observed phenomena could be explained with the existence of a planet in the debris disk. So far in one case, 55 Cancri, a planet and a dust debris shell have been observed around the same star. Future research may rely on the combined view of both, the dust shells around stars as well as their planetary companions.

For future studies of extra-solar planets, the NASA Terrestrial Planet Finder mission, TPF and the ESA Darwin mission are planned to make imaging observations of candidate habitable planets around other stars. Spectroscopic observations of the extrasolar planetary atmospheres are discussed to study the atmospheric composition in order to find evidence for biological activity (see [13]). Observations in the 8 to about 15 μm spectral range are expected to reveal the absorption features of H_2O, CO_2 and O_3. While the existence of H_2O is seen as an indication that a planet is habitable, O_3 indicates that photosynthetic activity takes place on it at a large scale. This is also the spectral regime where the thermal emission of the zodiacal dust cloud in our solar system has its maximum thermal emission. In Fig. 10 the cross-sectional at areas of the planets of our solar system, of the zodiacal dust and of the EKB dust are shown in comparison. While the dust makes up only a very small amount of the mass of the solar system, its geometric cross section exceeds that of the planets. An area of 0.3×0.3 AU observed at 1 AU would produce the same brightness from the zodiacal dust as from the Earth itself [2]. Knowledge of circum-stellar debris disks - even of relatively low density such as the zodiacal cloud - is needed for the study of extrasolar planets. In that sense the presence of dust shells may hamper the study of extra solar planetary systems. On the other hand dust disks are an important part of planetary systems and are connected in their origin and evolution to the planetary systems as a whole.

Acknowledgement

I would like to thank Dana Backman, Hiroshi Kimura, Alexander Krivov, Natalia Krivova, and James McKenzie for helpful discussions and comments on the text and Margit Steinmetz and Anita Brandt for their help in preparing the figures and the manuscript.

References

1. S. Amari, R.S. Lewis, E. Anders: Interstellar grains in meteorites. I - Isolation of SiC, graphite, and diamond; size distributions of SiC and graphite. II - SiC and its noble gases Geochimica et Cosmochimica Acta, **58**, 459 (1994)
2. D.E. Backman, L.J. Caroff et al.: L.J, Sandford, S.A., and Wooden, D.H., 1998 Exozodiacal Dust Workshop, Conference Proceedings, NASA/CP - 1998-10155
3. S.V. Beckwith, A.I. Sargent: Circumstellar disks and the search for neighbouring planetary system Nature, **383**, 139 (1996)
4. C.F. Bohren, D.R. Huffman: *Absorption and scattering of light by small particles* (Wiley, New York 1983)
5. J.A. Burns, P.L. Lamy, S. Soter: 1979: Radiation Forces on Small Particles in the Solar System, Icarus **40**, 1-48 (1979)
6. B.T. Draine, H.M. Lee: 1984 Optical properties of interstellar graphite and silicate grains Astrophysical Journal, Part 1, **285**, 89-108 (1984)
7. E. Gruen, H.A. Zook, H. Fechtin, R.H. Giese: Collisional balance of the meteoritic complex Icarus, **62**, 244 (1985)
8. M.S. Hanner, R.H. Giese, K. Weiss, R.H. Zerull: On the Definition of Albedo and Application to Irregular Particles, Astron. Astrophys. **104**, 42-46 (1981)
9. P. Kalas, D. Jewitt: 1995 Asymmetries in the Beta Pictoris Dust Disk Astronomical Journal **110**, 794 (1995)
10. H. Kimura, I. Mann: Electric Charge of Interstellar Dust in the Ksolar System Astrophys. J., **499**, 454 (1998)
11. J. Kissel, Sagdeev, R.Z., et al.: Kissel, J.; Sagdeev, R. Z.; Bertaux, J. L.; Angarov, V. N.; Audouze, J.; Blamont, J. E.; Buchler, K.; Evlanov, E. N.; Fechtig, H.; Fomenkova, M. N.; von Hoerner, H.; Inogamov, N. A.; Khromov, V. N.; Knabe, W.; Krueger, F. R.; Langevin, Y.; Leonasv, B.; Levasseur-Regourd, A. C.; Managadze, G. G.; Podkolzin, S. N.; Shapiro, V. D.; Tabaldyev, S. R.; Zubkov, B. V. Composition of comet Halley dust particles from VEGA observations Nature, **321**, 280-282 (1986)
12. A.-M. Lagrange, D.E. Backman, P. Artymowicz: *Planetary Material Around Main-Sequence Stars Protostars and Planets* IV Edited by Vince Mannings, A. P. Boss, and S.S. Russell, eds. Space Science Series (The University of Arizona Press, June 2000)
13. A. Leger: 2000, Strategies for Remote Detection of Life - DARWIN-IRSI and TPF missions Advances in Adv. in Space Research, **25**, 2209-2223 (2000)

14. Ch. Leinert, S. Bowyer et al.: S.; Haikala, L. K.; Hanner, M. S.; Hauser, M. G.; Levasseur-Regourd, A.-Ch.; Mann, I.; Mattila, K.; Reach, W. T.; Schlosser, W.; Staude, H. J.; Toller, G. N.; Weiland, J. L.; Weinberg, J. L.; and Witt, A. N. Leinert, Ch.; Bowyer, S.; Haikala, L. K.; Hanner, M. S.; Hauser, M. G.; Levasseur-Regourd, A.-Ch.; Mann, I.; Mattila, K.; Reach, W. T.; Schlosser, W.; Staude, H. J.; Toller, G. N.; The 1997 reference of diffuse night sky brightness Astronomy and Astrophysics Supplement, **127**, 1-99 (1998)
15. R.S. Lewis, E. Anders, B.T. Draine: Properties, detectability and origin of interstellar diamonds in meteorites Nature, **339**, 117 (1989)
16. J.-C. Liou, H.A. Zook, S.F. Dermott: 1996 Kuiper Belt Dust Grains as a Source of Interplanetary Dust Particles Icarus, **124**, 429-440 (1996)
17. I. Mann, H. Kimura: Interstellar dust properties derived from mass density, mass distribution, and flux rates in the heliosphere, J. Geophys. Res., **105**, 10317-10328 (2000)
18. G.W. Marcy, R.P. Butler: Planets Orbiting Other Suns The Publications of the Astronomical Society of the Pacific, **112**, Issue 768, 137 (2000)
19. M. Mayor, D. Queloz: A Jupiter-Mass Companion to a Solar-Type Star Nature, **378**, 355 (1995)
20. T. Mukai: 1981 On the charge distribution of interplanetary grains Astronomy and Astrophysics, **9999**, 1-6 (1981)
21. T. Mukai, H. Fechtig.: 1983: Packing Effects of Fluffy Particles, Planet. Space Sci., **31**, 655-658 (1983)
22. L.R. Nittler, C,M.O'd. Alexander, X. Gao, R.M. Walker, E. Zinner: 1997 Stellar Saphires: The Properties and Origins of Presolar AL 2O 3 in Meteorites Astrophysical Journal **483**, 475 (1997)
23. H.C. Van de Hulst: Zodiacal Light in the Solar Corona, Astrophys. J., **105**, 471-488 (1974)
24. K. Weiss-Wrana: Optical Properties of Interplanetary Dust: Comparison with Light Scattering by Larger Meteoritic and Terrestrial Grains, Astron. Astrophys., **126**, 240-250 (1983)
25. M. Witte, M. Banaszkiewicz, H. Rosenbauer: 1996 Recent Results on the Parameters of the Interstellar Helium from the Ulysses/Gas Experiment Space Science Reviews, **78**, Issue 1/2, 289 (1996)
26. S. Yamamoto, T. Mukai: Dust production by impacts of interstellar dust on Edgeworth-Kuiper Belt objects Astronomy and Astrophysics, **329**, 785-791 (1998)

Meteors, Meteor Showers and Meteoroid Streams

Iwan P. Williams

Astronomy Unit, Queen Mary, London E1 4NS, UK

Abstract. Meteors are phenomena that have been seen since time immemorial and many have been recorded in art and in literature. Recognition of what they actually are came somewhat later with a true understanding of meteor showers being no more than 150 or so years old. The recognition of the processes involved in the formation of a stream is even younger, starting only in the early fifties. Here, we wish to review our current state of knowledge and describe the physics involved in the formation and evolution.

1 Historical Background

Humans must have been aware of streaks of fire crossing the sky that we now call meteors ever since they started to notice their surroundings, though they might not have been aware of the true nature of the phenomenon. Indeed, in many of the early references to the observed phenomenon (for example in the book of Revelations in the Bible) meteors were likened to *falling stars like leaves falling off a fig tree* while in many ancient Chinese, Japanese and Korean records, mention can be found of *stars falling like rain*, or *many falling stars* (see [15]). It must have been the same general thought that gave rise to the English colloquial name for them of *Shooting Stars*. The usually spectacular display from the Perseids in early August was referred to by Irish country folk as "the burning tears of St Lawrence" (see [46]). From the eighteenth century onwards, meteors were often shown in the background in paintings of other events (see for example [31]). Within the Western Christian doctrine, there was a belief which lingered certainly until the begining of the nineteenth century, if not beyond, that the Universe was perfect. Hence, these displays could not of course be actually related to falling stars, or falling anything else for that matter, and for a very long time meteors were regarded as atmospheric phenomena. Indeed, the name *meteors* implies this.

The reference in the book of Revelations actually describes one of the signs that the end of the world is coming and the spectacular meteor displays of the nineteenth century associated with the Leonid stream gave rise to many predictions about this. [22] has an interesting discussion of this topic in connection with perhaps the most famous meteor shower engraving that exists which can be found in almost any discussion of meteor streams and is for example on the cover of [28]. [35] has discussed the possibility that part of *The Rime of the Ancient Mariner* by Coleridge was inspired by a display from the Leonid shower. Meteors have thus inspired art, religion and mythology and through this there exist good records of the appearances of meteors and meteor showers. These observations by themselves do not however give a scientific explanation of the phenomena, and neither will more modern observations *by themselves.* However, to understand the science, it is always wise to know the observational facts first. In their simplest form, these are quite straightforward. Meteors can be seen at any time of the year, appearing on any part of the sky and moving in any direction. Such meteors are called sporadic and the mean sporadic rate is very low – no more than a few per hour. This is not the stuff to inspire either Coleridge, religious fanatics nor artists. At certain well-determined times each year the meteor rate climbs by two or three orders of magnitude for a short interval. For example, around 12 August meteors are seen at a rate of one or two per minute. This is, of course, the Perseid meteor shower. In contrast to quiet periods, during such periods of high activity, the meteors do not appear uniformly distributed across the sky but appear to flow, or radiate out of a fixed point. Not surprisingly, this point is called the *radiant* of the shower and the radiant of the Perseid shower lies in the Constellation of Persius, hence the name. This behaviour is interpreted as implying that the meteoroids are moving on parallel courses and that the existence of a radiant point is due to parallax. In other words, there is a stream of meteoroids impinging the Earth and generating the meteor shower. The first persons on record to have noticed that shower meteors radiated from a point were [30] and [36]. A few years later, [20,21] pointed out that the annual showers were periodic on a siderial rather than a tropical year, in other words they were extra-terrestrial in origin. Some indication of this had come some thirty years prior to this when [8] had simultaneously observed the same meteors from two different locations and through parallax determined their height to be about 90 km. These annual showers may be spectacular enough to generate names for them in folklore, especially when they coincide with famous saint days, but again this is hardly the stuff to base predictions for the end of the world on. Fortunately, some showers appear to generate a very enhanced display at regular intervals. The most well-known of these is the Leonids, where truly awesome displays are recorded as having occurred. For example, in 1966, the rate was tens of meteors per second, a truly falling of the stars from the sky. such a display lasted for under an hour, but records show that such displays may be seen at intervals of time that are multiples of about 33 years. Two such recorded displays were in 1799 and 1833 and these helped [2], [26] and [34] to conclude that the orbit of the Leonid meteors were very similar to that of comet $55P/Tempel - Tuttle$ and that 33 years were very close to the orbital period of this comet. Since then comet-meteor stream pairs have been identified for virtually all recognizable significant streams, and a list of pairings was produced by [12]. Many of the gaps in Cook's list have since been filled. These simple facts allows a straightforward model of meteor showers and associated meteoroid streams to be constructed. Solid particles, which we shall call *meteoroids*, are lost from a comet. Since any relative speed between comet and meteoroid will be much less than the orbital speed, the meteoroids will move on orbits that are only slightly perturbed from the cometary orbit. If a large number of meteoroids are lost, this will of course form a cloud about the comet, co-moving with it. As the semi-major axes of each meteoroid will be slightly different, each will have a slightly different orbital period, resulting in a drift in the epoch of return to perihelion. After many orbits this results in meteoroids effectively being located at all points around the orbit. There will also be some reduction in the space density of meteorods with the passage of time for two main reasons, the small changes in semi-major axes of individual meteoroids due to gravitational perturbations will continue, so that the total volume occupied by a stream increase, while also meteoroids will be lost from the stream through collisions, radiation pressure and gravitational perturbations. A normal stream is thus middle-ages, with meteoroids all around the orbit so that a shower is seen every year. Gravitational perturbations from the planets will also cause a steady evolution of the mean orbit of the meteoroid stream, this being most noticeable in the longitude of the ascending node, that is the time of appearance of the associated shower. This drift is usually slightly under a day per century. Such evolution may also cause a change in the number of meteors seen as the heliocentric distance of the ascending node changes and the densest part of the stream moves away from the Earth's orbit. In a very old stream, the number density of meteoroids will be low so that the stream is never very noticeable, but again constant each year. A very young stream on the other hand will only show activity at certain years, and that at a much enhanced level whenever the Earth passes through the cloud of meteoroids which is still surrounding the cometary nucleus since insufficient time has passed for it to spread about the orbit. This picture of meteoroid stream evolution and the associated behaviour of meteor showers was firmly established by the 1950s. Indeed so firmly was it established that most astronomers came to the view (which is still widely held) that there was nothing much further to be gained from the study of meteor showers. Though I believe that the basic underlying physics implied in the above model is still true, I also believe that there is a considerable amount that we do not fully understand. I will discuss some of these in the following sections.

2 The Life of a Meteoroid Stream

The life of a meteor stream can be subdivided into three stages which in chronological order are as follows.

(i) The formation through ejection of small meteoroids from a parent body, usually assumed to be a cometary nucleus but may equally well be an asteroid. The end result of this process is a family of meteoroids all moving through the inter planetary space on fairly similar heliocentric orbits.

(ii) The evolutionary stage where the orbit of each individual meteoroid undergoes perturbations from the gravitational field of the planets and the effects of solar radiation. In general, this will increase any differences in orbital parameters so that the variation from the mean becomes larger – the stream is dispersing. This will also change the mean orbital parameters and, in particular, may change the position of the node.

(iii) The meteor shower, or observational, phase. This is when a meteoroid collides with the Earth's atmosphere and ablates, or burns up, to produce the familiar *shooting star*. To do this, the node of the meteoroid orbits must be at the Earth's orbital distance.

We shall consider these three phases.

2.1 The Meteoroid Ejection Process

As already mentioned, the generally accepted model for the formation of meteoroid streams was proposed by [37]. In this model, solar radiation causes sublimation of the nuclear ices and the resulting gas outflow drags with it small meteoroids. Since the gas outflow velocity will be considerably less than the orbital velocity of the cometary nucleus, this process results in meteoroids moving on orbits that are only marginally different from the original cometary orbit. Marginal differences in semi-major axis lead to a corresponding difference in orbital period and this leads to a drift in the mean anomaly of meteoroids relative to the parent, so that in the course of several hundred orbits, meteoroids are found at all values of mean anomaly. In other words, a meteor shower will be observed whenever the Earth intersects this orbit. In terms of the very basic physics, this model is almost certainly correct, though a number of authors (e.g. [13], [16]) have suggested minor modifications to the details which result in a higher ejection velocity than that given by Whipple.

The process of ejecting a meteoroid, expressed in the most basic form, simply changes both the energy per unit mass and the angular momentum per unit mass of the meteoroid relative to the parent nucleus.

Now, standard theory of Keplerian motion tells us that

$$E = \frac{-GM_\odot}{2a}, \tag{1}$$

and that

$$P^2 = a^3 \tag{2}$$

where a is the semi-major axis of the orbit in Astronomical Units and P the orbital period in years. Hence we can obtain

$$\frac{\Delta E}{E} = \frac{-\Delta a}{a} = \frac{-2\Delta P}{3P}. \tag{3}$$

Similarly, consideration of the angular momentum gives

$$\frac{\Delta h}{h} = \frac{\Delta a}{2a} - \frac{e\Delta e}{(1-e^2)} \tag{4}$$

Simple considerations also dictate that the orbit of the ejected meteoroid must pass through the point of ejection. In general, the small changes given by the above equations also imply that a small change in the argument of perihelion, ω, must also take place. If, further, the ejection is not within the plane of the cometary orbit, and in general there is no reason why this must be so, then the ejection process will also cause a small change in both the inclination, i, and the longitude of the ascending node, Ω.

The exact changes depend on the location of the ejection point and calculating these changes is easiest if the ejection takes place at perihelion. At first sight, it may appear to be a reasonable assumption that most meteoroids are ejected at perihelion since that is where the parent comet is most active. However, this only implies that the ejection RATE is highest there, the comet also spends less time at perihelion than at any other point on its orbit. If we assume that the ejection rate is proportional to the incident solar heating, then it is inversely proportional to the square of the heliocentric distance r, that is,

$$\frac{dM}{dt} = \frac{KL}{r^2} \tag{5}$$

where L is the solar luminosity and K is a constant. Also, assuming conservation of specific angular momentum, h,

$$\frac{d\nu}{dt} = \frac{h}{r^2}, \tag{6}$$

where ν is the true anomaly. From these two equations, it is clear that the number of ejected meteoroids in equal intervals of true anomaly is constant. Hence, we do have to consider a spread of ejection points and therefore a family of meteoroids moving on slightly different ellipses, slightly rotated in three dimensions relative to each other. The actual cross-section of this family will depend on the distribution of ejection points and on the ejection velocity. The cross-section as observed from Earth will also depend on where the Earth intersects this family of orbits, being small if the Earth passes through the family close to their perihelion and most meteoroids were ejected at this point, and large if the Earth passes through the stream far from the ejection region of most meteoroids, this region itself being well spread out. [14] has investigated this effect for a parent on a Geminid-like orbit, and figures can be found there illustrating the various possible cross-sections. Similar cross-sections are given by [41] for streams associated with Halley's comet. From these figures, it is clear that the spread in the cross-section in a direction orthogonal to the orbital plane is much less, and more constant, than in other directions, to a large extent, depending only on the ratio of ejection velocity to orbital velocity. This is a very useful result as far as our problem is concerned, for the activity profile that is obtained from the study of a meteor shower is, to a large extent, determined by the spread in the direction orthogonal to the orbital plane of the comet, most relevant comets having a reasonable inclination. To a first approximation, we can thus obtain an estimate of the ejection speed from such a study and indeed this has already been attempted by [23] for observations of the Leonids over the last few years. The distribution gets modified with the passage of time by evolutionary effects that we have already discussed and so application may only be meaningful for very young streams.

Observations of dust in comet Hale-Bopp suggested outflow speeds of the order of $300ms^{-1}$. This observed dust is smaller than conventional meteoroids and so may have been ejected with a higher speed than associated meteoroids. [38] reversed the argument, asking what the ejection velocity would be if all the dispersion found in meteor streams was due to the ejection velocity. This clearly gives an upper limit and it turned out to be in the range $160 - 880ms^{-1}$, in agreement with the above arguments but not of great help in narrowing the range. Theoretical considerations of the ejection process similarly do not help. Different models all produce values within the above range (e.g. [37], [13], [16]) but between them essentially cover the whole available range. Solving this problem is very important, It can be done either through a study of meteor showers, in which case we learn about cometary nuclei or through the study of cometary nuclei, in which case great advances will be made in the study of meteoroid streams.

The above discussion concentrated on the ejection of meteoroids from a cometary parent. It is also possible that small solid bodies can be ejected from asteroids, possibly subsequent to a collision. The actual physics of the process may be poorly understood, though again the range of values for the speed may well be in the same ball-park as for cometary ejection. The subsequent orbital evolution of the meteoroid will certainly be similar.

2.2 The Evolution of Streams

In terms of a physical understanding of the processes, this is the simplest, though of course the mechanics of keeping track of the effects of these processes on all the meteoroids that make up the stream may be an impossible task.

The most dominant force is solar gravity, expressed in the usual way as

$$F_{grav} = \frac{-GM_{\odot}m}{r^2} \tag{7}$$

where m is the mass of the meteoroid, r the heliocentric distance and $M_{\odot}$ is the mass of the Sun. To a reasonable approximation, meteoroids move on Keplerian ellipses about the Sun.

However, the meteoroids are in general small so that Solar radiation has some effect on their motion. There are a number of effects. Firstly the radiation pressure weakens gravity. The simplest way of dealing with this is to regard the gravitational constant G as being replaced by $G(1-\beta)$, where β is the ratio of the gravitational force to the radiation pressure. For a spherical meteoroid of radius b and bulk density σ,

$$\beta = 5.75 \times 10^{-5}/(b\sigma), \tag{8}$$

with b measured in centimetres and σ also in *cgs* units.

[25] showed that the effect is to cause the meteoroid to appear to move on an elliptical orbit with semi-major axis a_1 and eccentricity e_1 that are related to the Keplerian values a and e by

$$a_1 = ar(1-\beta)(r-2a\beta)^{-1}, \tag{9}$$

$$e_1 = 1 - (1-e^2)(r-2a\beta)(1-\beta)^{-2}r^{-1}. \tag{10}$$

Radiation also produces a drag, known as the Poynting-Robertson effect ([33]), through the radiation being absorbed in the solar rest frame but reemmitted in the moving frame of the meteoroid. Mathematical expressions for the rate of change of the orbital parametrs have been derived by [45], [40]. Those of use in the current discussion are

$$a\frac{da}{dt} = -\gamma(2+3e^2)(1-e^2)^{3/2} \tag{11}$$

and

$$\frac{de}{dt} = 2.5\gamma(1-e^2)^{-1/2}a^{-2}. \tag{12}$$

Here, $\gamma = \frac{GM_{\odot}\beta}{c}$, where c denotes the speed of light. For orbits in the inner solar system, this gives a timescale for significant change of the order of a^2/γ, or of the order of $b\sigma 10^7$ years. Major changes due to the Poynting-Robertson effect thus probably requires longer than the age of most streams. However, since the only relevant criterion for a meteoroid is whether or not it hits the Earth, then changes may not have to be that significant so that an important change may take a factor of order 1000 less than implied above.

The most important effect is, however, that due to the gravitational perturbations by the planets. In principle, the force on any meteoroid of known position, can be calculated at any instant provided we know all the planetary positions at the same instant. Hence the instantaneous change in the motion due to this force field can also be computed. Sufficient repetition of this calculation allows the meteoroid position to be obtained at any future time. [11] produced a mathematical algorithm based on this which was used, for example, by [44] to follow the evolution of the mean Taurid stream over an interval of 4700 years, demonstrating a similarity with the evolution of comet Encke's orbit. At this time, primarily due to a lack of available computing power, secular perturbation methods were popular and used for example by [32], [5,6].

With the improvement in both the speed and memory of computers, direct numerical integration methods gained in popularity. The first to use such a method was probably [17], where the motion of six Quadrantid meteoroids was investigated. Nearly two decade later, Hughes *et al.* had increased the number of test particles to over 200, while [14] increased this to 500 000. After this, the use of direct methods became widespread (see for example, [24], [19], [13] [4], [38], [43]).

With the computing power currently available, it is possible to follow accurately the evolution of millions of meteoroids over meaningful time scales. For this reason, following the evolution of meteor streams should not be regarded as a major problem. There is, however, one problem which is relevant, especially when the aim is to predict the occurrence of a meteor storm, namely that the number of meteoroids present in a stream far exceeds several millions, indeed numbers of the order of 10^{17} are more realistic. Hence a storm of 10^5 meteors seen for an hour is represented by not even a single particle hitting the Earth in the model.

2.3 The End, Observed Meteors

When a meteoroid enters the Earth's atmosphere, gas drag causes it to decelerate and the resulting energy loss causes heating of the meteoroid. This causes vaporization of the surface and ionization of the vapour. This leads to possible detection both of the trail optically and of the ionized material via radio waves. Unfortunately, the meteoroid can also fragment which makes the physics of the interaction with the atmosphere more complex and the determination of the ratio of mass to surface area more unreliable. The observations can, however, give accurate determination of both the influx and the orbital parameters. Though simple visual observations by naked eye or using binoculars does give some information, for example the influx rate at any given time or the colour and brightness, more accurate information is called for in order to obtain precise orbits for the individual meteoroids. An example is through the use of multiple observations with cameras. Such observations started in the thirties with the Harvard photographic program and the use of rocking mirrors to obtain photographs in Arizona. This development accelerated after the second World War with the use of cameras with a precisely timed occulting device to provide accurate velocity data for any meteor photographed from at least two locations. At least three major networks came into existence at about this time. The Prairie Network in central USA ran from 1964 to 1974, the MORP (Meteorite Observation and Recovery Program) project in Western Canada operated from 1971 to 1985 and the European Network in the Czech Republic, the Slovak Republic and Germany started in 1964 and is the only one of the three original nets that survives. An improvement of the photographic multi-station networks, at least in that visible meteors are recorded, is the use of low light level television methods. Active work known to the author in this field is underway in Canada, The Czech Republic, Japan, the Netherlands, Tajikstan and the USA. A review of this topic was given by [18] and a good example of what can be achieved using camcorders are the records of the Peekskill fireball over the Eastern USA in October 1992.

Following the development of radar and the radio wave band generally during the 1939-45 war, its value as a tool to investigate meteor trails was realized. Many radar systems were developed and, thankfully, a significant number are still operating. The systems, being essentially automatic, have the capacity to produce vast amounts of data with, for example, the AMOR system in New Zealand recording several hundred thousand orbits per year ([7]). One advantage of radar observations is that they can detect smaller meteoroids (by an order of magnitude or so) than visible observations. Since all mass distribution functions predict more small particles than large, this also means that radar has more meteors available for it to detect but this, in turn, can produce problems of confusion (being unable to distinguish between two meteor trails in the radar beam at the same time), a problem not encountered with visible observations.

At the present time, data are readily available on a large number of meteoroids belonging to streams. This data is wholly concerned with the meteor trail mostly its position in space, its length and its brightness. To calculate the velocity and position in space of the meteoroid on entry into the atmosphere, which allows a determination of energy and angular momentum, the height in the atmosphere is required. This can only be obtained from multiple observations from different locations, hence the setting up of the networks mentioned earlier. This information, together with the time of observation enables all the orbital parameters of the meteoroid to be calculated. Large numbers of orbits were obtained by these networks and thankfully many of these are now safely archived in the IAU Meteor Data Center at Lund (see [27]).

3 Outbursts in Meteor Showers

As already mentioned, we should expect young meteoroid streams to produce showers of uneven strength from year to year, being very strong when the parent is close to the Earth and much weaker otherwise. Such strong meteor displays we shall call outbursts. Outbursts are a regular feature of the Leonid stream, indeed it is these outbursts that have made the Leonids famous and in the early days helped towards our understanding of meteoroid streams. However, the situation within individual streams is not quite as simple as it looks at first sight, and in fact few streams behave like the simple picture when an outburst is observed. In this section we shall look briefly at three meteor showers that each illustrate a potential problem.

3.1 The Perseids

The Perseids is one of the few major and regular meteor showers that produces a display of roughly the same strength every year. It is also a fairly old shower with many record of it existing in ancient documents. Its parent is also well known, comet $109P/Swift-Tuttle$. In the early 90's, a second peak in the activity curve of the Perseids was noticed (The optical observations behind these activity profiles are summarised by [10]). Since the appearance of this peak was roughly coincident in time with the return of the parent comet to the Earth's locality, this peak was interpreted as being due to new meteoroids, recently ejected from the comet. Detailed models by [42] confirmed that this peak could be associated primarily with meteoroids ejected at the last (rather than current) apparition of the comet. This new peak in the Perseid shower activity profile is thus not such a mystery but its existence does remind us that meteoroid streams are not perhaps as static as had been thought. Not all the meteoroids may be of the same age and new meteoroids are added to the stream at each apparition of the comet. If significant changes in the cometary orbit occurs, then a meteoroid stream is not so much a single coherent stream as a number of similar filaments.

3.2 The Lyrids

In contrast to the Perseids where a strong display is seen each year, the Lyrids are almost non-existent in most years, but outbursts are seen which, in relation to the normal activity are quite strong, reaching a Zenithal hourly rate of several hundred, an increase over the norm of perhaps, a factor of 30 or so. Such an increase in the Perseid stream for example would lead to the event being labelled a major storm. [27] have chronicled all the recorded outbursts in the Lyrids and, not surprising perhaps in view of the actual weakness of the whole event, found that most records were recent with several being recorded this century. The parent comet of the Lyrids, comet Thatcher, has a period of order 400 years (it has actually only been observed once, at the epoch of discovery, so there is considerable uncertainty about the actual period of comet Thatcher, beyond the fact that it is long). These recorded outbursts can not therefore be associated with the return of comet Thatcher to the inner Solar System. [3] have produced a computer model of the stream, suggesting that the outbursts are caused by perturbations of stream filaments into an Earth-intersecting orbit by Jupiter. If this is correct, then the behaviour of the April Lyrids does not represent a mystery either, though it does illustrate that relative outbursts can occur for reasons other than a passage of the Earth through a dense cloud of recently ejected meteoroids.

3.3 The Leonids

The history of the developement of meteor stream science is peppered with observations of strong outbursts associated with the Leonids. Indeed, it is often claimed that it was the Leonid displays in 1799 and 1833 that gave birth to the study of meteor streams. At first sight, the Leonids display all the characteristics of a young stream that we described earlier, namely very strong displays whenever the parent comet is close to perihelion but very weak otherwise. Indeed, most modellers (e.g. [43], [9]) assume models based on this notion to try to predict the behaviour of the Leonids at the turn of millenium. Recently, [1], [29] had great success in predicting the behaviour of the Leonids in 1999.

Unfortunately, the Leonids have been observed for a long time (well over 1000 years) and can hardly be regarded as young. Also the change from an outburst to a non-outburst is very sharp, a gradual decline would be expected. [39] has suggested that perturbations due to Uranus are responsible for clearing meteoroids out of the stream in most parts of the orbit away from the parent. Hence, even in a well-studied stream there are still surprises to be found.

4 Conclusions

Some aspects of meteor studies are well understood, for example the evolution of the individual orbits under the effects of radiation and gravitational perturbations. Other, such as the ejection of meteoroids and their composition are less well understood. It is to be hoped that the continued studies of meteors from the ground and the study of comets both from the ground and through spacecraft, in space will in time lead to a clear understanding of all these aspects.

References

1. Asher, D.J. (1999): The Leonid meteor storms of 1833 and 1966. Mon. Not. R. astr. Soc. **307**, 919–924
2. Adams J. C., 1867, On the orbit of the November meteors, Mon. Not. R. astr. Soc, **27**, 247-252
3. Arter T.R. & Williams I.P., 1997, Periodic behaviour of the April Lyrids, Mon. Not. R. astr. Soc., **286**, 163-172
4. Asher D. Clube S.V.M. & Steel D.I. :1993, The Taurid Complex Asteroids, in *Meteoroids and their parent bodies*, Eds Stohl J. & Williams I.P., Slovak Academy of Sciences, Bratislava, 93-96
5. Babadzhanov P.B. & Obrubov Y.Y., 1980, Evolution of orbits and intersection condition with the Earth of Geminid and Quadrantid meteor Streams, in *Solid Particles in the Solar System* Eds Halliday I. & McIntosh B. A., D. Reidel, Dordrecht, 157–162

6. Babadzhanov P.B. & Obrubov Y. Y., 1983, Some features of evolution of meteor streams, in *Highlights in Astronomy*, Ed. West R.M., D.Reidel, Dordrecht, 411–419
7. Baggaley, W.J. Bennett, R.G.T. Steel, D.I. & Taylor, A.D., 1993, The Advanced Meteor Orbit Radar Facility: AMOR, *Quart. J. R. Astr. Soc.*, **35**, *293–320*
8. *Benzenberg J.F. & Brandes H.W., 1800, Versuch die Entfernung, die Geschwindigkeit und die Bahn der Sternschnuppen zu bestimmen, Annalen der Phys,***6***, 224–232*
9. *Brown P. & Jones J., 1996, Dynamics of the Leonid Meteoroid Stream: a Numerical Approach, in Physics, Chemistry and Dynamics of Interplanetary Dust Eds Gustafson B.A.S. & Hanner M.S., ASP Conf. Ser, 113-116*
10. *Brown P. & Rendtel J., 1996, The Perseid Meteoroid stream: Characterization of Recent Activity from Visual Observations, Icarus,* **124***, 414-428*
11. *Brouwer D., 1947, Secular variations of the elements of Encke's comet Astron.J.,* **52***, 190–198*
12. *Cook A.F., 1973, A working list of Meteor Streams, in Evolutionary and physical properties of Meteoroids, Eds Hemenway C. L. Millman P.M. & Cook A.F., NASA SP-319, Washington DC, 183-191*
13. *Gustafson B. A. S., 1989, Comet ejection and dynamics of nonspherical dust particles and meteoroids, Astrophys. Jl.,* **337***, 945–949*
14. *Fox K. Williams I.P. & Hughes, D.W., 1983. The rate profile of the Geminid meteor shower, Mon. Not. R. astr. Soc.,* **205***, 1155–1169*
15. *Hasegawa I., 1993, Historical records of meteor showers, in Meteoroids and their parent bodies, Eds Stohl J. & Williams I.P., Slovak Academy of Sciences, Bratislava, 209-223*
16. *Harris N.W. Yau K.K. & Hughes D.W., 1995, The True extent of the nodal distribution of the Perseid meteoroid stream, Mon. Not. R. astr. Soc.,* **273***, 999-1015*
17. *Hamid S.E. & Youssef M.N., 1963, A short note on the Origin and Age of the Quadrantids, Smithson. Cont. Astrophys.,* **7***, 309–311*
18. *Hawkes R.L., 1993, Television meteors, Meteoroids and their parent bodies, Eds. Štohl J. & Williams I.P., Slovak Academy of Sciences, Bratislava, 227–234*
19. *Hunt J. Williams I.P. & Fox K., 1985, Planetery perturbations on the Geminid meteor stream, Mon. Not. R. astr. Soc.,* **217***, 533–538*
20. *Herrick E.C., 1837, On the shooting stars of August 9th and 10th 1837, and on the probability of the annual occurrence of a meteoric shower in August, American Jl.Sci.,* **33***, 176-180*
21. *Herrick E.C., 1838, Further proof of an annual Meteoric Shower in August, with remarks on Shooting Stars in general, American Jl. Sci.,* **33***, 354-364*
22. *Hughes D.W., 1995, The World's Most Famous Meteor Shower Picture, Earth Moon and Planets,* **86***, 311*
23. *Jenniskens P., 1998, On the dynamics of meteoroid streams, Earth Planets Space,* **50***, 959–968*
24. *Jones J. & McIntosh B.A., 1986, On the structure of the Halley comet meteor stream, in Exploration of Comet Halley,* **ESA-SP 250***, Paris, 233–243*
25. *Kresák L., 1974, Bull. Astron. Inst. Czechos.,* **13***, 176*
26. *Le Verrier U.J.J., 1867, Sur les etoiles filantes de 13 Novembre et du 10 Aout, Comptes Rendus,* **64***, 94-99*

27. *Lindblad B.A. & Porubcan V., 1992, Activity of the Lyrid Meteoroid stream, in Asteroids Comets Meteors 91, Eds Harris A.W & Bowell E., Lunar and Planetary Institute, Tucson, 367-370*
28. *Littman M., 1998, The Heavens on Fire, Cambridge University Press, Cambridge UK*
29. *McNaught R.H. & Asher D.J., 1999, Leonid dust trails and meteor storms, WGN,* **27:2**, *85–102*
30. *Olmstead D.,1834, Observations on the meteors of 13 Nov.1833, American Jl. Sci.,* **25**, *354-411*
31. *R.J.M. Olson. & J.M. Pasachoff, Fire in the Sky, Cambridge University Press, Cambridge, UK, (1998)*
32. *Plavec M., 1950, Nature,* **165**, *362–XXX*
33. *Robertson, H.P., 1937, Dynamical effects of radiation in the Solar System, Mon. Not. R. astr Soc.,* **97**, *423–438*
34. *Schiaparelli G.V., 1867, Sur la relation qui existe entre les cometes et les etoiles filantes, Astronomische Nachrichten,* **68**, *331-332*
35. *Steel D., 1998, The Leonid Meteor showers and the genesis of the Ancient Mariner, A & G,* **13**, *20-23*
36. *Twining A.C., 1834, Investigations respecting the meteors of Nov.13th, 1833, American Jl. Sci.,* **26**, *320-352*
37. *Whipple F.L., 1951, A comet model II. Physical relations for comets and meteors, Astrophys. Jl.,* **113**, *464-474*
38. *Williams I.P., 1996, What can meteoroid streams tell us about the ejection velocities of dust from comets, Earth Moon and Planets,* **72**, *321-326*
39. *Williams I.P., 1998, The Leonid Meteor shower: why are there storms but no regular annual activity?, Mon. Not. R. Astr. Soc,* **292**, *L37-L40*
40. *Williams I.P.: 1983, Physical processes affecting the motion of small bodies in the Solar System and their application to the evolution of Meteor Streams, in Dynamical trapping and evolution in the Solar System, Eds Markellos V.V. & Kozai Y., D.Reidel, 83-87*
41. *Williams I.P. & Fox K., 1983, The evolution of meteor streams, in Asteroids, Comets, Meteors II, Eds Lagerkvist C.I. & Rickmann H., Uppsala Universitet Reprocentralen*
42. *Williams I.P. & Wu Z., 1994, The current Perseid meteor shower, Mon. Not. R. astr. Soc,* **269**, *524-528*
43. *Wu Z. & Williams I.P., 1996, Leonid meteor storms, Mon. Not. R. astr. Soc,* **264**, *980-990*
44. *Whipple F.L. & Hamid S.E., 1950, Sky and Telescope,* **9**, *248*
45. *Wyatt S.P. & Whipple F.L., 1950, The Poyntin-Robertson effect on Meteor orbits Astrophys. Jl.,* **111**, *134–141*
46. *Yeomans D.K., 1991, Comets: a Chronological History of Observations, Science, Myth and Folklore, J. Wiley and Son, New York*

Lecture Notes in Physics

For information about Vols. 1–542
please contact your bookseller or Springer-Verlag

Vol. 543: H. Gausterer, H. Grosse, L. Pittner (Eds.), Geometry and Quantum Physics. Proceedings, 1999. VIII, 408 pages. 2000.

Vol. 544: T. Brandes (Ed.), Low-Dimensional Systems. Interactions and Transport Properties. Proceedings, 1999. VIII, 219 pages. 2000

Vol. 545: J. Klamut, B. W. Veal, B. M. Dabrowski, P. W. Klamut, M. Kazimierski (Eds.), New Developments in High-Temperature Superconductivity. Proceedings, 1998. VIII, 275 pages. 2000.

Vol. 546: G. Grindhammer, B. A. Kniehl, G. Kramer (Eds.), New Trends in HERA Physics 1999. Proceedings, 1999. XIV, 460 pages. 2000.

Vol. 547: D. Reguera, G. Platero, L. L. Bonilla, J. M. Rubí(Eds.), Statistical and Dynamical Aspects of Mesoscopic Systems. Proceedings, 1999. XII, 357 pages. 2000.

Vol. 548: D. Lemke, M. Stickel, K. Wilke (Eds.), ISO Surveys of a Dusty Universe. Proceedings, 1999. XIV, 432 pages. 2000.

Vol. 549: C. Egbers, G. Pfister (Eds.), Physics of Rotating Fluids. Selected Topics, 1999. XVIII, 437 pages. 2000.

Vol. 550: M. Planat (Ed.), Noise, Oscillators and Algebraic Randomness. Proceedings, 1999. VIII, 417 pages. 2000.

Vol. 551: B. Brogliato (Ed.), Impacts in Mechanical Systems. Analysis and Modelling. Lectures, 1999. IX, 273 pages. 2000.

Vol. 552: Z. Chen, R. E. Ewing, Z.-C. Shi (Eds.), Numerical Treatment of Multiphase Flows in Porous Media. Proceedings, 1999. XXI, 445 pages. 2000.

Vol. 553: J.-P. Rozelot, L. Klein, J.-C. Vial Eds.), Transport of Energy Conversion in the Heliosphere. Proceedings, 1998. IX, 214 pages. 2000.

Vol. 554: K. R. Mecke, D. Stoyan (Eds.), Statistical Physics and Spatial Statistics. The Art of Analyzing and Modeling Spatial Structures and Pattern Formation. Proceedings, 1999. XII, 415 pages. 2000.

Vol. 555: A. Maurel, P. Petitjeans (Eds.), Vortex Structure and Dynamics. Proceedings, 1999. XII, 319 pages. 2000.

Vol. 556: D. Page, J. G. Hirsch (Eds.), From the Sun to the Great Attractor. X, 330 pages. 2000.

Vol. 557: J. A. Freund, T. Pöschel (Eds.), Stochastic Processes in Physics, Chemistry, and Biology. X, 330 pages. 2000.

Vol. 558: P. Breitenlohner, D. Maison (Eds.), Quantum Field Theory. Proceedings, 1998. VIII, 323 pages. 2000

Vol. 559: H.-P. Breuer, F. Petruccione (Eds.), Relativistic Quantum Measurement and Decoherence. Proceedings, 1999. X, 140 pages. 2000.

Vol. 560: S. Abe, Y. Okamoto (Eds.), Nonextensive Statistical Mechanics and Its Applications. IX, 272 pages. 2001.

Vol. 561: H. J. Carmichael, R. J. Glauber, M. O. Scully (Eds.), Directions in Quantum Optics. XVII, 369 pages. 2001.

Vol. 562: C. Lämmerzahl, C. W. F. Everitt, F. W. Hehl (Eds.), Gyros, Clocks, Interferometers...: Testing Relativistic Gravity in Space. XVII,507 pages. 2001.

Vol. 563: F. C. Lázaro, M. J. Arévalo (Eds.), Binary Stars. Selected Topics on Observations and Physical Processes. 1999.IX, 327 pages. 2001.

Vol. 564: T. Pöschel, S. Luding (Eds.), Granular Gases. VIII, 457 pages. 2001.

Vol. 565: E. Beaurepaire, F. Scheurer, G. Krill, J.-P. Kappler (Eds.), Magnetism and Synchrotron Radiation. XIV, 388 pages. 2001.

Vol. 566: J. L. Lumley (Ed.), Fluid Mechanics and the Environment: Dynamical Approaches. VIII, 412 pages. 2001.

Vol. 567: D. Reguera, L. L. Bonilla, J. M. Rubí (Eds.), Coherent Structures in Complex Systems. IX, 465 pages. 2001.

Vol. 568: P. A. Vermeer, S. Diebels, W. Ehlers, H. J. Herrmann, S. Luding, E. Ramm (Eds.), Continuous and Discontinuous Modelling of Cohesive-Frictional Materials. XIV, 307 pages. 2001.

Vol. 569: M. Ziese, M. J. Thornton (Eds.), Spin Electronics. XVII, 493 pages. 2001.

Vol. 570: S. G. Karshenboim, F. S. Pavone, F. Bassani, M. Inguscio, T. W. Hänsch (Eds.), The Hydrogen Atom: Precision Physics of Simple Atomic Systems. XXIII, 293 pages. 2001.

Vol. 571: C. F. Barenghi, R. J. Donnelly, W. F. Vinen (Eds.), Quantized Vortex Dynamics and Superfluid Turbulence. XXII, 455 pages. 2001.

Vol. 572: H. Latal, W. Schweiger (Eds.), Methods of Quantization. XI, 224 pages. 2001.

Vol. 573: H. M. J. Boffin, D. Steeghs, J. Cuypers (Eds.), Astrotomography. XX, 434 pages. 2001.

Vol. 574: J. Bricmont, D. Dürr, M. C. Galavotti, G. Ghirardi, F. Petruccione, N. Zanghi (Eds.), Chance in Physics. XI, 288 pages. 2001.

Vol. 575: M. Orszag, J. C. Retamal (Eds.), Modern Challenges in Quantum Optics. XXIII, 405 pages. 2001.

Vol. 576: M. Lemoine, G. Sigl (Eds.), Physics and Astrophysics of Ultra-High-Energy Cosmic Rays. X, 327 pages. 2001.

Vol. 577: I. P. Williams, N. Thomas (Eds.), Solar and Extra-Solar Planetary Systems. XVIII, 255 pages. 2001.

Vol. 578: D. Blaschke, N. K. Glendenning, A. Sedrakian (Eds.), Physics of Neutron Star Interiors. XI, 509 pages. 2001.

Vol. 579: R. Haug, H. Schoeller (Eds.), Interacting Electrons in Nanostructures. X, 227 pages. 2001.

Vol. 580: K. Baberschke, M. Donath, W. Nolting (Eds.), Band-Ferromagnetism: Ground-State and Finite-Temperature Phenomena. IX, 394 pages. 2001.

Vol.582: N. J. Balmforth, A. Provenzale (Eds.), Geomorphological Fluid Mechanics. X, 579 pages. 2001.

Monographs

For information about Vols. 1–27 please contact your bookseller or Springer-Verlag

Vol. m 28: O. Piguet, S. P. Sorella, Algebraic Renormalization. IX, 134 pages. 1995.

Vol. m 29: C. Bendjaballah, Introduction to Photon Communication. VII, 193 pages. 1995.

Vol. m 30: A. J. Greer, W. J. Kossler, Low Magnetic Fields in Anisotropic Superconductors. VII, 161 pages. 1995.

Vol. m 31 (Corr. Second Printing): P. Busch, M. Grabowski, P.J. Lahti, Operational Quantum Physics. XII, 230 pages. 1997.

Vol. m 32: L. de Broglie, Diverses questions de mécanique et de thermodynamique classiques et relativistes. XII, 198 pages. 1995.

Vol. m 33: R. Alkofer, H. Reinhardt, Chiral Quark Dynamics. VIII, 115 pages. 1995.

Vol. m 34: R. Jost, Das Märchen vom Elfenbeinernen Turm. VIII, 286 pages. 1995.

Vol. m 35: E. Elizalde, Ten Physical Applications of Spectral Zeta Functions. XIV, 224 pages. 1995.

Vol. m 36: G. Dunne, Self-Dual Chern-Simons Theories. X, 217 pages. 1995.

Vol. m 37: S. Childress, A.D. Gilbert, Stretch, Twist, Fold: The Fast Dynamo. XI, 406 pages. 1995.

Vol. m 38: J. González, M. A. Martín-Delgado, G. Sierra, A. H. Vozmediano, Quantum Electron Liquids and High-Tc Superconductivity. X, 299 pages. 1995.

Vol. m 39: L. Pittner, Algebraic Foundations of Non-Com-mutative Differential Geometry and Quantum Groups. XII, 469 pages. 1996.

Vol. m 40: H.-J. Borchers, Translation Group and Particle Representations in Quantum Field Theory. VII, 131 pages. 1996.

Vol. m 41: B. K. Chakrabarti, A. Dutta, P. Sen, Quantum Ising Phases and Transitions in Transverse Ising Models. X, 204 pages. 1996.

Vol. m 42: P. Bouwknegt, J. McCarthy, K. Pilch, The W3 Algebra. Modules, Semi-infinite Cohomology and BV Algebras. XI, 204 pages. 1996.

Vol. m 43: M. Schottenloher, A Mathematical Introduction to Conformal Field Theory. VIII, 142 pages. 1997.

Vol. m 44: A. Bach, Indistinguishable Classical Particles. VIII, 157 pages. 1997.

Vol. m 45: M. Ferrari, V. T. Granik, A. Imam, J. C. Nadeau (Eds.), Advances in Doublet Mechanics. XVI, 214 pages. 1997.

Vol. m 46: M. Camenzind, Les noyaux actifs de galaxies. XVIII, 218 pages. 1997.

Vol. m 47: L. M. Zubov, Nonlinear Theory of Dislocations and Disclinations in Elastic Body. VI, 205 pages. 1997.

Vol. m 48: P. Kopietz, Bosonization of Interacting Fermions in Arbitrary Dimensions. XII, 259 pages. 1997.

Vol. m 49: M. Zak, J. B. Zbilut, R. E. Meyers, From Instability to Intelligence. Complexity and Predictability in Nonlinear Dynamics. XIV, 552 pages. 1997.

Vol. m 50: J. Ambjørn, M. Carfora, A. Marzuoli, The Geometry of Dynamical Triangulations. VI, 197 pages. 1997.

Vol. m 51: G. Landi, An Introduction to Noncommutative Spaces and Their Geometries. XI, 200 pages. 1997.

Vol. m 52: M. Hénon, Generating Families in the Restricted Three-Body Problem. XI, 278 pages. 1997.

Vol. m 53: M. Gad-el-Hak, A. Pollard, J.-P. Bonnet (Eds.), Flow Control. Fundamentals and Practices. XII, 527 pages. 1998.

Vol. m 54: Y. Suzuki, K. Varga, Stochastic Variational Approach to Quantum-Mechanical Few-Body Problems. XIV, 324 pages. 1998.

Vol. m 55: F. Busse, S. C. Müller, Evolution of Spontaneous Structures in Dissipative Continuous Systems. X, 559 pages. 1998.

Vol. m 56: R. Haussmann, Self-consistent Quantum Field Theory and Bosonization for Strongly Correlated Electron Systems. VIII, 173 pages. 1999.

Vol. m 57: G. Cicogna, G. Gaeta, Symmetry and Perturbation Theory in Nonlinear Dynamics. XI, 208 pages. 1999.

Vol. m 58: J. Daillant, A. Gibaud (Eds.), X-Ray and Neutron Reflectivity: Principles and Applications. XVIII, 331 pages. 1999.

Vol. m 59: M. Kriele, Spacetime. Foundations of General Relativity and Differential Geometry. XV, 432 pages. 1999.

Vol. m 60: J. T. Londergan, J. P. Carini, D. P. Murdock, Binding and Scattering in Two-Dimensional Systems. Applications to Quantum Wires, Waveguides and Photonic Crystals. X, 222 pages. 1999.

Vol. m 61: V. Perlick, Ray Optics, Fermat's Principle, and Applications to General Relativity. X, 220 pages. 2000.

Vol. m 62: J. Berger, J. Rubinstein, Connectivity and Superconductivity. XI, 246 pages. 2000.

Vol. m 63: R. J. Szabo, Ray Optics, Equivariant Cohomology and Localization of Path Integrals. XII, 315 pages. 2000.

Vol. m 64: I. G. Avramidi, Heat Kernel and Quantum Gravity. X, 143 pages. 2000.

Vol. m 65: M. Hénon, Generating Families in the Restricted Three-Body Problem. Quantitative Study of Bifurcations. XII, 301 pages. 2001.

Vol. m 66: F. Calogero, Classical Many-Body Problems Amenable to Exact Treatments. XIX, 749 pages. 2001.

Vol. m 67: A. S. Holevo, Statistical Structure of Quantum Theory. IX, 159 pages. 2001.

Vol. m 68: N. Polonsky, Supersymmetry: Structure and Phenomena. Extensions of the Standard Model. XV, 169 pages. 2001.

Vol. m 69: W. Staude, Laser-Strophometry. High-Resolution Techniques for Velocity Gradient Measurements in Fluid Flows. XV, 178 pages. 2001.

Vol. m 70: P. T. Chruściel, J. Jezierski, J. Kijowski, Hamiltonian Field Theory in the Radiating Regime. VI, 172 pages. 2002.

FSC
www.fsc.org
MIX
Papier aus verantwortungsvollen Quellen
Paper from responsible sources
FSC® C105338

If you have any concerns about our products,
you can contact us on
ProductSafety@springernature.com

In case Publisher is established outside the EU,
the EU authorized representative is:
Springer Nature Customer Service Center GmbH
Europaplatz 3, 69115 Heidelberg, Germany

Printed by Libri Plureos GmbH
in Hamburg, Germany